Stirling and Thermal-La Engines

Motive power without the CO_2

ALLAN J ORGAN
Formerly of UNIVERSITY OF CAMBRIDGE, UK
and of KING'S COLLEGE LONDON, UK

Stirling and Thermal-Lag Engines

Motive power without the CO_2

World Scientific

NEW JERSEY • LONDON • SINGAPORE • BEIJING • SHANGHAI • HONG KONG • TAIPEI • CHENNAI • TOKYO

Published by

World Scientific Publishing Europe Ltd.

57 Shelton Street, Covent Garden, London WC2H 9HE

Head office: 5 Toh Tuck Link, Singapore 596224

USA office: 27 Warren Street, Suite 401-402, Hackensack, NJ 07601

Library of Congress Cataloging-in-Publication Data
Names: Organ, Allan J., author.
Title: Stirling and thermal-lag engines : motive power without the CO_2 /
Allan J Organ, formerly of University of Cambridge, UK and of King's College London, UK.
Description: London ; Hackensack, NJ : World Scientific Publishing Europe Ltd, [2023] |
The 2 in title is subscript. | Includes bibliographical references and indexes.
Identifiers: LCCN 2021048523 | ISBN 9781800611047 (hardcover) |
ISBN 9781800611054 (ebook for institutions) | ISBN 9781800611061 (ebook for individuals)
Subjects: LCSH: Stirling engines.
Classification: LCC TJ765 .O7375 2023 | DDC 621.4/2--dc23/eng/20211122
LC record available at https://lccn.loc.gov/2021048523

British Library Cataloguing-in-Publication Data
A catalogue record for this book is available from the British Library.

For any available supplementary material, please visit
https://www.worldscientific.com/worldscibooks/10.1142/Q0326#t=suppl

Desk Editor: Vishnu Mohan

Typeset by Stallion Press
Email: enquiries@stallionpress.com

Preface

This book has been written against the background of emerging recognition of the human contribution to climate change. There is finally open talk of a climate 'emergency' — even of a climate 'catastrophe'.

Recognition has not been matched by response. Such has been the pace of the former, however, that this preface has been rewritten three times while the book has been in production. It remains to be seen whether the publisher will agree to a last-minute change, but if so, readers may wish to check the recent Chatham House Climate Risk Assessment Report (https://www.chathamhouse.org/about-us) and/or a reaction to it in The Independent online dated 25 September 2021 (https://uk.yahoo.com/news/recent-climate-report-left-devastated-090454275.html).

COP-26 has been and gone. Billed as the last chance to save the planet, it might have done so — but only if it had resulted in unanimous agreement on the need, and the means, for instant and drastic reduction in reliance on fossil fuels.

With the benefit of hindsight — even of foresight — the expectation was naïve: Concentrated energy is muscle — and muscle is wealth (and military capability). A gallon of diesel oil does the day's work of ten fit laborers, allowing one to drive the tractor while the nine exercise their respective options. Transportable energy is the basis of the economy. The priority of any enterprise or administration is its own survival; Boeing and Airbus will continue to compete for aircraft sales (and are already comparing post-COP26 sales figures). A memo leaked from within UK government circles recommends prioritizing the economy over the environment. Those in high office and the well-heeled will continue taking short hops in long-range aircraft; the rest of us have no plans to trade in our 4-wheel drives for bicycles.

The closest to an unambivalent commitment during COP-26 was that of Pakistan, whose prime minister, Imran Khan, identified climate change as 'one of humanity's biggest crises', undertaking the planting of 'billions of trees' (Independent 3rd. Nov. 2021). Figures for carbon sequestration by trees are in terms of mature, broad-leaf woodland. Unfortunately, you can't plant mature trees — but the proposed measure is an essential complement to the indispensable initiative which must come from technology.

It is technology — born of the Industrial Revolution — which has bequeathed us this problem, together with the standard of living to which we have become addicted. Only technology can get us out of it without a precipitous drop in living standards. With no single technology poised to replicate the 45,000 kJ/kg of fossil fuel, it may be necessary to aggregate every single contribution, however insignificant in isolation.

When Stirling engine development was at its peak at the hands of NV Philips Gloeilampenfabrieken, 'pollution' meant oxides of nitrogen (NOx) and unburned hydrocarbons (Hcx). In that context, a Stirling engine with a sufficiently sophisticated combustion system was a low-pollution prime mover — although the feature failed to outweigh the projected cost of mass manufacture.

Things have changed: the pollutant of concern is now CO_2. When burning hydrocarbon fuel (oil, coal, wood), *all* prime movers — petrol engine, diesel engine, gas turbine, steam engine, and Stirling — emit comparable amounts of CO_2 per shaft-kW. But, in a crucial respect, the Stirling engine is fundamentally different: operation is *independent of the source of heat.* In other words, focused[1] sunlight will serve as 'fuel'. Table-top demonstrators have functioned convincingly in this mode, and there is nothing to suggest that they cannot be scaled up.

At first-draft stage, I was hesitant about promoting the modest role which the Stirling engine might have in a low-CO_2 economy. But, by many accounts, the situation is now dire — possibly even irreversible. Under the circumstances, it would be irresponsible to dismiss *any* potential contribution. But, first, there are prejudices and misunderstandings to be confronted.

[1]As, indeed, will unfocused sunlight, but in this latter case, the temperature difference driving operation is small, and thermal efficiency and specific power derisory.

There can be few concepts which have been innocently over-sold to the extent — and over the period of time (two centuries) — survived by the Stirling engine. The '*quiet, efficient, simple, affordable, burn-anything, environment-friendly Stirling engine*' does not exist — and has never existed.

This is not to claim that it will never exist. Indeed, this book has been written in the hope of showing the very opposite.

The cost of misplaced optimism is disillusionment, of which the Stirling engine has seen more than its fair share. Nothing has created greater backlash than claims for Carnot efficiency — to which the Stirling no more aspires than would a futile attempt to construct a 'real' Carnot engine.

So what of this brave new reality? By one respected measure of performance (specific cycle work or Beale number, N_B), the Stirling engine has never evolved beyond the concept disclosed in the original patent (Stirling 1816). This very same N_B can be used to characterize the reciprocating internal combustion engine — where numerical values are 50 to 100 times greater. Chapter 2 justifies and elaborates this point.

Rehabilitating the concept calls for abandoning illusions and reverting to unvarnished objectivity. Some 20 titles followed Walker's pioneering *Stirling Engines* of 1980, all echoing the claim of 'the perfect engine' first misleadingly put about by Professor Fleeming-Jenkin. *Stirling and Thermal-Lag Engines* is the first title whose material has developed out of recognition of the shortcomings rather than of the complacency induced by alleged unique strengths.

Chapter 2 pinpoints the underlying Achilles' heel — poor utilisation of the working fluid charge, thereby identifying the task ahead: improving cyclic distribution of the working fluid. This in turn calls for tools — appropriate tools.

The first of these arises out of strategy: calculating, simulating and displaying in terms of the loci over a cycle of individual fluid particles. This is the automatic outcome from formulation in Lagrange coordinates and results by default in a fluid particle trajectory map. The second is from a previously unexploited resource.

In 1993, a US patent was granted to one Peter Tailer for a 'thermal-lag' engine. An interpretation of the patent disclosure is that Tailer saw a performance limitation of the Stirling engine in terms of time lag — that between the establishment of a temperature difference and the corresponding flow of heat from hotter to cooler. All other things being equal, the penalty of such a time lag increases with increasing rpm.

Tailer's brilliant inventive step was to make a virtue of a necessity — to exploit (rather than to tolerate) the inevitable. The thermal-lag engine dispenses with the displacer, becoming what, on the surface, is the most ruthlessly simple heat engine ever conceived. From the inventor's perspective, it derived from the Stirling. But, that is not the point.

What matters here is that, if the thermal-lag engine were designed to the same criteria as the Stirling (maximum heat transfer for minimum flow resistance), *it could not work! The thermal-lag engine functions by virtue of irreversibilities, which it is a priority of Stirling engine design to eliminate*! (The same applies to the thermo-acoustic type. Perhaps it has been a mistake to call either a Stirling 'variant'. At a conference in Groningen, I queried an assertion that the (reversible) Schmidt analysis could guide the design of a thermo-acoustic cooler. I was howled down!)

The potential for insight here comes on recognizing that, despite fundamentally different design objectives, it is the same set of equations which defines the gas process interactions in both types, in all three, in fact — Stirling, thermal-lag and thermo-acoustic. There arises the intriguing possibility of a 'universal' numerical model of the gas process interactions, adaptable to any one of the three types merely by changing the input data defining the physical boundary conditions. *Stirling and Thermal-Lag Engines* explores and exploits the possibility.

An element of Zen comes into play: an enquiry advised by prior insight can have a different outcome from an enquiry conducted to see what turns up.

The Stirling remains poorly understood, a victim of unchallenged myth (Carnot efficiency), inappropriate physical data and unverified numerical tools together with pure, unadulterated complacency.

While no match for Robert Stirling's unprecedented intellectual leap of 1816, the present account also has a fresh starting point: it offers a novel view of internal gas process events, it devises and patents a kinematic mechanism to achieve those events and develops the concept step-by-step into a full-size, dimensioned, toleranced engineering design. As in 1816, an appropriate regenerator provision is a high priority. On the way, it estimates the embodied energy (and CO_2) in each component for eventual comparison with that of the competition (the p-v panel). A surprise awaits.

A study which stops short of construction and laboratory test remains an enquiry rather than a prescription for a viable product. The only successful Stirling engine is one which is working reliably and to the designer's performance predictions. This has to be left to a party having the necessary funding and machining resources.

I hope readers will approve of the departures from the mainstream, but constructive criticism will always be welcome at allan.j.o[at]btinternet.com.

By tradition, this is the point at which one takes personal responsibility for all outstanding defects. By tradition, however, one did not have to reckon with aggressive, meddling spellcheck software, recalcitrant too: resistant to recommended procedures for emasculation. Blame for bizarre spellings, malapropisms and other anomalies is hereby attributed to software which doesn't know its place.

Allan J Organ
Ullapool, Highland
March 2022

By the Same Author

Fiction
The Bridge — A Love Affair. CreateSpace Publishing/Amazon, 2014.

Non-fiction
Unravelling the Great Mull Air Mystery. FTRR, Inverness, 2015.

Murder in the Air — The Great Mull Air Mystery Re-visited. FTRR, Inverness, 2017.

The Hoax that Went Wrong — The Mull Air Mystery. Lucid Raccoon, Inverness, 2019.

Engineering/technology
Stirling Cycle Engines — Inner Workings and Design. Wiley, Chichester, 2014.

The Air Engine. Woodhead, Cambridge, 2007; Elsevier 2014.

Air Engines (jointly with Theodor Finkelstein). Mechanical Engineering Publishing (MEP) and Professional Engineering Publishing (PEP) 2001. Re-published with corrections by American Society of Mechanical Engineers (ASME) 2009.

Stirling and Pulse-Tube Cryocoolers. Mechanical Engineering Publishing (MEP) and Professional Engineering Publishing (PEP) 2005 Title subsequently acquired by Wiley, Chichester.

The Regenerator and the Stirling Engine. Mechanical Engineering Publishing (MEP) and Professional Engineering Publishing (PEP) 1997 Title subsequently acquired by Wiley, Chichester.

Thermodynamics and Gas Dynamics of Stirling Cycle Machines. Cambridge University Press, 1992. Re-published in paper-back 2010.

Acknowledgements

If this book has made a contribution, it is because many have helped in the process. Research student Claudio Takase has set a new standard in elegance and economy in technical writing. Peter Maeckel has challenged anything and everything I said or wrote, Geoff Vaizey and John McIntyre have demonstrated how strongly success depends on exemplary craftsmanship and uncompromising machining standards. Norris Bomford and the late Jimmy James inspired with their unquenchable enthusiasm and ingenuity, Jimmy's son Duncan, founder of, and driving force behind CADStudios, contributed a new dimension (all three of them, in fact!) to the work of mechanical design. An exchange with John Corey has invariably resulted in a more considered view of prospects, and in terms of time, no one has given more generously in person, by e-mail and by telephone than has Camille van Rijn.

The scholarly example of Ted Finkelstein remains in the background unsurpassed, pushing the algebra to its elegant limits before recourse to the computer.

Contents

Notation

Symbol		Units
A, A_{ff}	free-flow area — cross-sectional area of flow-passage e.g., of regenerative annulus	m^2
a	acoustic speed $\sqrt{\gamma RT}$	m/s
$a_{1,1}$, b_1, etc.	coefficients defined at point of use	–
c_p, c_v	specific heat at constant pressure/volume	J/kgK
D	total or substantial derivative $\partial/\partial t + u\partial/\partial x$	s^{-1}
D	outside diameter of cylindrical matrix stack also, abbreviation defined in text	m (as required)
D_{eq}	equivalent outside diameter of cylindrical matrix stack. Numerically equal to D_o for uniform cylinder, and equal to $\sqrt{(D_o^2 - D_i^2)}$ for annulus of radial thickness $D_o - D_i$	m
d_w	diameter of individual wire	m
f	cyclic frequency	s^{-1}
$f\{\ \}$	function of { }	–
Ge	dimensionless group $(4pr_h\sqrt{\gamma})/[\mu(T)\sqrt{RT}]$ allowing relationship between Re and Ma to be expressed $Re = Ge \times Ma$	–
h	coefficient of convective heat transfer	W/m^{2}K
	specific enthalpy	J/kgK

k	thermal conductivity	W/mK
L	overall length of flow passage	m
L_r	regenerator stack length in flow direction	m
L_{ref}	reference length, $V_{sw}^{1/3}$	m
M	mass (of working fluid charge)	kg
m'	mass rate, dm/dt	kg/s
m_w	mesh number — number of wires/m	m^{-1}
p	(absolute) pressure	Pa
p_{ref}	reference pressure (e.g., nominal charge pressure)	Pa
p_w	wetted perimeter	m
q'	heat rate per unit mass	W/kg
Q	heat quantity per cycle	J
Q_E, Q_C	heat per cycle added/rejected	J
R	specific gas constant	J/kgK
r_{hr}	hydraulic radius — (wetted volume)/(wetted area) = $\frac{1}{4}d_w¶_v/(1-¶_v)$ for cylindrical crossed rods	m
t	time	s
T	(absolute) temperature	K
T_E, T_C	respective (absolute) temperatures of expansion and compression spaces	K
ΔT_{sw}	cyclic temperature swing of matrix (= two times excursion either side of mean)	K or C
$\underline{u}$	mean mass speed or bulk speed, $m'/\rho A_{ff}$	m/s
u	specific internal energy	J/kgK
V_{dr}	matrix pore volume	m^3
V_{env}	regenerator housing volume	m^3
V	volume	m^3
V_E, V_C	respective swept volumes of expansion and compression spaces	m^3
V_{dr}	pore volume of regenerator	m^3

V_{sw}	swept volume	m^3
ν	cumulative volume reckoned from face of expansion piston of opposed piston machine. Independent variable aligned with x-axis of Lagrange coordinate system	m^3
W	work per cycle	J
x	linear distances in x coordinate direction	m
ε	perpendicular offset (horizontal in Fig. A-1) of axis of cylinder from axis of crank-shaft	m
ϕ	crank angle	rad
μ	coefficient of dynamic viscosity	Pa s
η_v	volumetric efficiency	–
ρ	density	kg/m^3
ρ_w	density of regenerator wire material	kg/m^3
ω	angular speed	s^{-1}

Dimensionless Variables

C_f	friction factor $\Delta p/(\frac{1}{2}\rho \underline{u}^2 L/r_h)$	–
Ma	Mach number $u/\sqrt{\gamma RT}$ (For ideal gas $Ma^2 = \rho\underline{u}^2/p\gamma$)	–
Re	Reynolds number $4\rho\|u\|r_h/\mu = 4m'r_{hr}L_r/\mu V_{dr} = 4SgMa^2$	–
r_p, r_v	ratio p_{max}/p_{min}, volume ratio V_{max}/V_{min}	–
Sg	Stirling number $pr_h/\underline{u}\mu = Re/(4\gamma Ma^2)$	–
St	Stanton number $h/\rho\|u\|c_p$	–
x	abbreviation for T_j/T_{j+1}	–
α	volume phase angle	–
δ_r	regenerator dead-space ratio V_{dr}/V_{sw}	–
η_T	thermal recovery ratio, e.g., $(T_{out} - T_{in})/(T_{target} - T_{in})$	–
ε_v	volume effectiveness, defined in text	–
κ	swept volume ratio V_C/V_E	–

μ	Finkelstein's dimensionless volume V/V_E	–
ψ	pV_E/MRT_C (Finkelstein defines $\psi = \frac{1}{2}pV_E/MRT_C$)	–
	dimensionless pressure pV_{sw}/MRT	–
σ	dimensionless mass m/M	–
σ'	dimensionless mass rate $m'/\omega M$	–
τ	dimensionless temperature T/T_C (Finkelstein's temperature ratio $T_C/T_E = 1/N_T$)	–
υ	dead-space parameter $(V_{de} + V_{dxe})/N_T V_E + V_{dr}/V_E \log_e N_T/(N_T - 1) + (V_{dxc} + V_{dc})/V_E$	–
υ_{adi}	υ in adiabatic analysis $= \upsilon - (V_{de}/N_T - V_{dc})/V_E$	–
υ_{TF}	Finkelstein's dead-space parameter $= 2\upsilon$	–

Dimensionless Parameters

a, b	coefficients of heat transfer correlation $StPr^{2/3} = a/Re^b$	–
c, d	coefficients of friction factor correlation $C_f = c/Re + d$	–
C	constant in expression for mean effective temperature	–
C_f	friction factor $\Delta p/(\frac{1}{2}\rho \underline{u}^2 L/r_h)$	–
n_g	number of gauzes in close-packed stack	–
N_{MA}	Mach parameter or dimensionless speed $\omega L_{ref}/\sqrt{RT_C} = \omega V_{sw}^{1/3}/\sqrt{RT_C}$	–
N_{Pr}, Pr	Prandtl number $\mu c_p/k$	–
N_{SG}	Stirling parameter $p_{ref}/\omega\mu$	–
N_{TC}	ratio of thermal capacity of matrix to that of gas with which it interacts at cyclic equilibrium.	–
N_{TCR}	thermal capacity ratio	–

$¶_v$	volume porosity — ratio of void volume to envelope volume = $1 - \frac{1}{4}\pi d_w m_w$ for cylindrical crossed rods	–
N_{MA}	characteristic Mach number	
N_T	characteristic temperature ratio T_E/T_C	–
N_{RE}	Reynolds parameter = $N_{SG}N_{MA}^2$	–
NTU	Number of Transfer Units StL/r_h	–
η_v	volumetric efficiency (of internal-combustion engine)	–
γ	specific heat ratio — isentropic index c_p/c_v	–

Subscripts

C, E	compression, expansion
d	un-swept or 'dead'
r, reg	relating to regenerator
ref	reference values

Supplementary Notation for Kinematics of 'Pseudo-Rhombic' Mechanism

r_p, r_{pin}	semi-throw of crank	m
x_p, x_{pin}	horizontal coordinate of crank-pin location	m
y_p, y_{pin}	vertical coordinate of crank-pin location	m
X, Y	coordinates of locations of pivot points Q and Q′	m
α	auxiliary angle defined in text	–
ϕ	angular position of crank-pin reckoned clockwise from vertical	–
γ	angular inclination to horizontal of link *RQ* in Fig. A.1	–
ς	angle set by designer to fix pivot points of coupler link	–
ξ	angular inclination to horizontal of a line through pivot points *R* and *T* of coupler in 'pseudo-rhombic' crank mechanism	–

1

Without the Rose-Tinted Spectacles

1.1. Mission Statement

There can surely be few instances where a concept having all the right credentials has attracted the full force of modern technology, has developed through so many dramatic advances — and failed to deliver a return in terms of significant commercial impact.

The Stirling engine is such an instance: The list of virtues routinely claimed includes silence, high thermal efficiency subject only to the Carnot limit, low moving-parts count — from which affordability and low maintenance are assumed to follow — low environmental impact and indifference to heat source. As a concept it ticks all the academic boxes, but how does it manifest in a world of commercial criteria?

The reality is that a Stirling engine embodying the full range of these attributes does not exist, never has existed — and never will exist without a drastic rethink. Competitive efficiency and specific power combined with quietness (this engine is not silent) call for technological and manufacturing sophistication, and sophistication comes with a daunting price tag. The 'Air-Independent Power/Propulsion' Unit (AIP — Fig. 1.1) reached batch-manufacture at the hands of the Swedish firm Kockums, but it does not take a production engineer to work out that the thermal efficiency, specific power and noise level which qualify it for its exclusive application (auxiliary power for stealth submarines) come at an eye-watering price.

The intermediate technological leaps merit a closer look. At a stroke '*... application of modern materials and modern theory of fluid flow and*

Fig. 1.1. Stirling-based air-independent propulsion unit (AIP) publicised by the Swedish firm Kockums.

heat transfer...' boosted specific power from 0.18 W/kg of the Rider 'Six-Inch' engine of 1906 (Fig. 1.2) by a factor of 55 to the 10 W/kg of the air-charged MP1002CA. At 250 W, the two engines were of comparable shaft output, so Fig. 1.2 illustrates the comparison to scale. The introduction of light gases, He and H2, brought improvement by a factor of 8 to 80 W/kg. Philips' invention of the rhombic drive allowed a single-cylinder engine to operate virtually vibration-free, and a problem of working fluid leakage was addressed by another mile-stone development, the oil-pressurised rolling-diaphragm seal.

Impressive brake thermal efficiencies were reported — and not just from laboratory 'one-offs'. On the basis of technology acquired from Philips, General Motors produced a batch of engines of nominally 3 kW

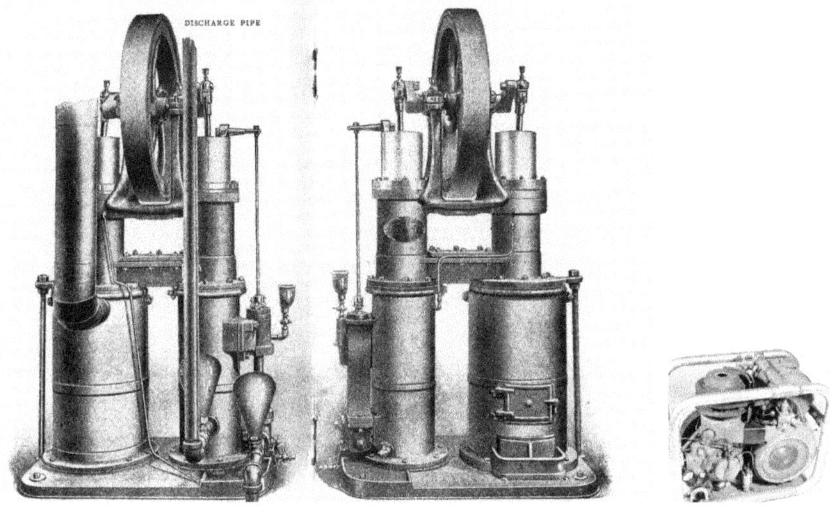

Fig. 1.2. Size and weight comparison (approximately to scale). Left: Rider 'Six-Inch'. Output: about 0.25 kW from a mass of one tonne. Right: Philips' 1950s air engine, the (rightly) celebrated MP1002CA. Same 0.25 kW from 12 kg (bare engine — estimated).

for the military. Designated the GPU-3 (**G**round **P**ower **U**nit **3** kW) it recorded thermal efficiency of 30%.

On the other hand, it remains the case that, of the millions of Stirling engines 'designed' and constructed world-wide, one-off or in factory production, the overwhelming majority have been of depressingly *low* thermal efficiency. Even after Philips had brought the concept into the 20th century, brake thermal efficiency of the MP1002CA overall (i.e., while driving the auxiliaries needed to make it self-contained) was measured at 5% (the author's final-year dissertation 1962).

Progress towards reality requires recognition that, to date, our perception of potential has been skewed.

1.2. A Seed of Doubt

Philips' early success breathed new life into the concept — but not commercial acceptance. The free-piston variant, essentially a 20th-century development not foreshadowed in the original invention, is enjoying commercial success in reversed-cycle form at the hands of Global Cooling. Where the criterion is reliability in unattended operation over indefinite service life, the same, free-piston configuration is proving itself in 'Stirling converter' units supplied by Sunpower to NASA Glenn for matching to nuclear heat sources.

From the point of view of this text, there remains an uncomfortable truth which, to date, has not been satisfactorily addressed. When the criterion is cycle work per unit charge pressure and per unit swept volume (or cycle work per unit mass of working fluid), numerical values are a risible fraction of those achieved by the internal combustion engine (I.C.E.). Worse still — *they have barely improved since Stirling's original of 1818!* Embodiments of the invention demonstrated to date make lamentably ineffective use of working-fluid charge *M*, delivering between 1/100th and 1/50th of the cycle work per unit mass achieved by the I.C.E.

A compression ratio of 2:1 combined with a temperature ratio one-third of that of the ICE goes part-way to explaining — but only part-way (see Chapter 2).

By obscuring this reality, the introduction of light gases and of pressurization may well have stifled long-term progress:

Three contributory factors stand out:

(1) Mal-distribution of working fluid: Only a faction — if any — of working fluid charge M [kg] undergoes a cycle taking in both nominal limits T_E and T_C. This alone invalidates glib comparisons with other cycles — the Carnot in particular.

For much of the expansion phase, the fraction m_c/M of gas occupying the compression space at T_C exceeds that in the expansion space at T_E — in contradiction of the basic function of a heat engine cycle.

(2) *Every* Stirling engine to date has been in the 'Low-ΔT' category. Expansion temperature T_E is about 600°C vs 2000°C or so for the gasoline engine.

(3) Inappropriate regenerator provision: if an instance exists of this crucial component's having been rationally *designed* for specific operating conditions, then it has yet to come to the notice of this writer.

Stirling science as it exists to date has done nothing to address the situation whereby Stirling's engine of 1818 and the lumbering Rider

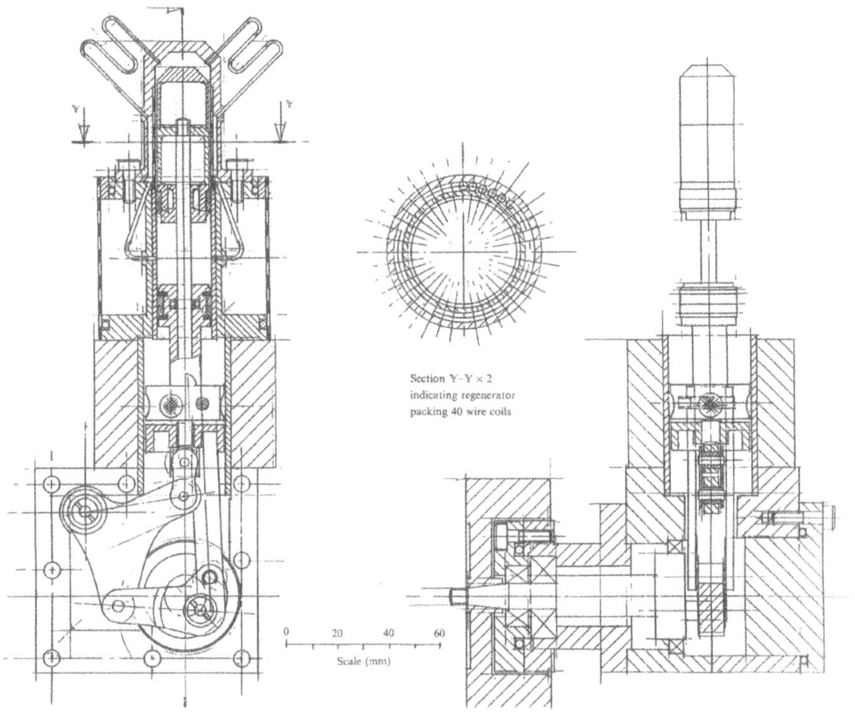

Fig. 1.3. The writer's reconstruction of Clapham's award-winning 5 cc He-charged engine.

'Six-Inch' of the 1890s extracted a somewhat greater amount of work per cycle from a kg of working gas[1] than does United Stirling's sophisticated P-40!

The claim might be called into question by the outcome of a competition held at the Model Engineering Exhibition. The worthy winner — by a substantial margin — was an engine of 5 cc displacement (Fig. 1.3), designed by a newcomer to the Stirling scene, Eddie Clapham. Pressurised with helium to 81 bar, brake output at 2000 rpm was measured at 78 W.

On the other hand, the figures convert to specific cycle work N_B of 0.058 — evidently a step backwards relative to the 1818 engine, to the Rider, to Philips' MP1002CA — and, indeed, to the P-40. As for power/

[1] Strictly per MRT_C, where R is specific gas constant. The comparison remains valid for hypothetical Otto cycles based on those gases (see Chapter 2).

weight ratio, the mass of the helium pressure vessel plus reducing valve has to be taken into account. Alternatively, the power required to pressurise to 81 bar and to maintain that pressure has to be deducted from the 78 W gross.

1.3. The Thermal-Lag Engine

The previous section has made a case for a radical rethink of prospects for the Stirling engine in terms of underlying shortcomings in implementation. The same is not possible for its younger cousin: there is no orthodox technology to be challenged, since none has been established. There is not even a coherent explanation of the principle of operation, so it is necessary to start from scratch.

Figure 1.4 shows the essential features of the drawing which accompanied Tailer's 1993 patent. It has been reversed left-to-right for consistency with a convention adopted here when displaying fluid particle trajectories and temperature-time histories, whereby the work-piston operates at the right-hand end of the gas path. The water jacket surrounding the working cylinder has been omitted for clarity, as has the burner and a feature which appeared to have been a flame-guard and/or exhaust collector.

Numerals are Tailer's originals, and include 112 which is a 'pressurised, annular rolling seal'. Numeral 101 is supposed to identify an

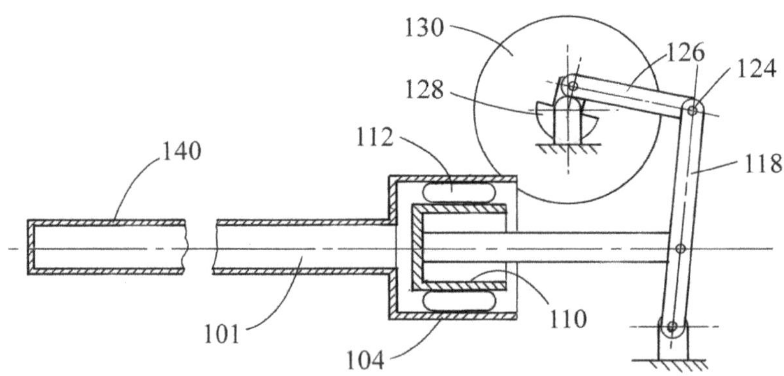

Fig. 1.4. Essential features of the 1993 patent drawing, but reversed left-to-right for consistency with illustrations to follow, whereby the work piston operates at the right-hand end of the gas path.

un-specified working fluid occupying cylinder 104 and 'chamber' 140 — possibly more readily understood as a pipe or tube.

What follows is quoted from the patent — verbatim except for interventions such as conversions from imperial to SI dimensions:

On finding that a slight time delay in pressure change was observable when a displacer was rapidly moved in a Stirling engine displacer cylinder (not shown), it was felt that this time delay could be harnessed or put to use. For this reason, an experimental test engine, as shown in FIG. 1 (the present Fig. 1.4), *was built. It had a cylinder 104 with a diameter of 9"* (228 mm) *and a piston 110 with a diameter of 7.5"* (190 mm). *Seal 112 was a 3.50/4.10×5 vehicle inner tube pressurized to 8 psig* (0.54 bar). *Crank 128 and link 118 provided a piston stroke of 0.75"* (19 mm). *The pressure change in engine volume 101 on turning flywheel 130 was between +2.5 psig to −2.5 psig* (±0.17 bar). *Heated chamber 140 was a capped 40"* (1016 mm) *length of nominal 11/4"* (100 mm) *pipe 100. Gas burner 144 was applied adjacent to flame guard 142 along chamber 140. The engine 100 showed no indication of running.*

On turning out burner 144, connecting rod 126 was removed from link 124. The heated chamber 140 had not appreciably cooled and the piston 110 was fortunately near bottom dead center. With amazement, it was observed that link 118 vibrated and the vibration continued. On relighting burner 144, the engine 100 ran in this manner as a free piston engine for half an hour with a piston 110 stroke of about 1/4" (6 mm) *at about 200 cycles/minute.*

The thermal lag engine 100 was operating, but only barely. It could not sustain a flywheel 130 even with a reduced stroke.

Links of chain (not shown) were placed in the heated chamber 140 and in the portion extending through the reservoir 108 to increase heat transfer. The 13/4" (100 mm) *nominal pipe heated chamber 140 held about 11 chain links per inch* (per 25 mm) *of length and each link had a surface area of about 1 sq in* (645 mm^2). *The chain increased heat transfer surface in chamber 140 from about 4 sq in/in of length to 15 sq in/in of length* (100 mm^2/mm to 380 100 mm^2/mm). *This was merely an experimental expedient and not a suggested way to increase heat transfer area.*

With the added chain (not shown) engine 100 then drove flywheel 130 at 450 rpm with a 3/4" (18 mm) *stroke of piston 110. This indicated that a thermal lag engine 100 could function and that further experimentation could lead to improved power and efficiency.*

This has hardly been the definitive explanation as to how the thermal-lag engine functions. However, there is now a connection between the 'chain' of Tailer's full-size engine and the regenerator of later desk-top scale devices which purportedly demonstrate the principle.

An indication of Tailer's depth of understanding is to be had from assertions later in the published patent:

While it is inexpensive to experiment with mean atmospheric pressure engines 100 with annular rolling seals 112, these experiments indicate superior working fluids such as helium or hydrogen under a high pressure, such as 2000 psi, the types of seals used in modern Stirling engines, and other known techniques could provide powerful external combustion engines with high efficiencies. In Stirling engines other factors being the same, engine power varies almost directly with the mean engine pressure. The same should hold true for thermal lag engines to the extent that they are analogous to Stirling engines.

It must be thought remarkable that, having identified thermal-lag as the driving mechanism, Tailer should now advocate working fluids well-known to minimise such effects. A later chapter of this book will recount an experiment in which the introduction of helium stopped dead in its tracks a thermal-lag engine which, on air, had run un-loaded at 3150 rpm. The proposal to design for 2000 psi (136 bar) will be looked at with the aid of dedicated computer simulation.

1.4. An Efficient, Affordable, Power-Producing Stirling Engine?

No conventional embodiment of the Stirling engine has surpassed by much of a margin a specific work value (Beale number) of 0.15. This affords a back-of-the-envelope estimate of likely power in terms of *rpm* and swept volume from a 1-atm. engine complying with 'traditional' Stirling science. The algebra is converted to graphical form in Fig. 1.5, in which the specimen 'solution' (oblique line) takes the 62 cc and 1500 rpm of the MP1002CA, predicting 22 W at $p_{\mathrm{ref}} = 1$ atm. (supposing those *rpm* can be achieved at 1 atm.).

The prognosis is not overwhelmingly attractive. But suppose there is scope, so far unexploited — perhaps not even identified, for persuading a larger fraction of working fluid mass M to participate in the full cycle

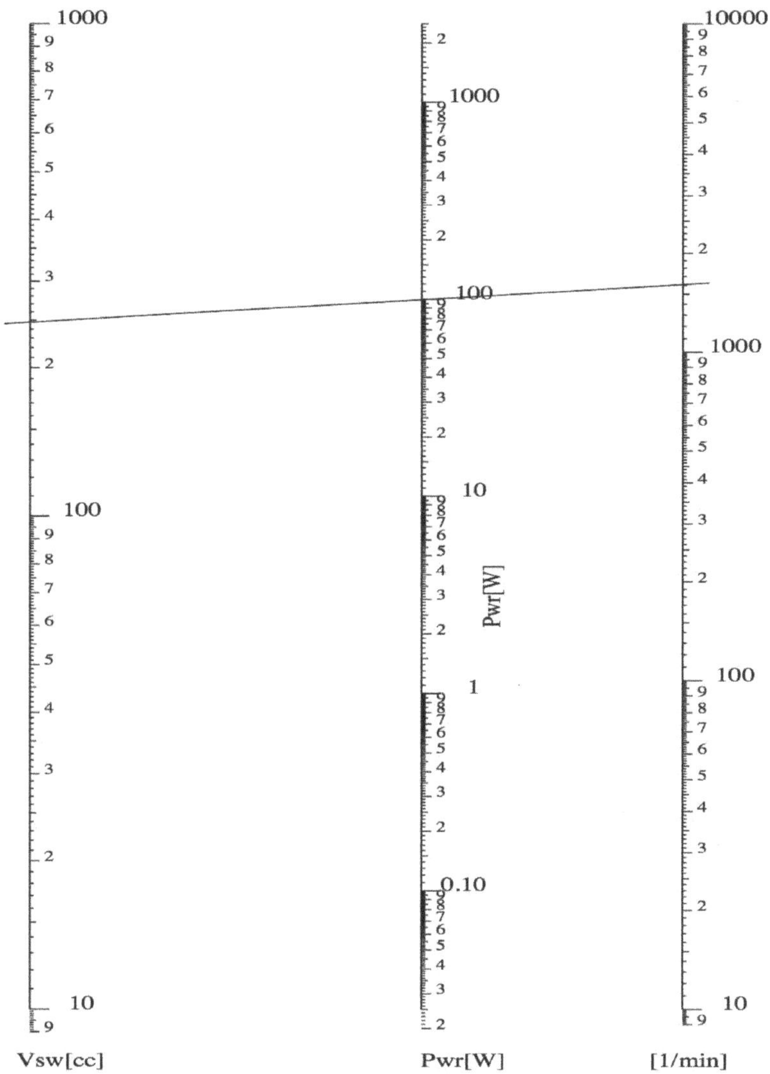

Fig. 1.5. Beale-number prediction of power in terms of swept volume and rpm when charge pressure restricted to 1 atm.

between T_E and T_C — say 5-times the usual fraction. In an era of LED lighting, 110 W is not to be sneezed at.

But can it really be the case that two centuries of technological development have left room for a fresh start — for innovation of any description?

With some advance knowledge of the outcome, the answer is 'yes', thanks to help from an unlikely source — from this author's sideline interest in forensic enquiry (see list of titles under '*By the same author*'). There is apparently an ABC of crime-scene investigation[2]: *Assume nothing; Believe nobody; Check everything.* It authenticates the bold title of this book: *one can go a long way to establishing a science simply by identifying and discarding what is not scientific.* It also determines that the route

Fig. 1.6. Size and weight comparison (approximately to scale). Left: Philips' 1950s air engine — the (rightly) celebrated MP1002CA. 250 W from 62 cc at 1500 rpm. Mass of bare engine 12 kg (estimated). Right: ED 'Hunter' two-stroke model aero engine. 197 W from 3.5 cc at 13,300 rpm. Mass 0.185 kg. — or 1.5% of mass of MP1002CA!

[2]Whether the origin is professional or an invention of the Worshipful Guild of Crime Fiction Writers remains to be established. According to Professor David Barclay, eminent forensic scientist, '... *a similar phrase was used in the very first official ACPO cold-case protocols published by National Crime Faculty around 1998. Of course it's a simple concept, so may have originated earlier, although Lausanne Uni attributes it to us. It probably got into crime fiction through Harrogate, where I frequently performed. And the title of my second crime novel was 'Believe No One'!*'

to the low-cost Stirling engine is backwards — to the pre-Philips era, to air-charging and to minimal reliance on pressurisation, to low parts-count, to simplicity, reliability and affordability.

The size/mass comparison of Fig. 1.6 indicates pictorially how far there is to go. Getting started will mean first of all taking a closer look at what is meant by working fluid utilisation. The exercise may be expected to open a can of worms!

2

An Uncomfortable Truth

2.1. Face-to-Face with Reality

It was asserted in Chapter 1 that the concept *per se* has not evolved since 1818. What appears to be an outrageous claim is merely a statement of the obvious: in order to achieve what the internal combustion engine (ICE) achieves from its single-atmosphere charge, the Stirling has to be pressurised to between 50 and 100 atm. The substantial performance gap has provided the primary motivation for the work reported in this book. Comparison with the ICE is implied throughout, so it is necessary to be sure that apples are not being unfairly compared to pears.

The Stirling engine, like the reciprocating internal combustion engine (gasoline and diesel) — and like the steam engine, and like the gas turbine — is a *heat engine*. Beneath the surface, the principle of operation is common to all: compression of an elastic medium (e.g., air) at low temperature, addition of heat followed by re-expansion at higher temperature.

The power which appears at the output shaft (or which is transmitted to the air stream in the case of the gas turbine) is the difference between the rate of energy supplied in the fuel, and the rate of energy rejection as heat. Efficiency in all cases is the ratio of power out to energy-rate in.

The various hydrocarbon fuels embody similar amounts of latent chemical energy per kg — about 46,000 kJ. This enables heat engines in general to be ranked in an informative pecking-order.

2.2. Internal Combustion Engine and Beale Number Limit

The heat released by combustion of unit mass of liquid hydrocarbon fuel with the working fluid (air) has an upper limit. Assuming the fuel to be 'pure' gasoline consisting of *n*-heptane and iso-octane, the calorific value LCV (lower) is known, and the heat release can be estimated by assuming that combustion is complete at the 'stoichiometric limit' of air/fuel ratio m_{air}/m_{fuel}, here denoted *af*.

From the ideal gas equation, mass of air m_{air} inducted/cycle = $\eta_{vol} p_{atm} V_{sw}/RT_{amb}$. η_{vol} is volumetric efficiency. Heat release/cycle is given by $Q_E = \eta_{vol} \times \text{LCV} \times p_{atm} \times V_{sw}/afRT_{amb}$. An expression for *work*/cycle follows on multiplication by thermal efficiency η_{th}.

Finally, dividing by $p_{atm} \times V_{sw}$ gives N_{BpV}:

$$N_{BpV} = \eta\,\text{th}\frac{\eta_v \times \text{LCV}}{af \times R \times T_C}. \tag{2.1}$$

In Eq. (2.1), N_{BpV} is Beale modulus (as in Stirling engine practice), the pV subscript indicating that the numerical value derives from the pV work. Temperature T_{amb} has been replaced by T_C for compatibility with Stirling engine usage. The relevant value of *af* is 14.7 and that of LCV is 46 MJ/kg. At assumed thermal and volumetric efficiencies of 30% and 85%, respectively, the upper limit of N_{BpV} is 9.17.

The alternative form of Beale modulus, here denoted N_{BmRT}, is due to Finkelstein (1960a, 1960b). It is obtained by dividing cycle work by mRT_C, and results in a slightly different numerical value. The difference in the algebra leading is the absence of the term η_{vol} for volumetric efficiency (because m is the mass of gas actually processed through the cycle). So in the present example N_{BmRT} is 9.17/0.85 = 10.78.

The yawning gap between these 'upper-bounds' and the 0.15 of the typical Stirling engine speaks for itself.

A first step in narrowing a gap is to quantify it. Acknowledging that Beale number N_B comes in two forms will reveal a yardstick common to all fossil-fuelled heat engines:

$$N_{BpV} = \frac{\text{work/cycle[J]}}{p_{ref}[\text{Pa}] \times V_{sw}[\text{m}^3]} = \frac{\text{Power[W]}}{p_{ref}[\text{Pa}] \times V_{sw}[\text{m}^3] \times \text{rpm}/60}, \tag{2.2a}$$

$$N_{BmRT} = \frac{\text{Work/cycle[J]}}{M[\text{kg}] \times R[\text{J/kgK}] \times T_C} = \frac{\text{Power[W]}}{M[\text{kg}] \times R[\text{J/kgK}] \times T_C \times \text{rpm/60}}. \tag{2.2b}$$

2.3. Beale Number and the Stirling Engine

Computation of N_{BpV} proceeds precisely as first carried out by Beale — except the choice of reference pressure p_{ref} is p_{min} for compatibility with the ICE calculation. (Where the engine is un-pressurised p_{atm} is assumed, the size of any error depending on the return-valve provision for leakage compensation.)

The corresponding value of N_{BmRT} is acquired by conversion with the aid of the Schmidt algebra — specifically Eq. (21) of the definitive reworking (Finkelstein, 1960a):

$$MRT_C/p_{min}V_E = \frac{1}{2}(1/N_T + \kappa + \upsilon)(1+\rho), \tag{2.3}$$

$$\rho = \frac{\sqrt{(1/N_T{}^2 + \kappa^2 + 2\kappa\cos\alpha/N_T)}}{1/N_T + \kappa + \upsilon},$$

$$V_{sw} = V_E\sqrt{(1+\kappa^2 + 2\kappa\cos\alpha)}.$$

Table 2.1 lists values of both forms of Beale number. The numerical value of N_{BpV} calculated for a specific combination of prime mover and operating conditions generally differs somewhat from the corresponding N_{BmRT}. In the case of the internal combustion engine the extent of the difference is due largely to volumetric efficiency η_{vol}. In the Stirling engine characteristically poor *volume effectiveness* ε_v is to blame. (A definition and picture of ε_v will emerge shortly.)

Table 2.1 covers the original engine of 1818[1] to the USS P-40 — and a couple of representative ICEs. What is particularly interesting is that, by this criterion, the lumbering Rider engine is up there with the best.

[1]The figure for the 1818 engine is derived not from the mythical 2 horsepower, but on the basis of extensive calculation, computer simulation and construction of full-size masonry, furnace and displacer cylinder according to the patent drawings of 1816 (Finkelstein and Organ 2001).

Table 2.1. Specific work comparison.

	$N_B = \dfrac{\text{brake power}}{p_{min} V_{sw} \text{rpm}/60}$	$\dfrac{\text{brake power}}{MRT_C \text{rpm}/60}$	Inferred volume effectiveness ε_v%
Equipment			
400hp/cyl	0.235	0.0935	0.39
GPU-3	0.182	0.0990	0.92
MP1002CA	0.128	0.0773	0.72
USS P-40	0.084	0.0412	0.38
V-160	0.094	0.0644	0.59
PD-46	0.076	0.0379	0.35
Stirling 1818	0.097	0.0482	0.45
Rider 'Six-Inch'	0.1044	0.0726	0.67
Honda 49 cc C-100 motorcycle engine (1958–67) 3.4kW @ 9500 rpm (4750 cpm)			
	5.47	***6.44***	(85.0)
OS-Max 2-stroke model aircraft engine 2½ cc, 0.4 hp at 17,000 rpm (17,000 cpm)			
	5.26	***6.19***	(85.0)
Rolls-Royce Trent 1000 gas turbine ***10.55*** (See final section of this chapter)			
(Volumetric efficiency for the ICE instances assumed to be 85%)			

Presumably 250 rpm gives all the time in the world for heat transfer while incurring minimal flow resistance. The above-average figure for the 400hp/cylinder Philips engine is in parentheses because it was a 'virtual' engine — an extrapolation on the basis of the brilliant concept of the 'part-engine'.

At first sight the open-cycle gas turbine threatens not to fit the picture. Estimating a Beale number is not straightforward because performance is quoted in terms of thrust rather than equivalent shaft power, for which an estimate has to be inferred from aircraft weight and performance. Figures and arithmetic at the end of this chapter for the Rolls-Royce Trent 1000 as fitted to the Airbus 380 lead to 10.55 as upper bound for N_B of the gas turbine — essentially as for the reciprocating ICE!

The power/weight disadvantage of the Stirling engine is routinely offset by pressurisation. If 50–100 atm. of hydrogen or helium is acceptable, then the power/weight problem (only) evaporates. On the other hand, specific work (Beale number by either definition) remains a depressingly low fraction of potential. Several factors contribute. One is the considerably lower temperature ratio of the *genre*. Quantifying additional contributory factors will be helped by first of all levelling the playing-field in this fundamental respect.

The algebraic expression for ideal cycle work is dominated by the multiplier $(1-1/N_T)$ (Finkelstein, 1960). Notional compensation for T_E increased to peak I.C.E. values (i.e., to ~ 2200 K) is to divide by $(1 - 1/N_T)$ with N_T calculated as T_E/T_C 873/313 and to multiply by the same term evaluated as T_E/T_C 2200/313. Values of N_{BmRT} adjusted in this way only marginally decrease the performance gap relative to the ICE. This focuses attention on working-fluid utilisation as the main culprit.

2.4. Volume Effectiveness ε_v

N_{BpV} figures for the ICE are lower than counterpart values for N_{BmRT} by a factor — *volumetric efficiency*. The latter recognises that the actual mass taking part in a cycle generally differs from nominal mass $p_{atm}V_{sw}/RT_C$. In the Stirling engine, working fluid utilisation is controlled largely by the volume events, suggesting use of a similar concept ε_v to denote 'volume effectiveness'. Multiplying the N_{BmRT} in the Stirling case by values of η_{vol} (less than unity) will inflate the numerical value. Experimenting with ε_v until values of N_{BmRT} compare with the ICE case will suggest the fraction of working fluid which is being used effectively. If the figure compares with the picture acquired by other, independent means (e.g., by inspection of the particle trajectory diagram), then useful insight is on the way.

2.5. A Changed Vision

Whether the comparison afforded by Table 2.1 is cause for pessimism depends on the hopes one holds out for the *genre*. A pipe-dream of this writer has been a kinematic Stirling engine to challenge the single-cylinder internal combustion engine for portable power — for electricity generation, pumping water, or a bit of unobtrusive Sunday-afternoon lawn-mowing.

Hopes had been pinned on the potential of the MP1002CA concept (recall Fig. 1.6) for revival and further development, until realisation

Fig. 2.1. Honda portable generator — early version. With written permission of an e-Bay advertiser who signs himself 'Terry'.

dawned that its 15 atm. pressurisation level would have to be raised by a factor of five or six before the Honda counterpart would feel threatened (Fig. 2.1). Supplying and managing pressure at between 50–100 atm. is an expensive business at ambient temperature, let alone when it involves components at red heat. Compounding cost in this way is not a promising way to compete.

It would be interesting to know what would be Robert Stirling's view were he to re-materialise. Evidently a thoroughly rational man, he might well have withheld comment until he had mobilized all available 21st-century resources in the search for corroboration of the diagnosis. Had he so done, he would have found some half-dozen features of implementation which are consistent with poor working fluid utilisation.

2.6. Corroboration

Real vs ideal cycle: Figure 2.2 is a fluid particle trajectory diagram constructed for the gas-path specification of the USS V-160 engine. Adjacent particle tracks separate equal masses of working fluid. Counting tracks within a given region indicates what fraction of total working fluid mass M occupies the region at that instant.

The expansion phase is where the total volume enclosed between opposed piston faces is on the increase. With crank angle elapsing

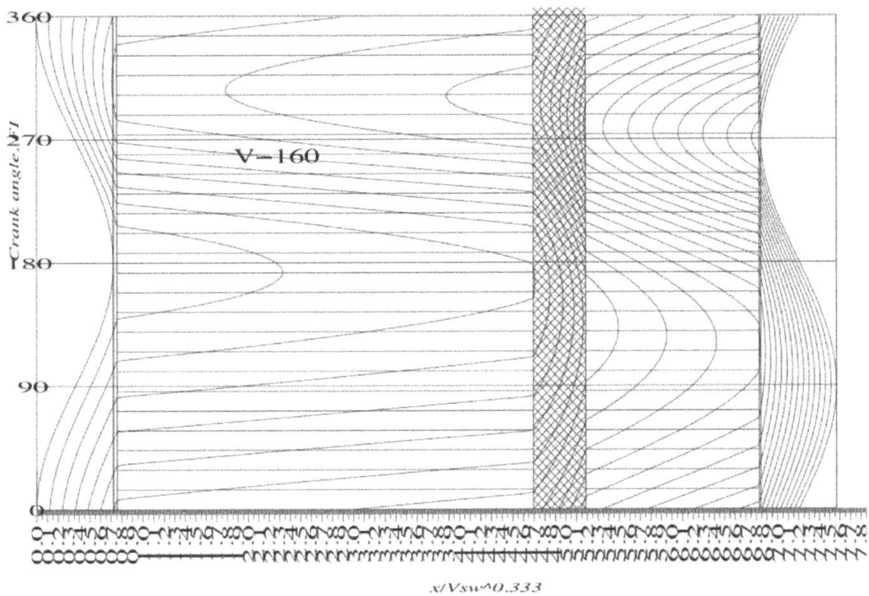

Fig. 2.2. Fluid particle trajectory map constructed for the gas-path specification of the USS V-160 engine.

vertically in the figure, expansion starts at about 270 degrees. Counting particle tracks confirms that, *during expansion there is a greater mass of working fluid at compression (i.e., sink) temperature than in expansion space and expansion exchanger combined*!

The principle of operation requires the opposite, so one could be forgiven for asking how the engine functions at all.

Cycle non-uniformity: The trajectory map has more to offer: tracking specimen particles over a complete cycle confirms that *only a fraction* complete a cycle between temperature limits T_E and T_C. In some configurations, notably the P-40, *no one working fluid particle experiences both limits*!

This is in stark contrast to the internal combustion engine cycles — Otto and Diesel — where, for practical purposes, the entire charge M is subjected to the same cycle.

Anomalous pressure-volume characteristic: In the context of reciprocating power cycles, there is an expectation that reduction in overall volume means increase in pressure.

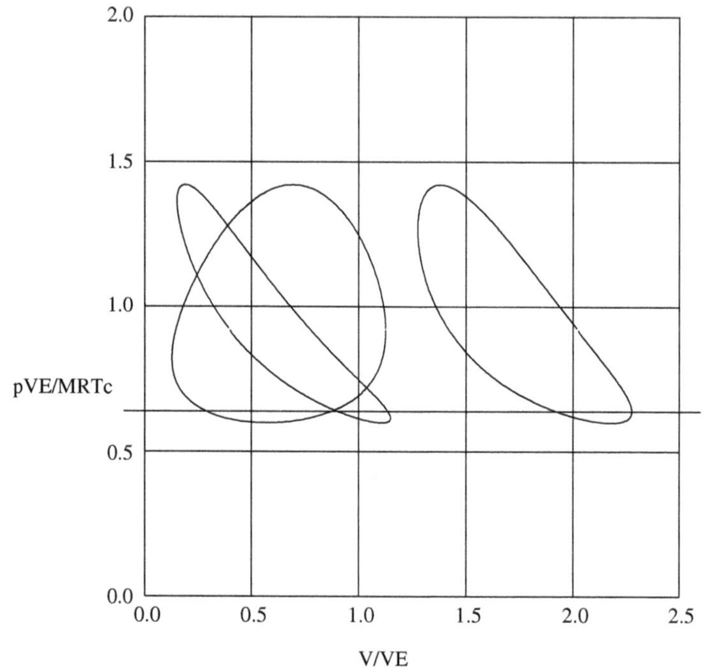

Fig. 2.3. Indicator diagrams generated for GM GPU-3 by 'adiabatic' analysis of Chapter 3.

The Stirling engine provides yet another exception to expectation. The right-hand trace of Fig. 2.3 is the indicator (net *p–V*) diagram generated for the specification of the GM GPU-3 by the 'adiabatic' cycle model (to be explored later). Tracing the diagram clockwise from the outer dead-centre position reveals that one-third of the initial volume reduction (which might be thought of as one-third of the compression stroke) is expended in merely restoring pressure to the value to which it had fallen at the outer dead-centre point.

The feature is readily explained: a substantial part of net pressure excursion is due to the transfer phases, which occur at nominally constant volume. The face-value interpretation remains: the feature is not consistent with maximum indicated work from a given working fluid charge.

A further expectation is that heat *input* to a power cycle occurs *after* compression and *before or during* expansion. Once again the Stirling defies expectation: Fig. 2.4 is the net *p–V* loop for the GPU-3 — again

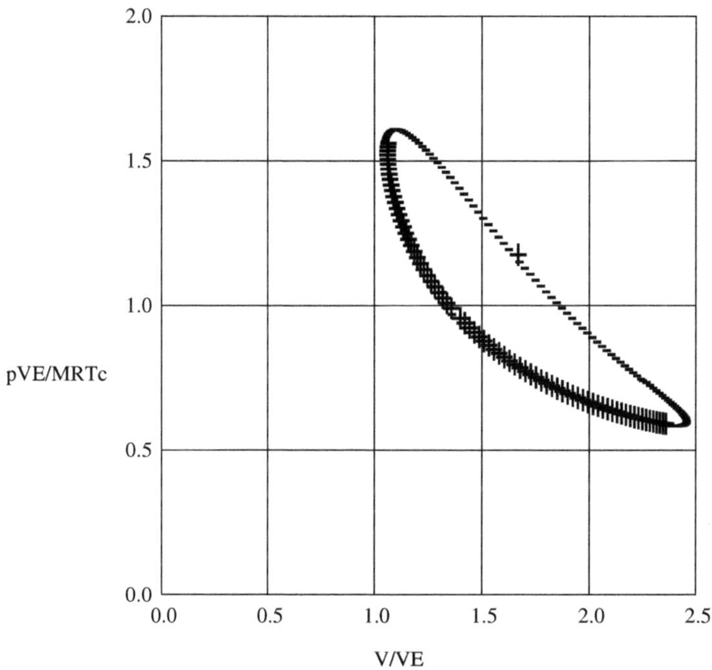

Fig. 2.4. A plus sign on the indicator diagram for the GPU-3 indicates where the adiabatic analysis considers that the expansion exchanger is taking in heat, a minus where it is rejecting.

generated by the adiabatic analysis, but this time with a plus sign (+) superimposed when heat is flowing *via* the expansion exchanger *into* the cycle, minus (–) when the expansion exchanger is rejecting heat. According to the simulation, heat *input* is predominantly during volume *decrease* phase — normally thought of as compression.

Instantaneous vs net exchanger heat flux: Every part of gas path — not just the regenerator — experiences a forward flow of heat once per cycle, and a flow in the reverse direction — also once per cycle.

The curves of Fig. 2.5 show heat flux rate at expansion and compression exchangers plotted over a cycle. *Net* useful heat flow driving the work-producing cycle is the *relatively small difference* between two relatively large quantities. Net useful heat flow is *a fraction of peak values* in both cases. Relative to the equivalent steady flow, the Stirling engine

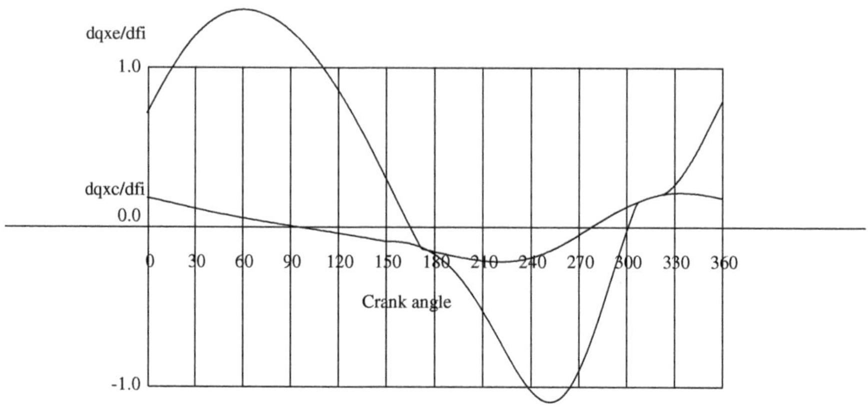

Fig. 2.5. Dimensionless exchanger heat rates. Upper: expansion; lower: compression.

makes far heavier demands on capacity of its exchangers. All things being equal, higher-than-necessary heat flow means higher than necessary pumping penalty.

Cycle temperature limits: The gasoline engine cycles its working fluid charge between temperature limits of ambient (say, 30°C) and about 2000°C. In Stirling engine terms, this is a temperature ratio N_T in excess of 7. The Stirling, by contrast, works between about 60°C and 650°C — an N_T of about 3.

The comparison places all Stirling engines *in the low-delta-T category*.

A heat engine by definition: A basic explanation for the working of a heat engine might be '*compress mass M of an elastic substance at low temperature, re-expand the same M at high temperature*'.

Compared to other reciprocating types, the SE has a massive ratio of clearance volume to swept volume. Volumetric compression ratios do not vary much either side of 2. The Stirling engine is heat engine. *If it doesn't do much compressing then it can't do much expanding*!

This does not, unfortunately, mean that increased compression ratio is the remedy: This writer has wasted a lot of his time — and that of other people — promoting the belief that pressurisation could be traded against compression ratio. This may be the definitive Catch 22 situation: raising compression ratio increases the adiabatic component of temperature swing and loads the heavily-loaded exchangers further to the point where performance falls off rather than improves.

2.7. 'Why Didn't Anybody Notice?'[2]

The aircraft gas-turbine and the engine of the modern automobile routinely perform close to the limits of their potential. The achievement reflects design and development using data and theory of direct relevance to the challenge.

The case of the Stirling engine could hardly be more different: at the centre of a 'first-principles' design approach is the regenerator — for which, however, experimental data of any direct relevance and generality have never been mobilized. The assertion has been made and justified previously. Indeed, steps have been taken to remedy the situation (e.g., Organ 2007, 2014) — although with no tangible effect to date.

A related concern under this heading was flagged two decades ago in '*The great Stirling cycle simulation fraud*' (Organ, 1996). Put briefly, authentic solution of the equations defining the unsteady, compressible gas process interactions of the Stirling gas path are not — and *cannot be* — achieved in the fixed-grid (Euler) coordinate framework routinely used for the purpose. If we provisionally accord the status of a science to Stirling studies, there is no other scientific context in which un-tried, *ad hoc*, do-it-yourself numerical discretization schemes would be tolerated.

But tolerated they are — which brings us back to Her Majesty's question: '*Why didn't anybody notice*?'. The answer is now to hand, and to both aspects — to the matters of irrelevant data and of inappropriate numerical analysis:

> *The abysmal working-fluid utilization obscures any benefit which would result from use of authentic data and/or of respectable numerical schemes.*

So the reason nobody notices is that, to date at least (i.e., in current embodiments of Stirling's cycle) regenerator design *is not critical*: the

[2]HM Queen Elizabeth II, referring to the 2008 trauma in the banking industry. The question was raised at the opening ceremony for the new building at the prestigious London School of Economics — and caught all concerned on the back foot.

practical cycle is *insensitive* to marginal improvement. It can be argued in support that, if it were otherwise, then there would be instances of 'grading' of the matrix — of end-to-end variation of hydraulic radius and/or porosity — to accommodate the considerable differences in flow conditions at the extremities.

A numerical example based on General Motors' GPU-3 will quantify. The engine returned Z_{pmin} at rated operating point of 0.274 [–] (Table 2.1 earlier). Now consider the Schmidt 'engine'. Its regenerator is *perfect*: zero flow loss and 100% thermal recovery. Fed with the specification and operating conditions of the GPU-3, the Schmidt 'engine' returns $Z_{pmin} = 0.439$ [–]. Allowing 85% mechanical efficiency the figure becomes 0.373 [–]. So the percentage improvement[3] relative to Z of the specimen internal combustion engine (Table 2.1) is a paltry 100 × (0.373 – 0.274)/5.47 — or 1.81%! Virtually undetectable!

It may now be clear why there has been no pressure to take a second look at reliance on incompressible, steady-flow data such as those of Kays and London (1964): With a bit of experience and/or background knowledge, regenerator 'design' can be a matter of guesswork.

2.8. The Thermal-Lag Engine in Context

One of several thermal-lag engines constructed by Vaizey (2020) was of 7.6cc swept volume and equipped with multi-channel heat exchangers — slots for the hot exchanger and parallel, 1 mm diameter holes (a total of 123!) for the cold. The engine was designed to accommodate 'dummy' cold exchangers with a view to evaluating the combined effect of 'pressure-lag' with thermal-lag. It will be described in some detail later, but for present purposes peak power was 1.6 W delivered at 1400 rpm with open-flame heating and with the 123-hole exchanger installed. Assuming mean cycle pressure to have been 1 atm., specific cycle work based on $p_{min}V_{sw}$ (Table 2.1) was 0.1128 [–].

Contrary to expectation, perhaps, the figure is comparable with those for Stirling engines. In other words, abysmal by comparison with the internal combustion engine, but remarkable for an engine for which design guidelines are non-existent, and which was equipped with

[3] Any such hypothetical improvement is due in part to further idealisations of the Schmidt 'engine'.

'wire-wool' regenerator of arbitrary wire size and of porosity which was (inevitably) unknown and non-uniform.

Vaizey's output of engines has included miniature gas turbines and has been prolific. All have been to exemplary standards of machining. However, it is safe to conclude that the performance potential of the thermal-lag type has yet to be demonstrated. The ruthless mechanical simplicity promise low manufacturing cost justifying developing a study of the cycle in parallel with that of the Stirling engine.

2.9. Matters Outstanding

This preliminary appraisal has identified a sufficient number of corroborating 'symptoms' to give confidence in the opening diagnosis: that the Stirling engine makes lamentably ineffective use of its working fluid charge M. The need to operate at elevated pressure (i.e., with large ratio of M/V_{sw}) is consistent with this.

The matter is evidently of little concern to operators of stealth submarines with their air-independent Stirling auxiliary power provision. It might, however, stand in the way of other potential applications for this prepossessing concept. What's to be done?'

It is never a bad thing to start by acknowledging reality: *the Stirling engine operates on a cycle in conflict with itself*. The unproductive — and ultimately wasteful — cyclic reversal of heat flows is one of several symptoms, all reflecting a basic problem of *timing* the distribution of working fluid in relation to compression and expansion phases.

Future prospects thus appear to hang on progress towards identifying some straightforward method of confining a much-improved percentage of M to the cold space for compression — and, equally, the majority of M to the hot space for expansion.

A priority will be to shake off dependence on the usual, simple-harmonic volume variations. A mechanism offering the appropriate kinematics has been identified (Chapter 12) and modelled in three dimensions and checked in AutoDesk. The target, however, is a *package* of changes — increased temperature ratio, reduced dead-space and enhanced exchanger performance, the latter made possible by acquisition of flow and heat transfer data pertinent to the unique task.

Stirling engine development over the past two centuries has been evolutionary, but has clearly faltered. It can be got moving again by '*raising the threshold*' from which study, analysis and design take off: The expression has been used before, but now urgently requires turning into action. There is scope for progress on four fronts:

(1) The internal combustion engine — reciprocating and gas turbine — could have reached an advanced stage of development on the assumption of instantaneously-uniform working-fluid properties. Even the crudest model of the Stirling cycle recognises that a thermodynamic event at one point in the gas path affects those at every other point — and not only at the moment, but at every other instant over a cycle.

    It must seem incomprehensible to an outside observer that half-a-century of digital computing has not led to a tool-box of standardised software for thermodynamic design. Given the benefits of interchangeability of research findings it is risible that quantitative comparison is not possible because each researcher improvises her/his own numerical analysis. An idea of what can be done with the aid of internet communication is the GNU open-source software project. It would be contrary to the principle of the GNU model, of Open Linux, etc., for an isolated researcher to inflict a way forward. On the other hand, a move to kick-start this long-overdue undertaking might be appropriate. Chapter 15 takes a nominal step by including an outline of a 'platform-independent' way of generating printable graphical output from within a basic Fortran program.

(2) If computer simulation is to speed development, it must take into account the way the regenerator functions in this unique environment. A factor suppressed by traditional regenerator theory is pressure swing, the influence of which can no longer be ignored.

(3) It is ludicrous that a misleading piece of 19th-century algebra — the Schmidt — should remain the default entry-point for study of the modern engine (and, incidentally, of thermal-lag and pulse tube types with which it has no connection[4]). There are vastly less restricted start points, arguably more approachable than Schmidt. A later chapter will elaborate.

(4) Long overdue is a critical look at the flow data universally relied-upon for regenerator calculations. Compressible flow is not just *subtly* different from incompressible: it is *fundamentally* different in that it embodies the possibility of *choking*. Data from tests below the compressibility threshold *cannot be used to explore the potential*

[4] If these complied with the Schmidt assumptions they would not function at all.

performance envelope of the air-charged engine. Doing so would be equivalent to designing the wing spar of an aircraft on the basis of static yield to the neglect of fatigue. Conversely, use of correlations derived from compressible flow permits *designing around* choking conditions — with benefits which do not need to be spelled out.

2.10. 'Beale Number' of Turbo-Fan Engine

Rolls-Royce Trent 1000-R

Rated thrust:	81,000 lbf/360 kN
Pressure ratio:	10:1
Dry weight:	6000 kg
Bypass ratio:	10:1
Air mass flow rate:	1100 kg/s
(i.e., engine core mass flow rate:	110 kg/s)

Airbus A-380

Max. take-off weight: 560 tonne
lift/drag ratio (presumably 'clean'): 20:1 *Sea-level power at 350 kt*:

$$u = 177 \text{ m/s}$$
$$Fu = 560 \times 10^3 \times 9.81 \times 0.05 \times 177/1000 \text{ kw}$$
$$= 48{,}600/4 \text{ kW/engine} = 12{,}150 \text{ kW/engine}$$

Fuel load: 200,000 kg – burned over 20 hours between 4 engines
$200{,}000/(20 \times 3600 \times 4) = 0.69$ kg/s per engine
@45MJ/kg = 31×106 J/s heat input rate
@ 40% thermal efficiency = $12{,}500 \times 10^3$ kW/engine

Max. thrust × velocity (say, shortly after take-off):

$$360 \text{ kN} \times 150 \text{ kt} \times 1.15 \text{ kt/mph} \times 8/5 \text{ km/hr/mph} \times 1/3600 \text{ km/s/km/hr} = 27{,}600 \text{ kW}$$

3

Raising the Threshold

3.1. Motivation

By tradition, the start-point for exploring the gas processes of the Stirling engine is the Schmidt analysis. The attraction usually cited is that it offers explicit algebraic expressions for indicated work W and for variation of pressure p with crank angle φ.

Not previously remarked is that it confirms the factor (25–50) by which specific cycle work in the Stirling engine falls short of that of the competition (I.C.E.). If cycle studies had focused on the latter feature rather than the first, they would surely have taken a different direction.

The Schmidt expressions are simple only because someone else has already done the algebra. If cycle pressure history, mass distribution and heat transfer phases are to be displayed, Schmidt still requires coding for digital processing. (Resulting heat rates will hopelessly misrepresent reality.) Schmidt is an analysis of an engine which *already works*: *It must not be used in reverse for design*, as it is insensitive to rpm.

An 'adiabatic' development may actually be easier to follow, since it does not require skill in manipulating trigonometric functions.

The present exercise is not the first attempt to gain acceptance for such a replacement: The adiabatic assumption is a limiting case of Finkelstein's *Generalized thermodynamic analysis* of 1960. His algebraic *tour de force*, however, differs from the present approach in two respects: Finkelstein specified clearance volume to be zero at inner dead-centre positions. The simple-harmonic volume variations thus passed through zero ($\mathrm{d}V/V = \mathrm{d}m/m = 0/0$), calling for extensive supplementary algebra to work around the singularities.

He chose not pursue opportunities afforded by algebraic abbreviation and substitution. The choice cost him dearly in terms of computational effort: '... *the solutions are very sensitive near the discontinuities* (i.e., *singularities*) *so that very small computational steps and great numerical accuracy are required. One single solution corresponds to about six weeks' work on an electric desk machine, which warrants the use of an electronic computer*'. (In hindsight, an under-statement — but this was 1960.)

Given the computational labour, his choice of worked example is bizarre — a specification (numerical values of κ, α, v) '... *typical of many old engines of symmetrical design* ...' and working on air — but as a heat-pump with temperature ratio $T_E/T_C = ½$. Interpretation is not helped by a schematic illustration suggesting a modern configuration, nor by the change in notation for $\varepsilon = T_C/T_E = 1/N_T$ between his nomenclature and the tabulated specification of the 'engine'.

Electronic digital computing was well-established by the time (1968) Qvale and Smith (1968) tackled the same cycle description, although their advantage over Finkelstein was not exploited.

Walker (1980) reports that Lee's re-visiting (1976) of the Finkelstein algebra resulted in a '... *fast, well-behaved computer program in the Fortran IV language*.' Temperature solutions illustrated by Walker are not informative, and the original is inaccessible.

Urieli and Berchowitz (1984) address the adiabatic problem *via* an independent formulation and numerical integration using a Runge–Kutta scheme. Specimen results reflect input data for the GPU-3 — including volume variations deriving from the kinematics of the rhombic drive. Relating the resulting solutions to present findings and those of Finkelstein requires account to be taken of the 240 degrees or so between respective crankshaft datum positions.

An adiabatic formulation by Rios (1969) was acquired wholesale by Martini (1983) in the form of a copy deck of Hollerith punched cards. Like the Finkelstein prototype, it addressed the reversed cycle machine. By the time Martini had tweaked and annotated the Fortran code it extended to 710 lines, i.e., to that number of punched cards. (This compares to the 180 or so lines of Fortran coding in the formulation offered below.)

3.2. Defining Equations

With two exceptions, the starting assumptions are those of the Schmidt analysis (1871). The crucial exception is that heat exchange in the variable-volume spaces is zero. Temperatures T_e and T_c of cylinder contents m_e and m_c thus vary with crank angle φ, but uniformity is ensured by instantaneous and complete mixing. A minor departure is that, while volumes V_e and V_e of expansion and compression spaces still vary as simple-harmonic functions, clearance volume is included:

$$V_e = \tfrac{1}{2}V_E(1 + \cos\varphi) + V_{de},$$
$$V_c = \tfrac{1}{2}\kappa V_E(1 + \cos(\varphi - \alpha)) + V_{dc}.$$

Not only is this more realistic (clearance volume of a real machine is *never* zero!), but the threat of indeterminacy at inner dead-centre which taxed Finkelstein is avoided.

With this provision, the formulation can also handle any alternative functions $V_e(\varphi)$, $V_c(\varphi)$ because derivatives $dV_e(\varphi)/d\varphi$, etc. can be evaluated numerically as $[V_e(\varphi + \tfrac{1}{2}d\varphi) - V_e(\varphi - \tfrac{1}{2}d\varphi)]/d\varphi$, etc.

The ideal gas law is applied to either working space both before and after an increment $d\varphi$ of crankshaft rotation during which mass dm enters or leaves:

Before: $pV = mRT$,
After: $(p + dp)(V + dV) = R(m + dm)(T + dT)$.

Subtracting the former from the latter gives, for either space:

$$pdV + Vdp = R(mdT + Tdm) \qquad \{-dpdV + RdmdT\}. \tag{3.1}$$

Retention of higher-order term $-dpdV + RdmdT$ in Eq. (3.1) by iteration achieves the accuracy needed to prevent the solution 'drifting' over successive cycles.

Under the adiabatic assumption the energy equation applied to a variable-volume space is

$$-dW = du - dh.$$

Work increment $\mathrm{d}W$ is $p\mathrm{d}V$. Enthalpy increment $\mathrm{d}h$ is $\mathrm{d}mc_pT$ during outflow and $\mathrm{d}mc_pT_x$ during inflow, where T_x is exchanger wall temperature. Using $T|T_x$ to denote the alternatives:

$$\mathrm{d}mc_pT|T_x = c_v(m\mathrm{d}T + T\mathrm{d}m) + p\mathrm{d}V, \qquad \{+ c_v\mathrm{d}m\mathrm{d}T\}. \tag{3.2}$$

The higher-order term in parentheses succumbs to iteration. Substituting the expanded gas law (Eq. (3.1)) into Eq. (3.2):

$$mc_p\,\mathrm{d}T = V\mathrm{d}p + \text{higher-order terms (outflow)}, \tag{3.3a}$$

$$mc_p\,\mathrm{d}T - c_p\,\mathrm{d}m(T_x - T) = V\mathrm{d}p + \text{higher-order terms (inflow)}. \tag{3.3b}$$

With m_d for mass instantaneously occupying the spaces not swept by either piston:

$$m_e + m_c + m_d = M. \tag{3.4}$$

From Eq. (3.4):

$$\mathrm{d}m_e + \mathrm{d}m_c + \mathrm{d}m_d = 0. \tag{3.5}$$

By assumption, pressure p at any instant is common to the entire gas path. Corresponding to each increment of crankshaft rotation there are accordingly five unknowns: $\mathrm{d}m_e$, $\mathrm{d}m_c$, $\mathrm{d}T_e$, $\mathrm{d}T_c$ and $\mathrm{d}p$. There are two equations of type (3.2), one for each variable volume space, and likewise two Eqs. (3.3) – a or b depending as $\mathrm{d}m$ is positive or negative), i.e., five equations in all.

The final section of this chapter shows how the five are initially reduced to four by eliminating $\mathrm{d}p$ in favour of $\mathrm{d}T_e$, $\mathrm{d}T_c$, $\mathrm{d}m_e$ and $\mathrm{d}m_c$, then further reduced to two by expressing $\mathrm{d}T_e$ and $\mathrm{d}T_c$ in terms of $\mathrm{d}m_e$ and $\mathrm{d}m_c$, finally reaching an explicit expression for each of $\mathrm{d}m_c$ and $\mathrm{d}m_e$ in terms of independent variable crank-angle φ. The $\mathrm{d}p$, $\mathrm{d}T_e$ and $\mathrm{d}T_c$ each follow in a single step by back substitution. The values are used to update p, m_e, m_c etc. by simple addition, as $p = p + \mathrm{d}p$, $m_e = m_e + dm_e$, etc.

Instantaneous heat exchanger loading follows by application of the energy equation to fixed volumes V_{dx}, viz, $\mathrm{d}q_x = \mathrm{d}u_x - \mathrm{d}h_x$.

For flow from exchanger *into* adjacent cylinder ($\mathrm{d}m > 0$)

$$\mathrm{d}q_x = c_v\mathrm{d}m_xT_x - c_p\mathrm{d}m_xT_x.$$

Change in mass content of the exchanger $\mathrm{d}m_x$ follows by differentiating the ideal gas law applied to the fixed volume, giving $\mathrm{d}m_x = V_{dx}\mathrm{d}p/RT_x$. Recalling that $c_p - c_v = R$:

$$\mathrm{d}q_x = -V_{dx}\mathrm{d}p \qquad (\mathrm{d}m > 0). \tag{3.6}$$

When flow is outward from the cylinder ($\mathrm{d}m < 0$), $\mathrm{d}q_x$ is reduced by any excess of entering enthalpy corresponding to $T > T_x$ viz, by $(-\mathrm{d}m)c_p(T - T_x)$:

$$\mathrm{d}q_x = -V_x\mathrm{d}p - (-\mathrm{d}m)c_p(T - T_x)\ (\mathrm{d}m < 0). \tag{3.7}$$

3.3. Evidence of Appropriate Functioning

The analysis has been coded in Fortran and exercised on the data of Table 3.1. Replacing isentropic index γ by unity should give results indistinguishable from those of the Schmidt analysis applied to the same data. This is confirmed: working-space temperatures T_e/T_C and T_c/T_C are constant over the cycle at N_T and at unity, respectively. Net cycle heat fluxes for hot and cold exchangers, Q_{xe} and Q_{xc}, are zero to within fractions of a percent. Conversely, increments in working-space heat exchange $\mathrm{d}q_e$ and $\mathrm{d}q_c$ sum over a cycle to Q_E and Q_C, respectively, such

Table 3.1. Gas path specification.

Temperature ratio $N_T = T_E/T_C$	3.0
Swept-volume ratio $\kappa = V_C/V_E$	1.0
Volume phase angle α	90 deg
Clearance-volume and dead-volume ratios:	
expansion space clearance $\mu_{de} = V_{de}/V_E$ (variable)	0.125
expansion exchanger $\mu_{dxe} = V_{dxe}/V_E$	0.092
regenerator $\mu_{dr} = V_{dr}/V_E$	0.316
compression exchanger $\mu_{dxc} = V_{dxc}/V_E$	0.092
compression space clearance $\mu_{dc} = V_{dc}/V_E$ (variable)	0.125
Isentropic index $\gamma = c_p/c_v$	1.4

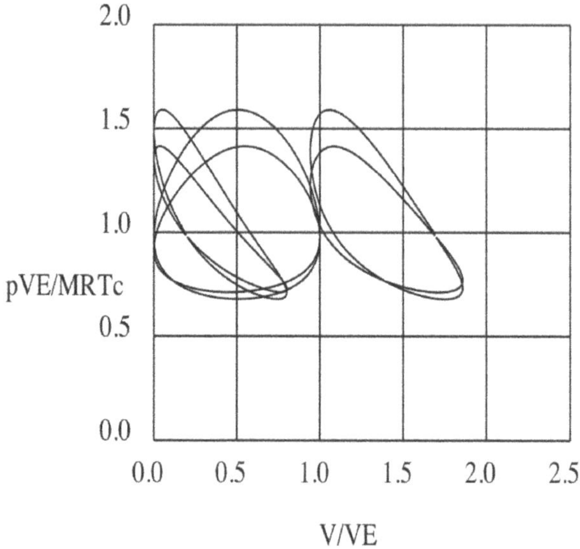

Fig. 3.1. *p-V* diagram comparison. Loops of lesser amplitude generated by Schmidt analysis fed with gas-path specification of Table 3.1.

that $W = Q_E + Q_C$ (NB Q_C negative). With γ retained at unity, the variation in (uniform) pressure *p* superimposes over the Schmidt counterpart.

Figure 3.1 is the statutory set of *p-V* loops, here shown superimposed over corresponding loops from the Schmidt analysis fed with the same data. As might be anticipated, the adiabatic assumption gives a wider pressure excursion than the isothermal.

3.4. Specimen Results

Figure 3.2 is generated from the data of Table 3.1. It appears cluttered, but the super-position will lead to an insight which appears not to have emerged from previous studies. It is potentially controversial, so it is worth keeping in mind that this chapter recounts what the present, specific adiabatic formulation predicts — not necessarily what intuition, or experience, or this writer, or another formulation might anticipate.

The uppermost curve is expansion-space temperature T_e.

The majority of heat inflow *via* the expansion exchanger dq_{xe} is seen to be in phase with *decreasing* $V_e + V_c$ — i.e., during *compression*. This is

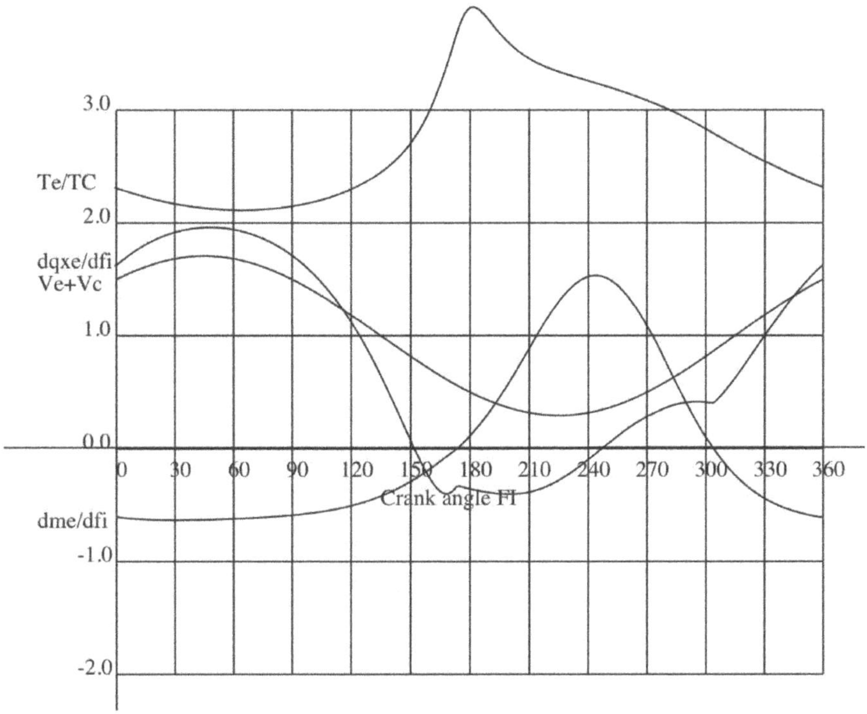

Fig. 3.2. Cyclic variation of temperature, mass rate and exchanger heat flux for expansion space superimposed over instantaneous volume total $V_e + V_c$.

nevertheless consistent with outflow ($dm_e/d\varphi$ negative) in combination with T_e below exchanger temperature T_E over the same interval of crank-shaft rotation.

Urieli and Berchowitz (1984 — their Figs. 4.2, 4.4 and 4.5) also have dq_{xe} peaking during outflow ($dm_{xe} < 0$) and with T_e/T_E close to minimum — just as in Fig. 3.2 above. In other words, heat flow *into* the expansion *exchanger* peaks during mass *outflow.*

Taking into account the angular phase-shift between the present plots and theirs, their curve of dq_{xe} in particular matches that of Fig. 3.3, lacking only the sharp discontinuities of the latter.

Walker (1980) notes from Lee's graphs (1976) that $T_e(\varphi)$ '*... was less than the heater temperature for much of the time.*' Unfortunately no plot of $dV/d\varphi$ is available to allow confirmation.

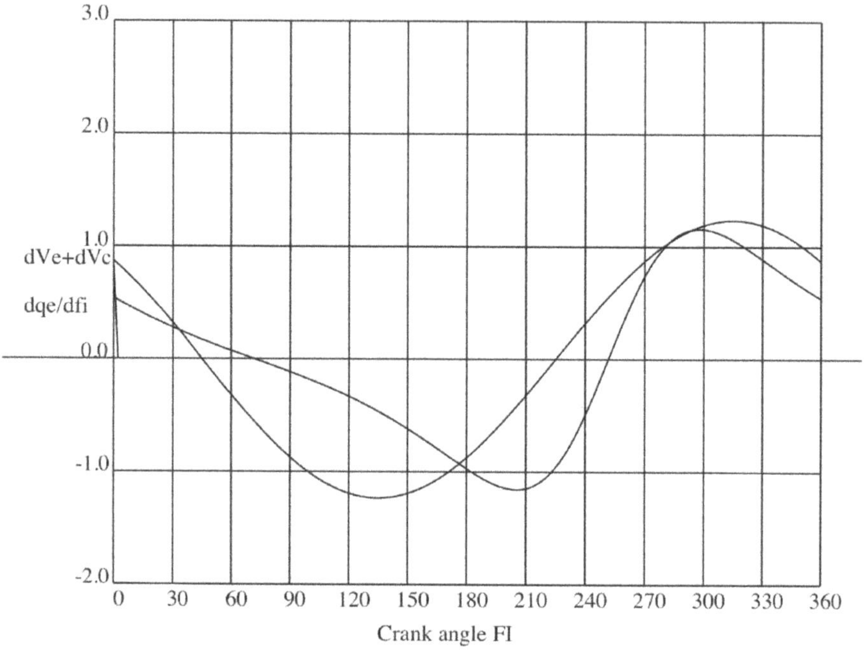

Fig. 3.3. Schmidt analysis fed with the data of Table 3.1 has heat addition to the hot space substantially in phase with expansion ($dV_e + dV_c > 0$).

It is an expectation that a heat engine cycle should function by adding heat during and/or shortly before the expansion phase (having rejected heat during and/or immediately before the compression phase). The irrelevant 'ideal' or 'text-book' Stirling cycle achieves just this, but only because it envisages the entire working fluid mass M at T_E for expansion and at T_C for compression. Schmidt at least acknowledges that M is distributed throughout the entire gas path, but still has $dq_e/d\varphi$ more or less in phase with expansion ($dV_e + dV_c > 0$) — Fig. 3.3.

The adiabatic analysis thus suggests a fundamental distinction between the Stirling and other power cycles: The usual picture whereby increasing pressure accompanies decreasing volume does not apply. Neither, on present findings, do heat reception and rejection lie in phase with volume increase and decrease respectively. Figure 3.4 illustrates. Uniquely in the Stirling, a major contribution to the rise and fall in

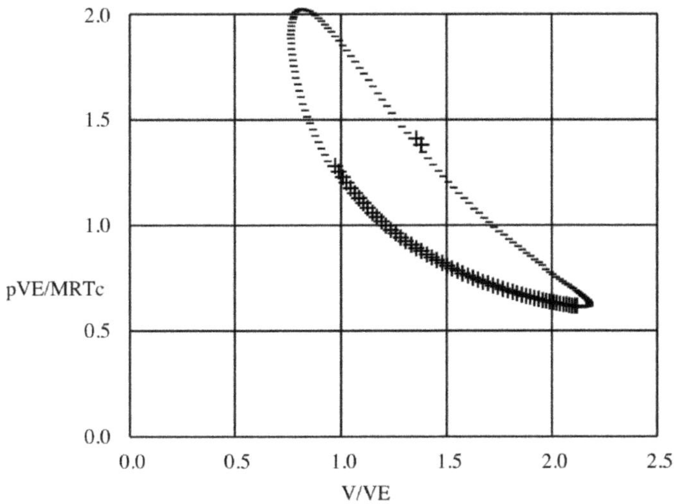

Fig. 3.4. Heat reception at expansion exchanger marked by plus + and rejection by minus –.

pressure is from the transfer phases during which mechanical compression and expansion are intentionally minimal.

The Schmidt analysis is a triumph of elegance over relevance. The built-in assumption of infinite heat transfer capability in the variable-volume spaces turns reality on its head and gives a thoroughly misleading picture of the function of crucial gas-path elements — the exchangers.

The fact that it offers algebraic expressions for per-cycle quantities W, Q_E and Q_C is no justification for continued reliance on a resource which is unhelpful for first-principles design: The reality is that the variable-volume spaces are essentially adiabatic (Organ, 2014). This has called for a different analytical approach — but one which has turned out to be more transparent than the Schmidt. When it comes to exploring the gas processes, it is no more demanding of coding.

The adiabatic formulation leads to a fundamental distinction between the 'real' Stirling engine and other idealised cycle models: operation of the Stirling goes against expectation that heat rejection is induced by decreasing volume, heat reception by the reverse.

3.5. Extension to the *Generalised Thermodynamic Analysis*

With the solution algorithm established, extension to Finkelstein's pioneering cycle model is achieved in a single, elementary step. The equation for the energy increment of the variable-volume space is enhanced by heat exchange term dq to become $dq - pdV = du - dh$. In terms of instantaneous wetted area A_w, coefficient of convective heat transfer h and enclosure temperature T_w (assumed uniform and constant):

$$dq = hA_w(T_w - T)dt. \tag{3.8}$$

Equation (3.8) is substituted into Eq. (3.2) and the substitution carried forward to Eqs. (3.3), where insight is from normalisation: this yields terms $hA_w(T_w/T_C - T/T_C)d\varphi/\omega Mc_p$, one for each working space. The group $hA_w/\omega Mc_p$ is dimensionless and is evidently a Stanton number. With the appropriate respective cyclic variation of A_w in each case (expansion and compression) the term is simply added wholesale to the right-hand side of Eqs. (3.3), adding to coefficients b_3 and b_4 in the solution matrix.

Having referred to the Stirling machine exclusively as a prime mover, Finkelstein unaccountably illustrates in terms of operation in reversed-cycle mode — as a heat pump. Attempting to replicate his results involves reversing direction of rotation to maintain the expansion space at the higher temperature.

For Fig. 3.5, the cycle variation of expansion- and compression-space temperatures T_e/T_C and T_c/T_C is plotted against crank angle φ. The parameter is Stanton number, which covers the range of Finkelstein's own graphs, viz from 0 (adiabatic) via 0.1, 0.5 and 1.0 to infinity. The curves are not labelled, being identified by the fact that the largest excursions correspond to the lowest values of 'Stanton' number. Specifying small unswept volumes ($0.015V_E$) has gone some way to mimicking Finkelstein's use of zero. Taking account of the different vertical scales, respective sets of curves for T_e/T_C vs φ share the same idiosyncrasies — the 'blip' close to inner dead-centre for the adiabatic case and the feature whereby all curves cross twice per cycle.

The family resemblance is not matched in the curves for the compression space — surprising considering the symmetry of the algebraic equation sets and of the corresponding matrices (Section 3.6). The matter remains under investigation.

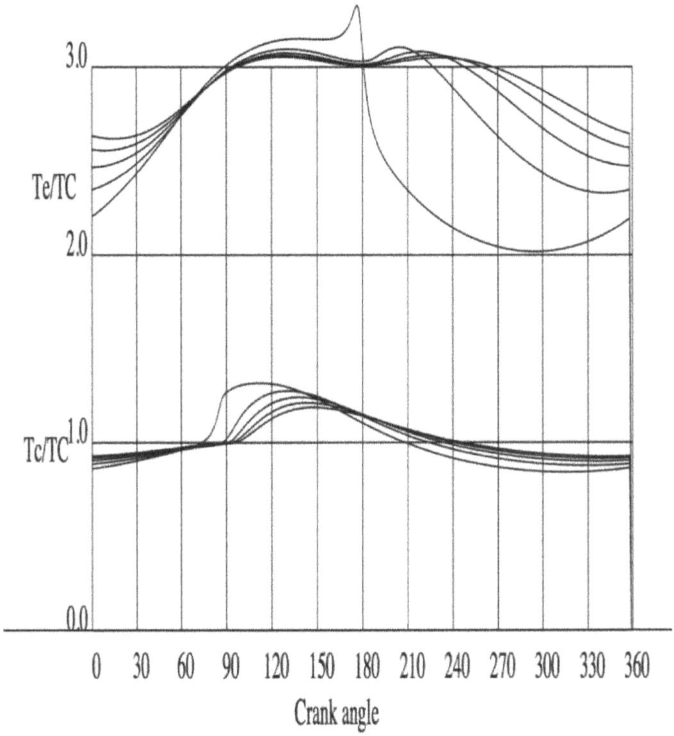

Fig. 3.5. Working space temperature histories as a function of Finkelstein's 'Stanton' number. (Key: largest excursion, smallest Stanton number.)

3.6. Algebraic Reduction Prior to Coding

The development picks up at Eq. (3.5) of the main text, applying the ideal gas law to individual spaces at respective temperatures:

$$m_{\mathrm{d}} = p/R\{(V_{\mathrm{de}} + V_{\mathrm{dxe}})/T_{\mathrm{E}} + V_{\mathrm{dr}}/T_{\mathrm{C}}\log_e N_{\mathrm{T}}/(N_{\mathrm{T}} - 1) + (V_{\mathrm{dxc}} + V_{\mathrm{dc}})/T_{\mathrm{C}}\}$$
$$= (pV_{\mathrm{E}}/RT_{\mathrm{C}})\upsilon,$$
$$\upsilon = (V_{\mathrm{de}} + V_{\mathrm{dxe}})/N_{\mathrm{T}}V_{\mathrm{E}} + V_{\mathrm{dr}}/V_{\mathrm{E}}\log_e N_{\mathrm{T}}/(N_{\mathrm{T}} - 1) + (V_{\mathrm{dxc}} + V_{\mathrm{dc}})/V_{\mathrm{E}}. \quad (3.9)$$

Differentiating Eq. (3.9) gives

$$m_{\mathrm{d}} = (\upsilon V_{\mathrm{E}}/RT_{\mathrm{C}})\mathrm{d}p.$$

Using the result to eliminate dp from the earlier Eq. (3.5) leaves four equations in four unknowns:

$$a_{1,1}\mathrm{d}m_\mathrm{e} + a_{1,2}\mathrm{d}m_\mathrm{c} + a_{1,3}\mathrm{d}T_\mathrm{e} + 0 = b_1,$$
$$a_{2,1}\mathrm{d}m_\mathrm{e} + a_{2,2}\mathrm{d}m_\mathrm{c} + 0 + a_{2,4}\mathrm{d}T_\mathrm{c} = b_2,$$
$$a_{3,1}\mathrm{d}m_\mathrm{e} + 0 + a_{3,3}\mathrm{d}T_\mathrm{e} + 0 = b_3,$$
$$0 + a_{4,2}\mathrm{d}m_\mathrm{c} + 0 + a_{4,4}\mathrm{d}T_\mathrm{c} = b_4.$$

From row 3, dT_e = $(b_3 - a_{3,1}\mathrm{d}m_\mathrm{e})/a_{3,3}$ and can be eliminated from row 1. From row 4, dT_c = $(b_4 - a_{4,2}\mathrm{d}m_\mathrm{c})/a_{4,4}$ and can be eliminated from row 2. The result is two equations simultaneous in dm_e and dm_c:

$$a'_{1,1}\mathrm{d}m_\mathrm{e} \quad a'_{1,2}\mathrm{d}m_\mathrm{c} = b'_1,$$
$$a'_{2,1}\mathrm{d}m_\mathrm{e} \quad a'_{2,2}\mathrm{d}m_\mathrm{c} = b'_2,$$

in which

$$a'_{1,1} = a_{1,1} - a_{1,3}a_{3,1}/a_{3,3},$$
$$a'_{1,2} = a_{1,2} - a_{1,3}a_{3,2}/a_{3,3},$$
$$b'_1 = b_1 - a_{1,3}b_3/a_{3,3},$$
$$a'_{2,1} = a_{2,1} - a_{2,4}a_{4,1}/a_{4,4},$$
$$a'_{2,2} = a_{2,2} - a_{2,4}a_{4,2}/a_{4,4},$$
$$b'_2 = b_2 - a_{2,4}b_4/a_{4,4}.$$

Coding for solution is in terms of dimensionless variables. Finkelstein's system (1960b) is followed closely:

$$\psi = pV_\mathrm{E}/MRT_\mathrm{C} \text{ (Finkelstein defines } \psi = \tfrac{1}{2}pV_\mathrm{E}/MRT_\mathrm{C}\text{)},$$
$$\sigma = m/M,$$
$$\tau = T/T_\mathrm{C}.$$

With Δ to denote the determinant of the coefficients:

$$\Delta = a'_{1,1}a'_{2,2} - a'_{2,1}a'_{1,2},$$

$$\mathrm{d}\sigma_{\mathrm{e}} = (b'_1 a'_{2,2} - b'_2 a'_{1,2})/\Delta,$$
$$\mathrm{d}\sigma_{\mathrm{c}} = (a'_{1,1} b'_2 - a'_{2,1} b'_1)/\Delta.$$

From Eq. (3.5)

$$\mathrm{d}\psi = -\,(\mathrm{d}\sigma_{\mathrm{e}} + \mathrm{d}\sigma_{\mathrm{c}})/\upsilon.$$

From Eq. (3.1)

$$\mathrm{d}\tau_{\mathrm{e}} = -\,(\psi \mathrm{d}\mu_{\mathrm{e}} + \mu_{\mathrm{e}} \mathrm{d}\psi)/\sigma_{\mathrm{e}},$$
$$\mathrm{d}\tau_{\mathrm{c}} = -(\psi \mathrm{d}\mu_{\mathrm{c}} + \mu_{\mathrm{c}} \mathrm{d}\psi)/\sigma_{\mathrm{c}}.$$

4

Raising the Threshold Further

4.1. Scope

Graduating from isothermal to adiabatic has failed to address a serious limitation: both formulations are independent of time, and might be better described as *thermostatic* than thermodynamic.

Heat transfer in its relevant forms — conduction, convection — takes time. And as a later chapter will demonstrate, it is also necessary to deal with unsteady cyclically-reversing flow in the regenerator in terms of impedance rather than of steady-flow resistance.

This means that neither the isothermal nor the adiabatic formulation can be used unconditionally for first-principles design. Conversely, both offer scope for probing the inner workings *of an already developed engine*. This, taken in conjunction with the relative algebraic simplicity, justifies keeping them in the repertoire.

This book was undertaken partly as a result of disconcerting insights from the *fluid particle trajectory map*. Provided such a map is generated from data defining an engine *known in advance to be viable*, it can hardly be grossly in error. This justifies presentation here of a simplified approach to map construction.

4.2. Constructing the Map

The gas path has been illustrated so many times 'straightened-out' into a horizontal, opposed-piston ('alpha') layout that the picture is taken for granted here. However, Fig. 4.1 represents a point in the gas path where

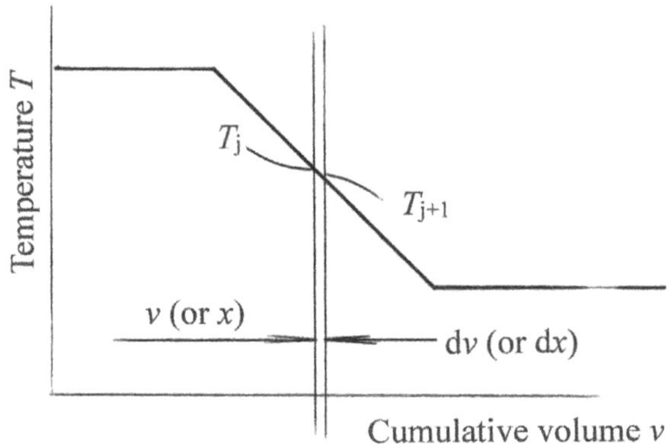

Fig. 4.1. Working fluid element d*m* between adjacent particle paths, indicating representative temperature distribution.

the temperature across the planar gas element — assumed thin — varies linearly between T_j and T_{j+1}. Working fluid occupies *volume* rather than length, so the horizontal coordinate is cumulative volume V rather than linear distance, x. Conversion of temperature gradient between dT/dx and dT/dV is elementary.

From the ideal gas equation, mass dm_j contained between the plane faces is:

$$dm_j = \frac{p}{RT_{j+1}} \frac{\log_e\left(T_j/T_{j+1}\right)}{T_j/T_{j+1} - 1}\left(V_{j+1} - V_j\right). \tag{4.1}$$

Under the 'isothermal' assumptions local fluid temperature T_{gj} is equal at all locations to the temperature T_w of the adjacent enclosing surface. Calculation can therefore proceed in terms of T_w. The latter varies across the regenerator with linear gradient (invariant):

$$dT_w/dV = (T_C - T_E)/V_{dr}.$$

Here two adjacent values of cumulative volume V_{j+1} and V_j enclose gas element dm_j:

$$T_{wj+1} = T_{wj} + (dT_w/dV)(V_{j+1} - V_j), \tag{4.2}$$

where V is cumulative volume reckoned from the left-hand entry to the regenerator.

Working fluid mass M is divided into n sub-masses $\mathrm{d}m_j$. An option is to have all $\mathrm{d}m_j$ equal, but this is not essential. One approach to map construction is to fix crank position and to proceed from left to right computing the location of each successive mass boundary j + 1 in terms of that of preceding boundary j.

On reaching the right-hand piston face, the crank-shaft is advanced by a small angular amount, the new pressure p calculated corresponding to the new piston-face position and the earlier process re-started.

Combining Eqs. (4.1) and (4.2):

$$\mathrm{d}m_j\{T_{g,j} + (\mathrm{d}T_w/\mathrm{d}V)(V_{j+1} - V_j)\} = pR^{-1}(V_{j+1} - V_j)\log_e N_{Tw}/(N_{Tw} - 1),$$

$$N_{Tw} = T_{wj}/T_{wj+1}.$$

Assembling like terms:

$$V_{j+1} = V_j + \frac{\mathrm{d}m_j T_{wj}}{pR^{-1}\mathrm{flog}(N_{Tw}) - \mathrm{d}m_j(\mathrm{d}T_w/\mathrm{d}V)}, \tag{4.3}$$

$$\mathrm{flog}(N_{Tw}) = (\log_e N_{Tw})/(N_{Tw} - 1).$$

Equation (4.3) is not explicit in V_{j+1} because T_{wj} is a function of V_j, *viz* $T_{wj}(V_j)$ and likewise $T_{wj+1}(V_j)$. However, under iteration Eq. (4.3) converges rapidly.

4.3. Specimen Maps

In the map of Fig. 4.2, the upper numerals are labels for vertical lines outlining individual gas-paths elements. So the expansion exchanger extends from 2 to 3. The regenerator lies between 3 and 4, the compression exchanger between 4 and 5. In order that all maps plot to comparable physical size, cumulative volume (horizontal coordinate) is normalised by swept volume V_{sw}.

The specimen is constructed for the gas-path specification of General Motors' GPU-3 engine, but provisionally using the equivalent volume ratio κ and phase angle α rather than volume variations generated by the rhombic drive.

The assumption of uniform $\mathrm{d}m_j$ means the distribution of working fluid at any crank angle can be estimated by counting particle paths. Expansion starts at crank-angle $\phi = 180°$ as plotted, and it is obvious without counting that the mass at T_C (between 4 and the right-hand piston face)

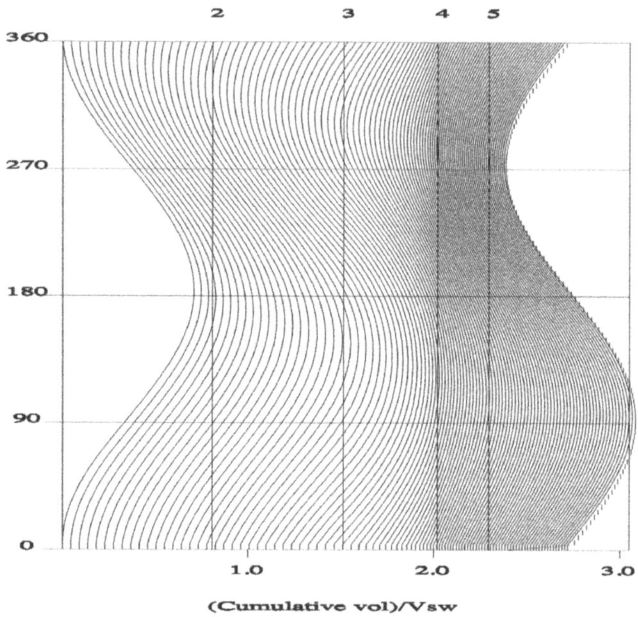

Fig. 4.2. Trajectory map for the gas-path specification of the GPU-3.

remains considerably in excess of that at T_E (between left piston face and 3) for the duration.

Where the individual elements of the gas path (expansion exchanger, regenerator, compression exchanger) are of uniform free-flow area,[1] no additional insight is gained by re-expanding the map from the cumulative-volume coordinate back to physical length x, since a particle located, say, half-way through the regenerator is at the same half-way point in either coordinate system.

The construction method is not tied to any particular volume variations, and that specified for Fig. 4.3 is discontinuous. It is obvious that the volume history sometimes perversely referred to as the 'ideal' has done little to improve the poor working fluid utilisation achieved by simple-harmonic motion.

Previously unexplored insight from particle trajectories is illustrated by Fig. 4.4, which examines the possibility of T_E in the

[1]As is almost invariably the case.

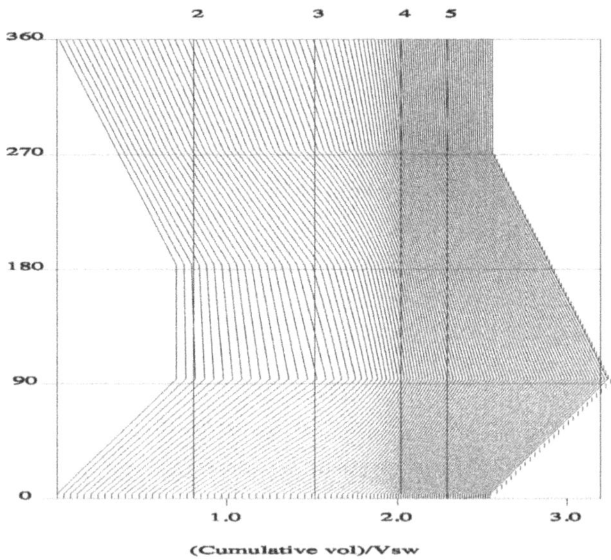

Fig. 4.3. Volume variations discontinuous corresponding to the inappropriately-named 'ideal' cycle.

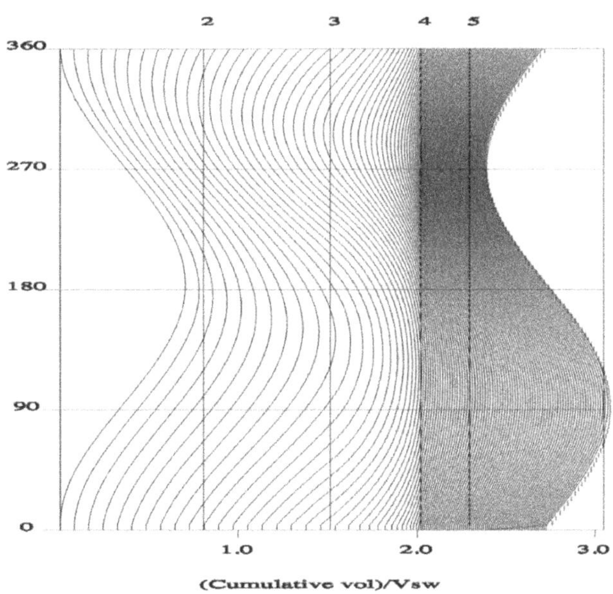

Fig. 4.4. Gas path specification unchanged from that of Fig. 4.2, but temperature ratio N_T increased to reflect $T_E = 2000°C$ of solar heating study.

internally-focusing, solar-heated variant increased to the 2200°C or so of the internal-combustion engine. If the solar version embodies conventional gas-path proportions and volume variations, then the expectation of increased in thermal efficiency and specific power is going to be thwarted by the fact working fluid distribution is even less favourable (fraction of mass at T_C during expansion even greater!).

The observation has implications for *scaling*, a tool of potential use in first-principles gas-path design. This derives from the principle of Dynamic Similarity. When applied to the Stirling engine, this has, to date, been restricted to temperature ratios N_T close to the conventional value of 3.

4.4. Extending the Scope of Scaling

A step in established methods (Organ, 1997, 2014) is to achieve a distribution of working fluid mass in the 'derivative' design which is the same over a cycle as that in the 'prototype'. A precondition has previously been that the two should share a common value of ratio N_T.

According to Finkelstein's 1960 re-working of the Schmidt analysis, pressure variation over a cycle is the following function of the cycle minimum p_{min}:

$$p = p_{min}\frac{1+\rho}{1+\rho\cos(\phi-\theta)}, \tag{4.4}$$

$$\rho = \frac{\sqrt{(1/N_T^2 + \kappa^2 + 2\kappa\cos\alpha/N_T)}}{1/N_T + \kappa + \upsilon}, \tag{4.5}$$

$$\tan\theta = \frac{\kappa\sin\alpha}{1/N_T + \kappa\cos\alpha}. \tag{4.6}$$

Numerator and denominator of Eqs. (4.5) and (4.6) can be multiplied through by temperature ratio N_T without affecting the outcome:

$$\rho = \frac{\sqrt{[1 + (\kappa N_T)^2 + 2\kappa N_T\cos\alpha]}}{1 + \kappa N_T + \upsilon N_T}, \tag{4.5'}$$

$$\tan\theta = \frac{\kappa\, N_T \sin\alpha}{1 + \kappa\, N_T \cos\alpha}. \tag{4.6$'$}$$

All engines having the same numerical values of κN_T and υN_T will, by definition, have the same (ideal) cyclic variation of pressure. Taking the matter further calls for expanding the term υN_T:

$$\begin{aligned} \upsilon N_T = {} & (V_{de}+V_{dxe})/V_E \\ & + N_T\{V_{dr}/V_E(\log_e N_T)/(N_T - 1) \ldots N_T(V_{de}+V_{dxe})/V_E\}. \end{aligned} \tag{4.7}$$

This introduces the additional similarity requirement that prototype and derivative designs should share the same respective numerical values of individual terms of the dead-space parameter, i.e., of Eq. (4.7). The only arithmetic of any consequence is that required to revise fractional regenerator dead-space V_{dr}/V_E. Let the original value be denoted by μ_{dr} and the replacement to be determined by μ^*_{dr}. Similarly, denote original temperature ratio by N_T and the proposed replacement value by N^*_T. Then equating unknown and knowns of the relevant term of Eq. (4.7):

$$\mu^*_{dr} = \mu_{dr} \times N_T/N^*_T \times \ln N_T/\ln N^*_T \times (N^*_T - 1)/(N_T - 1).$$

Figure 4.5 shows the 'engine' of the earlier Fig. 4.2 scaled in this way to operate with $T_E = 2000°C$. While the process has changed the boundaries of individual gas-path elements, similarity has been achieved to the extent that, at any chosen crank angle, the original cyclic distribution of the working fluid mass throughout the gas path has been preserved, as confirmed in terms of numbers of particle tracks. (It has been necessary to reduce the number to retain legibility at the compression end. However, the sceptical reader can confirm the identical *ratio* of tracks between respective gas-path elements.)

That Eq. (4.4) is unaffected means that cyclic pressure excursions are unchanged. This means in turn (all other things being equal) that adiabatic temperature swing remains the same whereas it would otherwise have increased (in the present example) to the detriment of performance.

The reader familiar with the Finkelstein treatment will recall the expression (his Eq. (30a)) for dimensionless cycle work which derives from Eq. (4.4). Multiplying numerator and denominator by N_T as for the

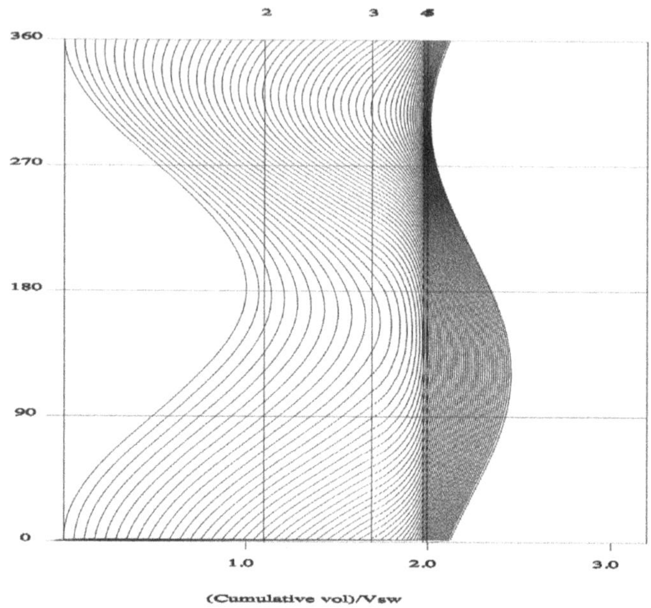

Fig. 4.5. Gas-path specification scaled so that cycle pressure history and working fluid distribution remain unchanged at T_E = 2000°C from the picture at 600°C. (Number of tracks reduced to retain legibility at compression end.)

earlier Eqs. (4.4)–(4.6) leaves a term in which the product κN_T fails to appear, namely $1 - 1/N_T$. Transposing this to the left-hand side:

$$\frac{W}{p_{\min} V_E (1 - 1/N_T)} = \frac{2\pi\kappa\, N_T \sin\alpha}{1 + \kappa\, N_T + \upsilon\, N_T} f\left(p_{\min}(\rho)\right), \tag{4.8}$$

$$f\left(p_{\min}(\rho)\right) = \frac{\sqrt{(1+\rho)}}{\sqrt{(1-\rho)}\left[1 + \sqrt{(1-\rho^2)}\right]}. \tag{4.9}$$

A change to κ which leaves the value of κN_T unchanged not only promises unaltered cycle pressure excursion and unchanged adiabatic temperature swing, but now keeps the value of Eq. (4.8) constant. The only way this can happen as temperature ratio N_T increases (i.e., as the numerical value of the term $1 - 1/N_T$ increases) is for specific cycle work $W/p_{\min} V_E$ to increase in sympathy — in line with expectation.

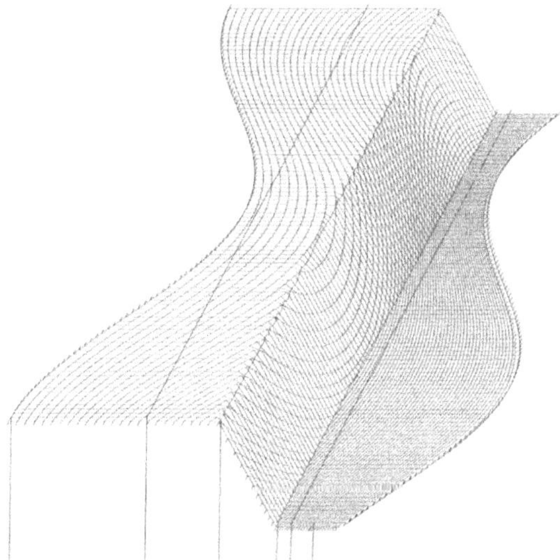

Fig. 4.6. Trajectory map displayed to the format of regenerator temperature solution. Acquired by superimposing over (invariant) distribution of enclosure temperature.

Design by scaling is not the focus of this chapter. This, however, is the cue for a reminder that the cycle model underlying map construction to this point is Schmidt.

This is indifferent to the realities of heat transfer and flow friction, so heat-exchanger and regenerator have yet to be scaled. However, the multi-channel exchanger and wire-mesh regenerator offer considerable scope for accommodating the required hydraulic radii r_h (and, in the case of the regenerator, volume porosity $¶_v$) within the changed values of dead-space.

The approach may increase confidence in the performance achievable when designing for elevated pressure ratio.

The trajectory map is the coordinate framework for acquiring the regenerator temperature-time history. Figure 4.6 is the map of Fig. 4.2 re-plotted in the format of a representative solution, i.e., as a temperature relief. The transformation is purely in the display, with no change to the code which generates the trajectory map.

4.5. Application to Design

The preoccupation of Chapter 2 was the low Beale modulus of the Stirling engine, and its relationship to the pattern of working-fluid distribution

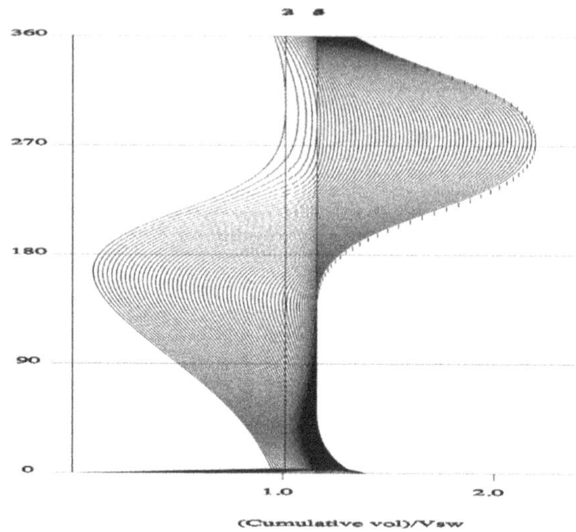

Fig. 4.7. Trajectory map generated by novel crank mechanism proposed for internally-focusing solar Stirling engine. The map is the opposed-piston equivalent of the coaxial ('beta') configuration of the actual mechanical layout, clearly showing working fluid excluded from compression space over majority of expansion phase — and from the compression space for expansion.

over a cycle. The tool for visualising that pattern is the particle trajectory map.

A thread running through this book is the design of an internally-focusing solar Stirling engine which it is hoped will deliver 100−150 W on unpressurised air. The task is justified in terms of the dramatically improved picture of working fluid utilisation afforded by a novel drive mechanism (Organ, 2019) which allows bespoke volume variations. The mechanism will be fully described in the course of a staged exposition of the design in chapters to follow. It drives displacer and piston in axial (beta) configuration, but the working-fluid distribution achieved is shown in Fig. 4.7 transformed to opposed-piston layout.

5

Thermal-Lag Engine

5.1. Rationale

The thermal-lag (T-L) engine functions by virtue of processes which one attempts to design out of the Stirling engine — pressure drop and thermal bottle-necks. It is one of life's paradoxes that it takes *precisely* the same set of equations to model either type! To understand the T-L is therefore to understand the finer points of the practical Stirling cycle — or at least to understand why the latter does not perform to expectation. Since that is most of the time, this chapter is not a digression but essential background.

It might be a searching test of a cycle model that it should simulate either type, Stirling or T-L, with no modification other than to the pointer to the respective data file.

5.2. Context

If one digs deep enough, all heat engines work by the same principle: an elastic substance — frequently air — is compressed. Heat is then added, raising the internal energy and also, in the case of a gas, the temperature. The substance is then re-expanded — usually to the original volume. As Peter Tailer then says in US patent 5414997A *Thermal lag machine* (1993): ... *the working fluid is at a lower average temperature and pressure during compression and a higher average temperature and pressure during expansion (the difference manifesting as) net work.*

Completing the sequence of events in readiness for a repeat cycle calls for rejection of sufficient internal energy to reduce the start temperature to the original value.

The difference between the internal combustion engine (ICE), the Stirling, the thermal-lag, the thermo-acoustic and Ericsson variants is merely in terms of the mechanism for bringing about the heat addition and heat rejection. Further discussion will consider only closed-cycle, regenerative variants, i.e., the former three. These share the common features of heated region, cooled region and regenerator (in the thermo-acoustic context frequently referred to as the ‘stack’).

In the Stirling, the required displacement of working gas is brought about by out-of-phase cyclic variation of the volume of heated and cooled regions. ‘Biased’ is more appropriate than ‘displaced’, since only a fraction of the working fluid inventory is fully involved. Moreover, net bias achieved is due only partly to intentional displacement. A further component is the incidental result of compression and expansion.

In the T-L, volume changes and displacement merge completely. Differential heating is now achieved in terms of the time lag between displacement to a region of different temperature and the response of the fluid — the ‘thermal-lag’ time. Tailer’s patent abstract sheds light:

A thermal lag engine has a cooled cylinder and piston connected to a heated chamber. A working fluid in the cooled cylinder and heated chamber is alternately compressed and expanded by the piston. There is a thermal lag or time interval before the working fluid entering the hot chamber is heated and the working fluid expanding back into the cooled cylinder is cooled.

The confusion which prevails with the thermo-acoustic variant is difficult to account for: there is no physical similarity, and operating frequency (which admittedly depends on choice of working gas) is at least an order-of-magnitude higher. Gas displacement must be thought of in terms of pressure waves: for given temperature T the mass of gas over given length of pressure wave is greater than that over the same length of rarefaction wave. Acoustic waves travel at the sum of local acoustic speed $a = \sqrt{\gamma RT}$ and local particle speed u, viz, at $a \pm u$. Moreover, they reflect, particularly from closed and open ends of the containing duct. Forward and reflected components interact, and at certain frequencies — including the ‘natural’ frequency of the system — can be additive, giving larger amplitudes than those associated with an acoustic wave radiating in free-air. The net result can be 'standing waves', with pressure p oscillating at a point between maximum and minimum. It is therefore possible to envisage compression and expansion exchangers so spaced as to give a cycle having two extremes: (a) peak p in the former when p in the latter is a minimum and (b) vice versa. The excitation frequency at which this occurs might be called the 'natural' frequency of the system.

For the arrangement to convert heat to work now only requires the rarefaction wave to reach the piston (or diaphragm) during volume reduction and the pressure wave to arrive during volume increase.

Returning to the T-L type, a given machine heated and cooled to a specific temperature difference will likewise settle to a particular no-load cycle speed. This, however, does not reflect a 'natural frequency' in the sense that characterises the thermo-acoustic counterpart.

5.3. What Has This to Do with the Stirling Engine?

Everything! Thermodynamics and heat transfer are studied under (at least) four separately-identifiable headings: convective heat transfer; viscous fluid flow; thermal conduction (more usefully thermal *diffusion*) and the interchangeability of work, heat and internal energy. (The latter can manifest as 'adiabatic temperature change' — the bicycle tyre pump gets hot during inflation.) Thermal radiation plays a role, but neither this nor the elastic-wave phenomenon which predominates in thermo-acoustic types requires further consideration.

To say that the four phenomena interact in the T-L machine would be an understatement: they merge in different proportions over a cycle, and each point in the gas path experiences a different cycle of merging.

At an early stage of attempting to make sense of the T-L, this writer fell to wondering whether Tailer had designed the engine from first principles. If so, he must have been a genius or a psychic — or both. His own account under *Background* in the patent clarifies:

Experimenting with Stirling engines using air at mean atmospheric pressure as a working fluid, it was found that a main factor limiting power for a given size of engine was the time required for heat to transfer to the working fluid. For this reason, an experimental engine that was not a Stirling engine was built to take advantage of — and actually use — this

Table 5.1. Outline specification of Vaizey's impressive single-cylinder thermal-lag engine.

Swept volume $V_{sw} = V_{ref}$	7.6 cc
Charge pressure $p_{ref} = p_{min}$	0.66[a] atm./10^5 Pa
rpm	variable
T_C (= T_{ref}), T_E	30°C, 600°C
Volumetric compression ratio, V_{max}/V_{min}	1.645
Pressure ratio at rated temperatures p_{max}/p_{min} (simulated)	2.037

(Reference length $L_{ref} = V_{ref}^{1/3}$: 19.66 mm, Reference area $A_{ref} = V_{ref}^{2/3}$: 386 mm^2)

Regenerator:					
Packing length	65 mm				
Diameter	10.0 mm				
Material	'wire wool' volume porosity $¶_v$: 0.85 (assumed)				
Pulse-tube:					
Outer diameter 17.5 mm	inner diameter 10.0 mm				
Heater:					
Length	2 × 20 mm (reverse-flow)				
No. flow channels	30 rectangular 2.0 mm × 2.0 mm				
Cooler:					
Length	20 mm				
No. flow channels	123 holes 1.0 mm dia.				
	exp. exch.	regen.	comp. exch.	comp. cyl.	pulse tube
Dead-space ratio V_d/V_{ref}	0.188	0.4418	0.0548	0.100	0.766
Length L_x/L_{ref}	1.474	2.947	0.736	0.100	2.947
Area/A_{ref}	0.128	0.149	0.074	1.000	0.259
Hyd. rad L_x/L_{ref}	0.0184	0.021	0.007	0.250	0.300

[a]The figure for p_{min} is arrived on the basis that, without provision to the contrary, mean cycle pressure probably settles to atmospheric. Pressure-ratio of about 2 then suggests $p_{max} = 2 \times p_{min} = 1.33$, giving $p_{mean} = 1/2(0.66 + 1.33) =$ unity.

time required for heat transfer, which is referred to as thermal lag. An ultimately simple engine using only a cooled cylinder and piston connected to a heated chamber ran both as a free piston engine and as a kinematic engine. (This writer's punctuation.)

While it reflects a remarkable thought process, the patent account suggests experimental verification of an enlightened hunch rather than first-principles design. A remarkable picture emerges nevertheless: an engine operating by virtue of *effects which one attempts to design out of a Stirling engine*!

This account will proceed by reference to an engine demonstrated by Vaizey. The choice reflects the fact that it out-performs all comparable Stirling derivatives (un-pressurised, table-top 'hot-air' engines) known to this writer. No-load rpm was 2370. In Fig. 5.1, the upper image is a screen-grab of Vaizey's YouTube video showing the engine driving a 7-inch × 6-inch APC propellor and producing a healthy slip-stream. The lower image is a cross-section — schematic but more or less to scale.

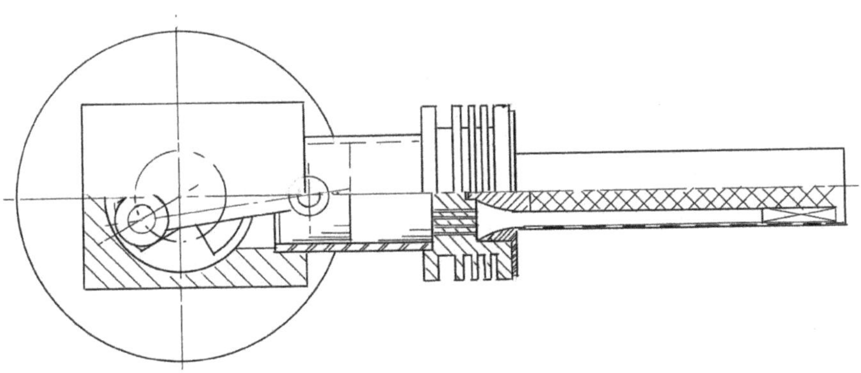

Fig. 5.1. 20 cc thermal-lag engine. One of three built to the impeccable machining standards of Geoff Vaizey. Reproduced with permission granted by e-mail.

5.4. The Role of Simulation

A little-discussed strength of cycle simulation is the possibility of displaying, in isolation, gas processes which, in reality, overlap and interact. The adiabatic component of temperature swing has already been mentioned. Suppressing heat exchange between gas and enclosure throughout (by setting local, instantaneous *NTU* to zero) means that *any* change in temperature can be attributed to pressure change alone. Moreover, the distribution of gas temperature at any point in the cycle is due exclusively to the combined effect of wholesale translation of the original temperature profile plus that of pressure swing.

The cycle model which generated the illustrations which follow employs precisely the equations used to set up a Stirling cycle simulation. It is relevant to the bigger picture that only two modifications are required to adapt to present purposes: *a* an appropriate temperature distribution to the gas-path enclosure and *b* expansion space volume is held at zero.

The start-up cycle is initiated with gas temperature distribution set equal to that of the adjacent enclosure. Figure 5.2 is a relief of temperature vs location and crank angle over a start-up cycle. Horizontal location is cumulative volume, so a region of reduced free-flow area (such as the compression exchanger) plots to less-than-natural length. Crank angle elapses from lower left to upper right. For the purposes of display, this and temperature reliefs to follow are plotted using a reduced number of angular and distance subdivisions, resulting in some apparent inaccuracies.

A particle-displacement diagram generated from a simulation of one-dimensional ('slab') flow is one of several essential checks of appropriate coding. In Fig. 5.3, region 1–2 outlines the regenerator, 2–3 the high-temperature exchanger, 3–4 the pulse-tube section, 4–5 the cooler (and/or flow restriction). Adjacent trajectories enclose equal subdivisions of total working-fluid mass, so horizontal spacing is proportional to local (absolute) temperature.

The T-L engine functions with piston motion continuous (e.g., pure simple-harmonic), discontinuous[1] or a number of possibilities between.

[1]The reversed-cycle pulse-tube cooler evolved around discontinuous motion achieved by switching valves (hence, presumably, 'pulse'). The pulse-rate of large coolers can be as low as 1–2 Hertz!

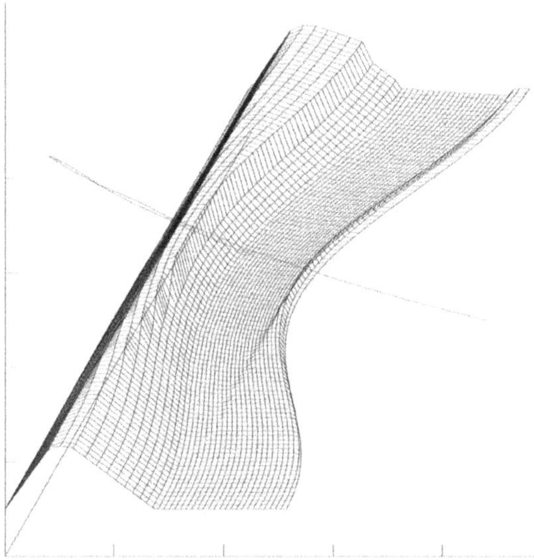

Fig. 5.2. Temperature-crank-angle history for the Vaizey engine over a start-up cycle.

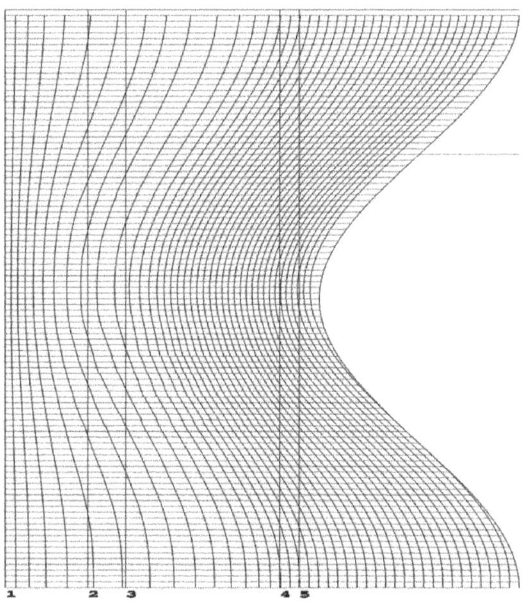

Fig. 5.3. Fluid particle trajectory map corresponding to Fig. 5.2. (Apparent mismatch at piston face at extreme right disappears when subdivisions increased to realistic number.)

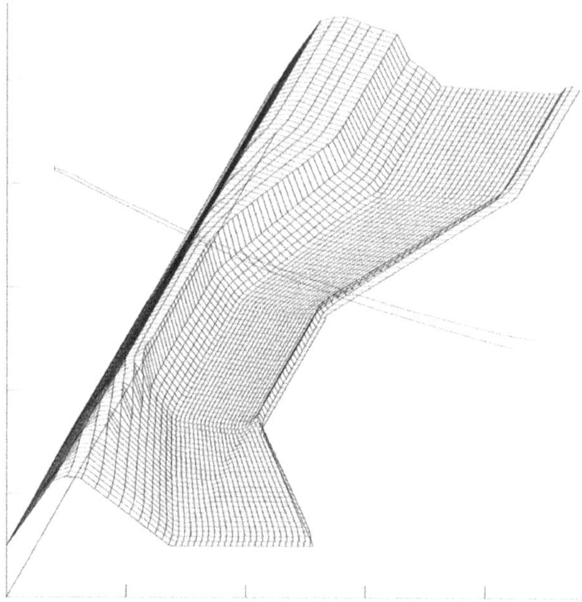

Fig. 5.4. Temperature relief of start-up cycle generated by discontinuous piston motion.

Discontinuous motion may be the closest it is possible to get to separating the overlapping phases. Figure 5.4 is the discontinuous counterpart of Fig. 5.2.

A qualitative description of a cycle of operation will assist interpretation of the quantitative results which follow.

Compression phase 1–2 shifts the entire temperature profile leftward, while the accompanying reduction in volume brings about a pressure rise overall. The latter causes a tendency for an accompanying temperature rise. Net temperature change at any given location will depend on the sign (+ or −) of local difference $T_w - T_g$, and the intensity of any corresponding heat exchange.

Constant-volume phase 2–3 might thought of as a period of temperature 'relaxation'. The tendency of gas temperature T_g to adjust to the T_w of the adjacent enclosure is accompanied by further, slight motion of the gas. The extent of the temperature adaptation depends on location: it can be almost complete in regenerator and heat exchangers, but substantially incomplete in pulse-tube and cylinder. In order for the expansion stroke which follows to provide a surplus of work, the relaxation phase must bring about an increase in overall internal energy. This shows up as an increase in pressure above that achieved at the end of compression stroke 1–2.

Expansion phase 3–4 causes wholesale rightward translation of the 'relaxed' temperature profile. The profile is modified in the process (a) by the tendency for temperature to drop overall in sympathy with falling pressure and (b) heat exchange in regions of small hydraulic radius — regenerator and exchangers. The fact that effect (b) is only slight in pulse-tube section and cylinder results in the development of a temperature 'front' prominent in Figs. 5.2 and 5.4.

A second 'relaxation' phase 4–1 follows, in which regions of small hydraulic radius (regenerator, exchangers) see relatively rapid reduction in local temperature difference $\Delta T = T_w - T_g$, while ΔT in pulse-tube and cylinder adjusts only slightly. There is a slight redistribution of working fluid. Pressure falls as net internal energy of the gas reduces.

On the initial cycle the *p–V* traces (superimposed on temperature reliefs) do not close to loops. The number of cycles to closure depends strongly on *rpm*, with some cases requiring only 3 cycles.

5.5. A Closer Look

Figure 5.5 (upper) amounts to a 'slice' taken from the temperature relief between the start of the cycle and the end of compression — a superposition of temperature distributions over the first quarter of the cycle. Numbers of crank-angle increments per cycle and working gas subdivisions has been reduced to 100 and 40, respectively, with the advantage of avoiding a completely black image — but the disadvantage of jerky outlines. It shows the original trapezoidal profile shifted bodily leftward and simultaneously raised by the adiabatic element of temperature response to rising pressure. Evidence of the shift is greatest in the pulse-tube section and least in the regenerator, reflecting contrasting heat transfer provision.

Figure 5.6 (upper) shows the re-expansion phase. The temperature profile at any instant is the combined outcome of rightward bodily shift, of adiabatic cooling due to expansion, and local heat transfer. The leading profile has never left the pulse-tube, and returns more or less to the starting distribution. Gas re-entering the pulse-tube from left to right emerges from the exchanger close to T_E regardless of the temperature at which it entered during compression. The resulting temperature front advances little-affected by contact with the pulse-tube, but very much under the influence of falling pressure, as seen in terms of the steady fall.

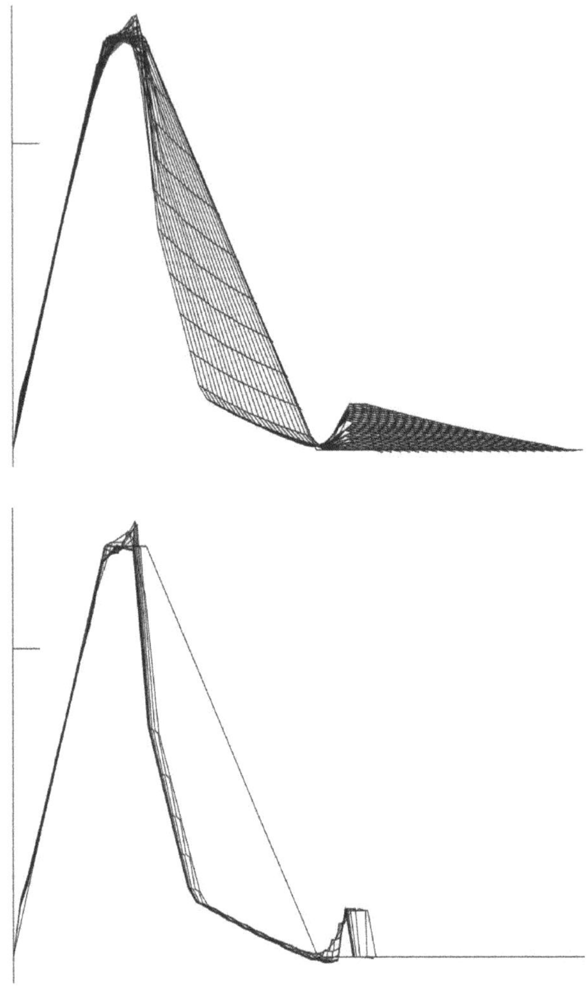

Fig. 5.5. Upper: Initial compression phase not only causes an overall rise in pressure (and temperature), but shifts the starting temperature profile bodily leftward. Effective heat transfer in the regenerator and the opposite in the pulse-tube have the obvious outcome. Lower: Temperature peak relaxes towards T_E of the hot exchanger, while the profile in the pulse-tube inclines towards the wall temperature distribution of that section. Overall, internal energy and pressure increase at constant overall volume.

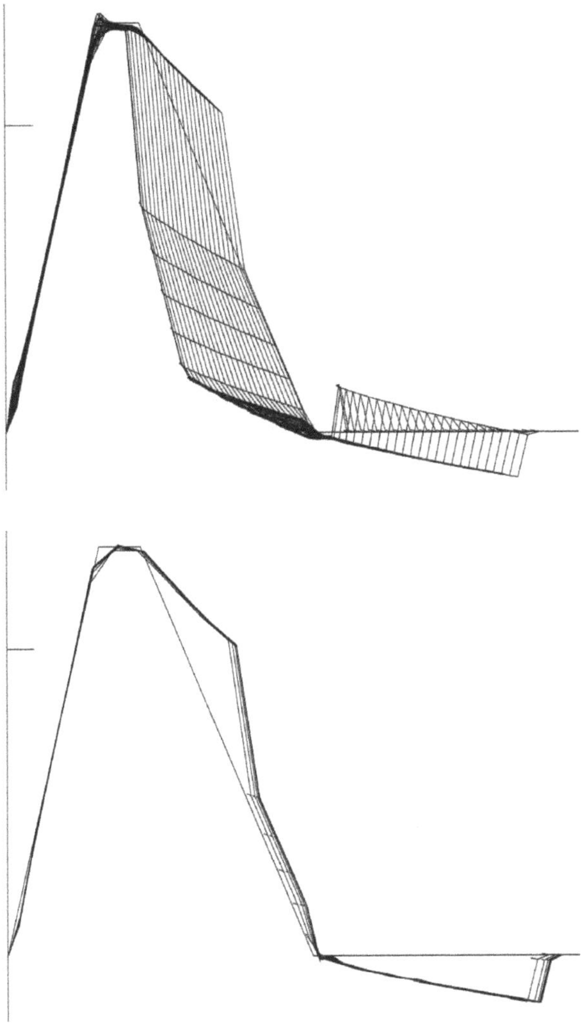

Fig. 5.6. Upper: Expansion phase: temperature profile displaced progressively rightward. There is everywhere a tendency to fall in sympathy with the reduction in pressure, either enhanced or opposed by heat exchange with the adjacent wall. Lower: A slight redistribution of the working fluid accompanies relaxation of the temperature profile during the dwell phase at outer dead-centre. Temperature peak adjusts towards T_E of the hot exchanger, while the profile in the pulse-tube inclines towards the wall temperature distribution of that section. Net internal energy and pressure increase at constant overall volume.

A further, slight re-distribution follows arrival of the piston at outer dead-centre (Fig. 5.6 (lower)). In the supposed absence of leakage, pressure falls to the start value.

5.6. Exchanger Pressure Drop

In the context of Stirling engines, flow friction is to be avoided at all cost: '*maximum heat transfer per minimum pumping power*' — or so has been the creed to date.

The 'rule' may not apply to the thermal-lag type: Resistance by the ambient exchanger to flow outward from the compression space suggests increased mass at compression temperature relative to the lossless case. Just what the doctor ordered — up to a point! And that is a break-even point, where the gain (if any) is outweighed by the pumping penalty incurred. (A beautiful problem in optimisation for some optimist!)

Effective heat exchange is paid for in terms of pressure drop whether one likes it or not. The question which arises is whether to design for a larger drop than necessary, i.e., than if the sole criterion were maximum heat transfer rate.

While the matter is pondered, Fig. 5.7 shows pressure-crank-angle history over a start-up cycle. The relatively small end-to-end gradients do not show up well at operating *rpm* in prospect, so the result for 5500 rpm is shown. (The superimposed p–V loop cannot be expected to close after a single cycle.)

A more focused picture emerges from plotting local, instantaneous pressure gradient $\partial p/\partial x$. Figure 5.8 is for the more realistic 2550 no-load rpm approached by the Vaizey engine. Excursions above the reference plane indicate positive gradient. Leftward flow is down a pressure gradient, which is positive (decreasing right-to-left), so at least the simulation gets the sense correct!

5.7. Discussion

This writer became aware of the *genre* on being offered a contract[2] to attempt computer simulations of pulse-tube and thermo-acoustic coolers.

[2]The prospect was attractive on two counts: (a) it was accompanied by offers from two independent organisations to supply full specifications and performance data and

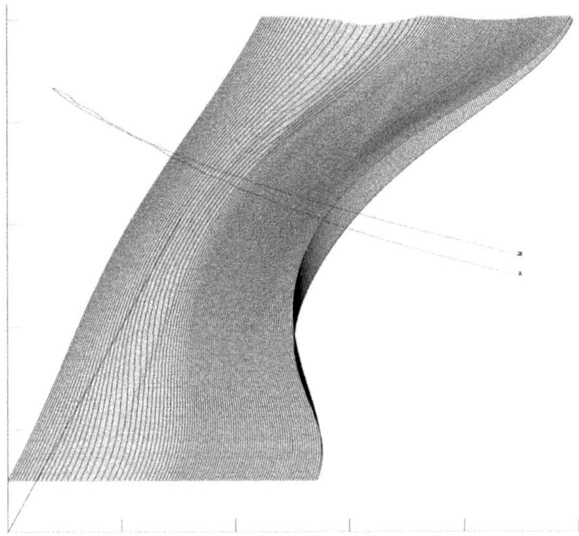

Fig. 5.7. Relief of pressure over start-up cycle. RPM exaggerated to 5500 to give clearer display.

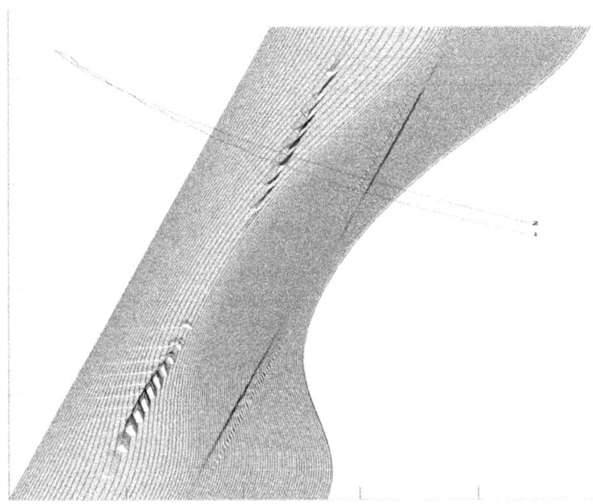

Fig. 5.8. Relief of pressure gradient $\partial p/\partial x$ over start-up cycle at 2550 rpm.

(b) *carte-blanche* to publish. (With hindsight the conflict should have caused alarm bells to ring.) There was the exciting prospect of a counterpart to *The Regenerator and the*

Attention was attracted back to the topic more recently through an internet post by Mayes, highlighting mechanical simplicity and opening discussion on possible design criteria/optimisation.

The Vaizey design is inspired, and benefits from exemplary machining, a facility in CAD/CAM and a background in the manufacture and development of sophisticated prototypes (miniature gas turbines, sub-contract to Formula One, high-performance hydro-foils, etc.). As far as is known, however, it was not the outcome of experiment with the thousands of combinations and permutations of volumes, areas, hydraulic radii, wire diameters, porosities, etc., which would constitute practical 'optimisation'. In other words, the possibility remains that the impressive performance demonstrated goes no way to pre-figuring potential.

This in turn prompted recollection of the remarkable performance of the engine demonstrated by Vaizey. When simplicity is the over-riding criterion, might not the thermal-lag phenomenon be preferable to the mechanically-actuated displacer?

Computer simulation is the much-touted alternative to the time and eye-watering cost of experimentation. Its reputation is tarnished somewhat by early abuse of computing power spent 'optimising' the Schmidt cycle. An ideal nevertheless remains: a simulation sensitive to all proposed design changes, recalibrated continuously against measured performance as changes are implemented.

The simulation which generated the illustrations for this chapter comes nowhere close. To date, there is only one benchmark indicating the realism achieved: Changing from air to He stopped Vaizey's thermal-lag engine dead in its tracks. A change in the simulation input data from the 287 J/kgK of air to 2082 for He also stopped the simulation in its tracks. (In the Stirling engine context, substituting helium for air typically results in an increase in the rpm at which work/cycle peaks. Conversely, the thermal diffusivity of He is 10-times that of air — evidently reducing the lag time essential for functioning of the thermal-lag type.

Validating a computer model without counterpart laboratory experiment would be a challenge in any area of endeavour. In the present case, however, a searching test offers itself. Some thermal-lag variants have functioned without crank and flywheel, but with piston (or water column)

Stirling Engine (1997). In the event, neither organisation responded to repeated reminders of their undertakings, and work proceeded in an academic vacuum (Organ, 2005).

alone oscillating at some 'natural frequency' of the system. To be credible, any cycle model must do likewise — and at life-like frequency! The simulation featured here is being modified to check that it copes with this case.

It was developed for Stirling engine use. Adaptation for present purposes has been solely through changes to the input file — gas-path geometry and distribution of enclosure temperature. It may be a further critical test of a Stirling simulation that it should predict pulse-tube behaviour with no more than a change to input data.

6

So How Does It Work?

6.1. The Stirling Engine, that Is?

Certainly not according to the account in any textbook on engineering thermodynamics.

If Fig. 6.1 rings a bell, it does so for the wrong reasons. It has no link to any hypothetical 'ideal' cycle. Volume variations are shown discontinuous for the reason given earlier in connection with the thermal-lag engine: to achieve a degree of separation between the overlapping processes purely for purposes of explanation.

A cycle starts with the gas at rest and carrying the temperature distribution of the adjacent solid enclosure. It can be considered in four phases, each comprising a primary 'intentional' process accompanied by at least one secondary effect. It is assumed throughout that temperature excursions of the regenerator matrix are slight.

1–2. Right-hand piston moves inwards, reducing total volume and achieving the primary purpose of compression. However, compression is not confined to gas at T_C, but affects the entire working fluid mass M. The relationship between p and V does *not* follow a process path $pV = RT_C$. A secondary effect is a biasing of M, together with its impressed temperature profile, leftward in the direction of the expansion space. This precedes nominal transfer phase 2–3, and meets two forms of resistance:

(a) flow friction and (b) 'back-pressure' as the leading elements of gas expand in response to higher wall temperature. The rising

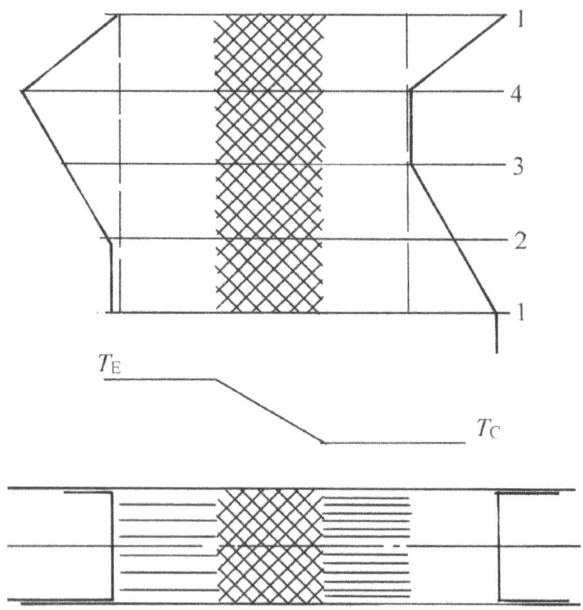

Fig. 6.1. Gas-path 'straightened out' to opposed-piston configuration. Discontinuous piston motion does *not* indicate the notoriously misleading 'ideal' cycle: it merely offers a degree of artificial separation between otherwise inextricably overlapping processes.

pressure is accompanied by an adiabatic component of temperature rise. This works in favour of the biasing process, but adds to the thermal load on the compression-space exchanger.

2–3. Both pistons move leftward in synchronism. This goes some way to achieving the primary purpose of transfer from cold space to hot. On the other hand, no more is achieved than a *biasing* of M towards higher temperature. Secondary effects are flow resistance and a rise in pressure throughout the gas-path. The latter contributes a further component of adiabatic temperature rise.

3–4. Compression-space piston is held stationery at inner dead-centre position while the expansion piston continues leftward. This achieves the primary purpose of expanding a fraction of M at T_E but incurs the secondary effect of expansion throughout the system across the full range of temperatures between T_E and T_C. While this is happening, transfer of fluid left to right continues.

4–1. Both pistons return in synchronism to respective start-points, biasing M towards temperatures which are lower but not, in general, close to T_C. The primary process is subject to flow friction.

6.2. The Important Part

The interactions loosely described above are but a small part of the story: changing volume-ratio and/or phase-angle and/or temperature ratio and/or any of the ratios of dead-space to swept volume results in a changed interplay.

Consider the matter of volumetric compression ratio r_v. The characteristically low value suggests treating it as a parameter of Stirling engine operation. This would be unhelpful, since the same numerical value may be achieved by different means.

Low r_v is associated with low specific cycle work (low N_B). This does not mean that raising it is necessarily a good idea, since the outcome — improvement or deterioration — depends upon just how the increase is achieved.

Take a single-cylinder, parallel-coaxial engine operating vertically, equip it with an hypothetical drive mechanism allowing the displacer to be held stationary at top dead-centre over the entire range of movement of the compression piston, with compression-space volume initially at maximum, now begin reducing it, the majority of working-fluid charge M being in the compression space, the *initial* stage of compression takes place substantially at T_C. In the typical high-performance engine, the combined dead-space of regenerator and expansion exchanger considerably exceeds that of the cooler. This means that, if volume reduction is continued past the point at which the transfer phase[1] would normally begin, the fraction of M exchanging heat at T_C is less than that exchanging at higher temperatures — up to T_E. Not forgetting that intuition is a treacherous tool in the context of Stirling engines, this strongly suggests not raising compression ratio beyond this point.

If low specific work (low N_B) goes along with low compression ratio (low r_v) and if low r_v is inevitable, is there any scope for the much-needed raising N_B by manipulating volume variations?

[1]The compression phase has already given rise to considerable transfer.

Depending upon the proposed application, yes! A later chapter will revisit in considerable detail the internally-focusing solar concept invented and demonstrated by Finkelstein (1961). The expansion exchanger in this case is integral with the regenerator, reducing dead-volume at T_E to near-zero. The small thermal capacity of a regenerator *appropriately designed* for single-atmosphere on air results in a matrix having minimal dead space. Referring to the earlier explanation, this permits the compression stroke to proceed further before the fraction of M being compressed at $T > T_C$ becomes significant.

The matter is left here until picked up again in context.

7

The Inscrutable Regenerator

7.1. Context

The Schmidt and adiabatic analyses share (at least) one severe limitation: formulation is independent of *time*. Performance predictions therefore do not reflect the effect of *speed*, or rpm.

Meaningful performance forecast calls for gas-process modelling of a different order. Traditionally, this has involved use of 'correlations' to evaluate heat transfer rates and flow resistance — experimentally-acquired relationships between Stanton number *St* and Reynolds number, *Re* and between friction factor C_f and *Re*.

Bearing in mind that it is some eight decades since Philips made claims for the '*application of modern methods of fluid flow and heat transfer*', a welcome development would be correlations not merely applicable to the unique flow environment (compressible, unsteady, cyclically-reversing), but appropriate also to exploration of the *potential* performance envelope.

The matter has been addressed before in the writer's 2007 text — and at some length. But that was before a bit of simple arithmetic drew attention to the fact that even *steady* flow through the wire gauze cannot be adequately characterised by flow at ambient temperature. Indeed, a correlation acquired in this way will generally be different from that resulting from tests at 650°C — the temperature of the hot end of the regenerator. Confirmation will sound the death-knell for correlations to traditional format.

7.2. 'Friction Factor' C_f

Studies of flow through wire gauzes fall under (at least) two distinct headings. (The fact that the one does not acknowledge the existence of the other might be cause for concern in its own right!)

(1) Tests under conditions where compressibility is the dominant influence. Pressure drop is correlated to Mach number *Ma* with *Re* (or equivalent) as parameter. The 1967 account of Pinker and Herbert stands out.
(2) With compressibility discounted and pressure drop correlated to Reynolds number *Re* via the concept of friction factor C_f with volume porosity as (geometric) parameter. The work of Kays and London (K & L, 1964) may be the best-known in the category.

Despite obvious overlap, the two approaches have never been reconciled. Nor — as far as is known — have the findings of Pinker and Herbert been applied in the modelling of Stirling engines. Conversely, the correlations of Kays and London are widely relied upon, despite involving the user in the risks of gross violation of Dynamic Similarity principles.

The legitimate context of friction factor C_f is viscous shear parallel to an enclosing wall in turbulent flow. It is thanks to Kays and London that it has been hijacked to the alien environment of the regenerator. The literature on flow through porous media does not make use of C_f. Neither do Pinker and Herbert nor the studies on which they draw.

The Kays and London correlations *do not apply* with any generality to conditions in the Stirling engine:

(a) They were not acquired from tests on regenerators.
(b) High Reynolds numbers were attained not at high speed, u, but by massive hydraulic radius r_h achieved in turn by use of rods of up to 9.5 mm diameter. With $r_h = 0.04$ mm of the typical Stirling regenerator, achieving the Kays and London maximum *Re* of 10^5 calls for $u = 9375$ m/s — or Mach 27!
(c) They were acquired at ambient temperature. Suitability for use at the 900 K temperature of the hot end of the Stirling regenerator appears never to have been questioned. The matter is explored below, where implications for the reversed-cycle cooler promise to be even more disconcerting.

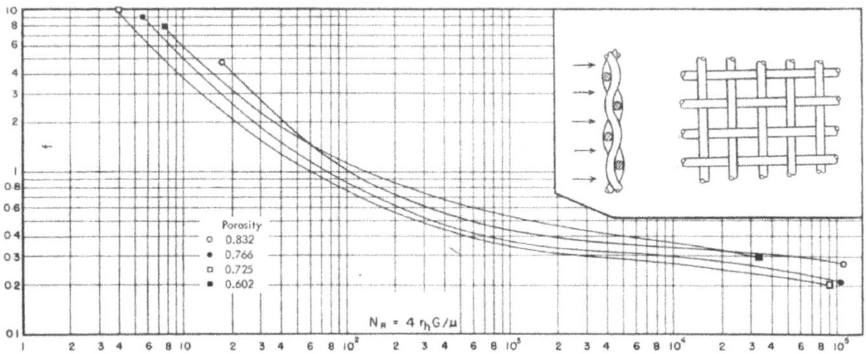

Fig. 7.1. Use of crossed bars of 9.5 mm (3/8-inch) diameter allowed Kays and London to achieve *Re* of 10^5. In the Stirling context (wires of 0.04 mm diameter — 0.0016 inch) attaining the same *Re* at ambient temperature calls for a Mach number of 27! Reproduced with permission of Krieger Publishing Co.

Ambient temperature is not representative of 700°C! Prof. William M Kays was tracked down at Stanford University and the matter drawn to his attention. The dismissive reply (*A Reynolds number is a Reynolds number* — Fig. 7.1) is included as an instance of thinking which prevails — and which must change if the Stirling engine is to benefit (Fig. 7.2).

A definition of *Re* is the ratio between viscous and inertia forces. That ratio changes with temperature: $Re = 10^2$ at 300 K does not mean the same thing as $Re = 10^2$ in the same fluid at 1000 K. In the present context, C_f is demonstrably a function of Mach number and, as the good professor would no doubt insist, a Mach number is a Mach number.

Consider the definition of *Re* and extract *u:*

$$Re = 4\rho u r_h/\mu(T) = 4pur_h/RT\mu(T),$$
$$u = RT\mu(T)Re/4pr_h.$$

Now retrieve *u* from the definition of Mach number, $Ma = u/\sqrt{\gamma RT}$:

$$u = Ma/\sqrt{\gamma RT}.$$

Equate the two expressions for *u*:

$$Re = \frac{4pr_h\sqrt{\gamma}}{\mu(T)\sqrt{RT}}Ma. \tag{7.1}$$

Department of Mechanical Engineering
Thermosciences Division
Stanford University
Stanford CA 94305

Phone: (415) 723 - 5657
Date: May 20, 1997

Dr. Allan J. Organ
UNIVERSITY OF CAMBRIDGE
Department of Engineering
Trumpington Street
Cambridge CB@ 1PZ
England

Dear Dr. Organ:

I hope I understand your problem.

The crossed-rod matrix data in Figures 10-90 to 10-92 in the 1964 edition of Compact Heat Exchangers was obtained by a transient technique that has nothing to do with the Schultz paper. Rather large rods were used for the tests, but there is no reason why the rods cannot be scaled down to 0.04 mm, or whatever, since the data are already in non-dimensional form. Thus, when scaled down to the size you are talking about the correlations in these figures should be perfectly applicable to your application, providing only that the Reynolds numbers are the same.

If you will note on Pages 135 and 136, the crossed-rod matrices were tested using the same technique as we used for tube banks. The other technique was only used for the *screen matrices* shown on Figs. 7-8 and 7-9, and also for the stacked sphere matrix shown on 7-10. So I don't think you have a problem unless I don't understand your problem.

Sincerely yours,

William M. Kays
Professor

cc: A. L. London

Fig. 7.2. Written response of Professor W M Kays.

In the case of air, coefficient of dynamic viscosity $\mu(T)$ increases markedly with temperature. Over the temperature range of interest (60°C – 700°C) the combined effect of product $\mu(T)\sqrt{T}$ amounts to a factor of nearly 7. According to this algebra, the relationship between the viscous contribution to pressure drop and the contribution of compressibility is strongly temperature-dependent.

Now suppose that the K&L tests on air at ambient pressure and temperature had been confined to a single size of gauze, i.e., to fixed hydraulic radius r_h. Variation of *Re* would then have been achieved by varying *u*. Under present assumptions, *u* imposed by *Re* fixes the value of *Ma*, thereby obscuring the rôle of the latter as an independent parameter — and at the same time making the resulting 'correlation' a function not only of *T* (Eq. (7.1)), but *also of pressure p, gas constant R and specific heat ratio* γ (Eq. (7.1) again).

In contrast, steady-flow tests carried out with due regard for Dynamic Similarity principles yield results as shown in Fig. 7.3, applicable to any temperature and gas. (A Reynolds number in air is a Reynolds number in H2 — as likewise is a Mach number!) Plotting is to Pinker and Herbert format, the independent variable being approach Mach number, Ma_{app}

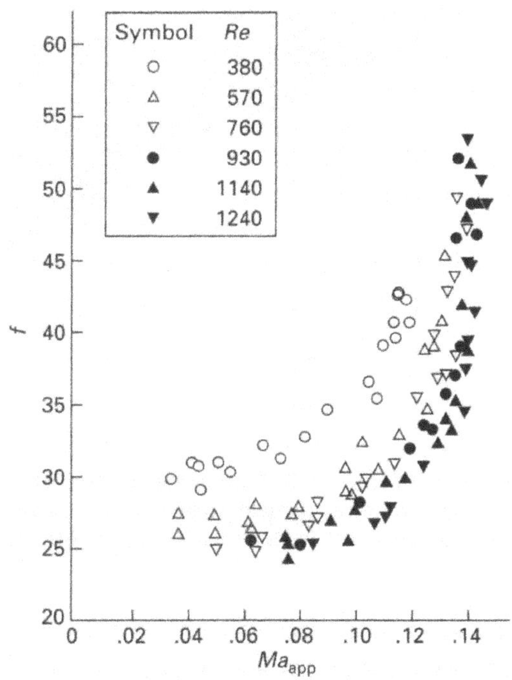

Fig. 7.3. Pressure coefficient vs approach *Ma* for stack having volume porosity $¶_v$ = 0.689. After Su 1986. The results corroborate those of Pinker and Herbert.

with *Re* as parameter. Note the influence of *Ma* down to 0.02, and its *increasing* influence at decreasing *Re*. The latter reflects the reduction in effective flow area of the individual gauze aperture due to increasing encroachment of the boundary layer.

Figure 7.3 shows that for $Ma_{app} \geq 0.14$ flow is choked for all *Re*. A realistic upper working limit of *Ma* might therefore be 0.10. Setting the right-hand side of Eq. (7.1) to this value yields maximum of *Re* in terms of r_h at specified *T*, *p* and *R*. It can therefore be used to identify the Re_{max} above which Kays' and London's correlations (Fig. 7.1) are vulnerable to *Ma and, to this extent, un-safe*.

Figure 7.4(a) replots the original C_f–*Re* curves over the 'safe' working range. On the basis that the original heat transfer (*St–Re*) correlations were acquired by the same short-cut method, these are replotted over the same truncated range of *Re* (Fig. 7.4(b)).

Figure 7.5 is constructed for air or N2 as working fluid, and allows instant readout of Re_{max} in terms of *p* and r_h. Two values of *T* are represented viz, 300 K and 900 K.

The focus on C_f involves an irony: it is of no *direct* relevance to regenerator design! In contrast, a figure for pressure-drop ratio d*p*/*p* of, say, 1% has the same significance for a helium-charged engine at 4000 rpm as for an air-charged machine at half that speed. The rest of this chapter proceeds accordingly.

7.3. A Replacement for C_f

A function form to supersede $C_f = C_f(Re, ¶_v)$ might be:

$$dp/p|_{screen} = f(Re, Ma, ¶_v). \tag{7.2}$$

Equation (7.2) can be displayed to advantage by introducing Stirling number *Sg*:

$$Sg = pr_h/\underline{u}\mu, \text{ where } Sg = Re/(4\gamma Ma^2),$$
$$dp/p|_{screen} = f(Sg, Ma, ¶_v). \tag{7.2'}$$

Figure 7.6 maps experimental results by Su (1986) into that format. The range of *Sg* acquired experimentally is insufficient as it stands for simulation use, but can be extended *ad infinitum* with the aid of a numerical model (Organ, 2014) of flow through the gauze stack.

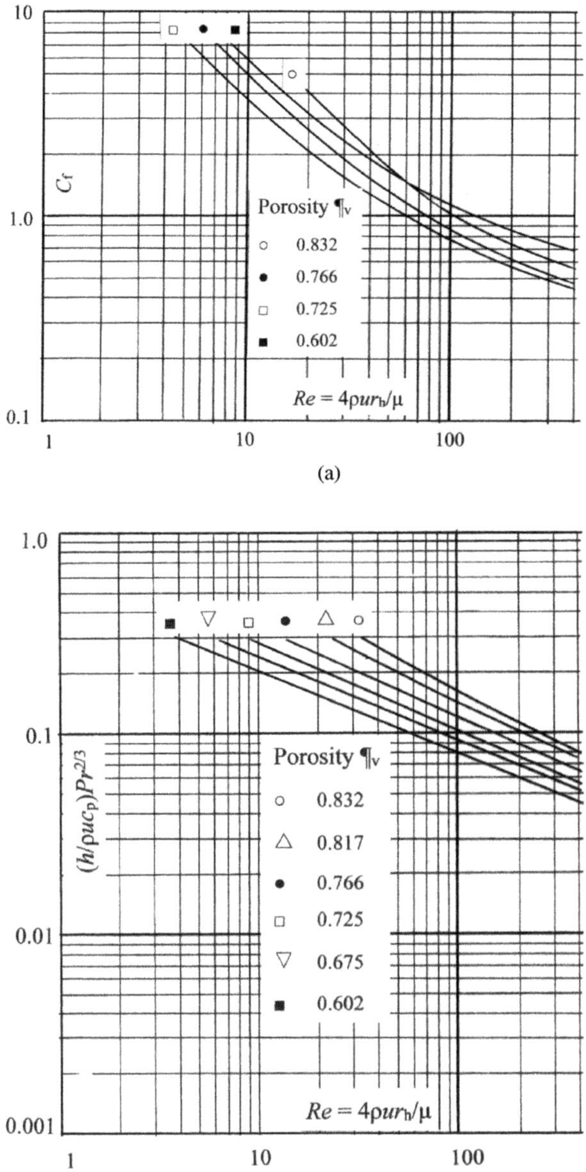

Fig. 7.4. Kays' and London's correlations truncated beyond the 'safe' range. (a) C_f vs *Re*. (b) *StPr*2/3 vs *Re*.

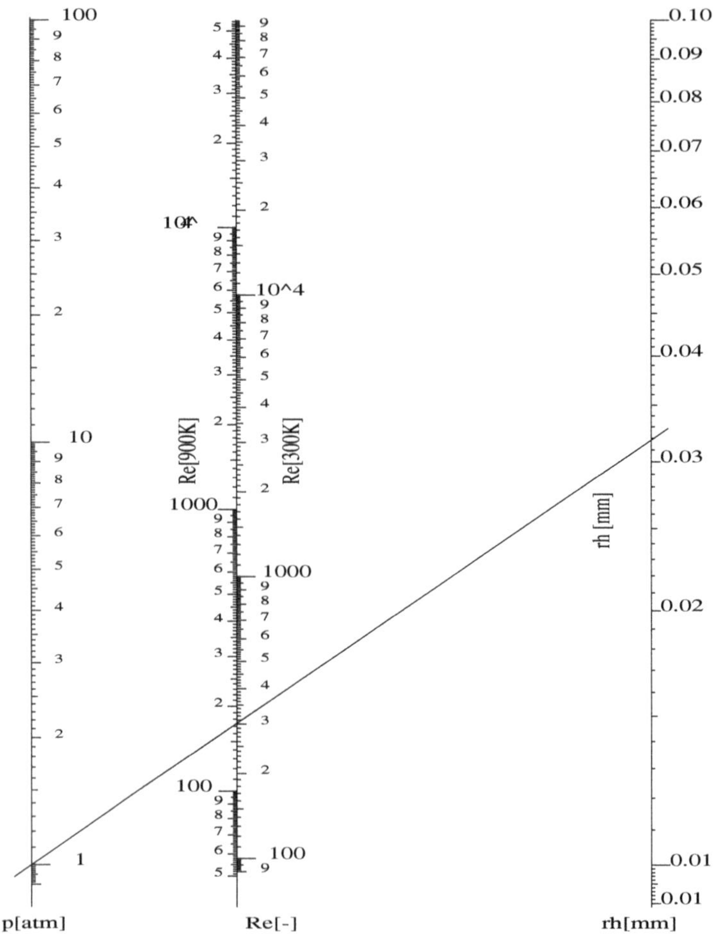

Fig. 7.5. Equation 7.3 in nomogram form for Ma = 0.1. Connecting nominal charge pressure p [atm.] to value for r_h gives Re on centre scales which is upper 'safe' limit of Kays' and London's correlations of C_f–Re at temperatures indicated.

Figure 7.7 shows the correlation generated by the steady-flow numerical model over the range of *Ma* and *Sg* covered by the same experiments by Su. There is discrepancy at the point at maximum *Ma* and minimum *Sg* but the model captures all essential trends of the experimental curves.

Simulating dp/p at each of the 350 or so individual wire screens of the typical stack threatens a heavy computational overhead. At the development and verification stage, an empirical equivalent (Eq. (7.3) below) can be used. Figure 7.7 is the corresponding graphical display:

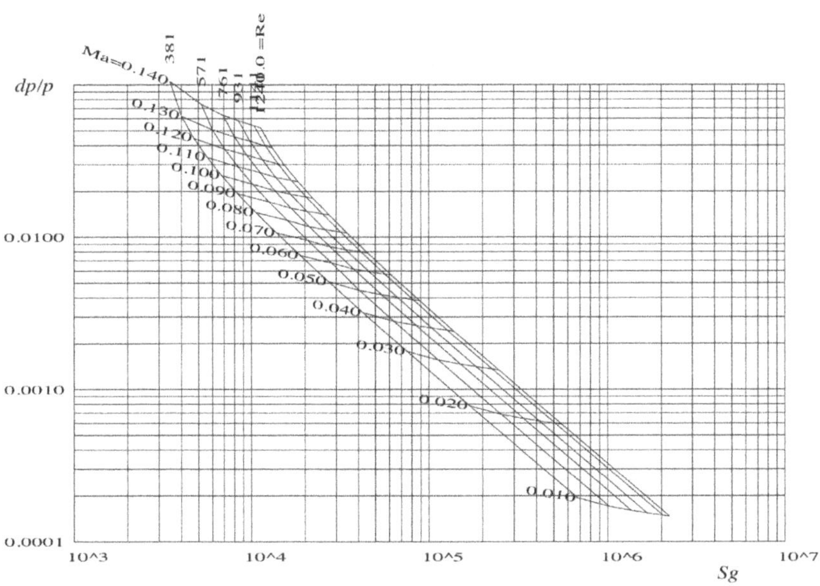

Fig. 7.6. Su's experimental results of Fig. 7.2 remapped in the function form of Eq. (7.2).

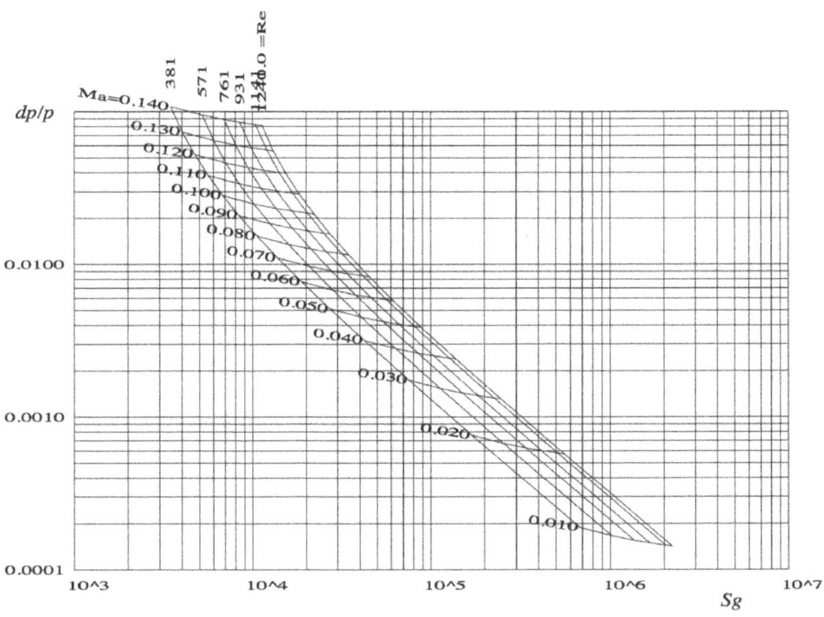

Fig. 7.7. $dp/p|_{gauze}$ predicted by steady-flow model (Organ, 2014) and plotted over the range of *Sg* and *Ma* covered by Su's experiments (cf. Fig. 7.5).

$$dp/p\big|_{\text{gauze}} = \frac{50 + 0.267Re}{Sg - 6.33Re}. \tag{7.3}$$

A passable replica of Su's original results is achieved, and the formula mimics the predictions of the flow model over all *Sg* and *Ma* explored to date (Fig. 7.8).

Presentation in terms of *Sg* reveals (as C_f does not) the criterion for minimum d*p*/*p* — which is maximum *Sg*. From the definition of *Sg* as $pr_h/u\mu$, minimum d*p*/*p* is achieved at maximum *p* and minimum *u*. All other things being equal, this calls for maximising charge pressure p_{ref} and minimising *rpm* — a tentative design rule, perhaps.

The agreement achieved between experiment and steady-flow simulation is over a limited range, and not proof of wider applicability. In search of further corroboration, Fig. 7.9(a) shows the low-*Re* range of a Kays and London plot 'reverse-engineered' to the proposed new format. Figure 7.9(b) is the prediction by the steady-flow model for the same range of *Re* and *Ma*, and for the volume porosity of the Kays and London

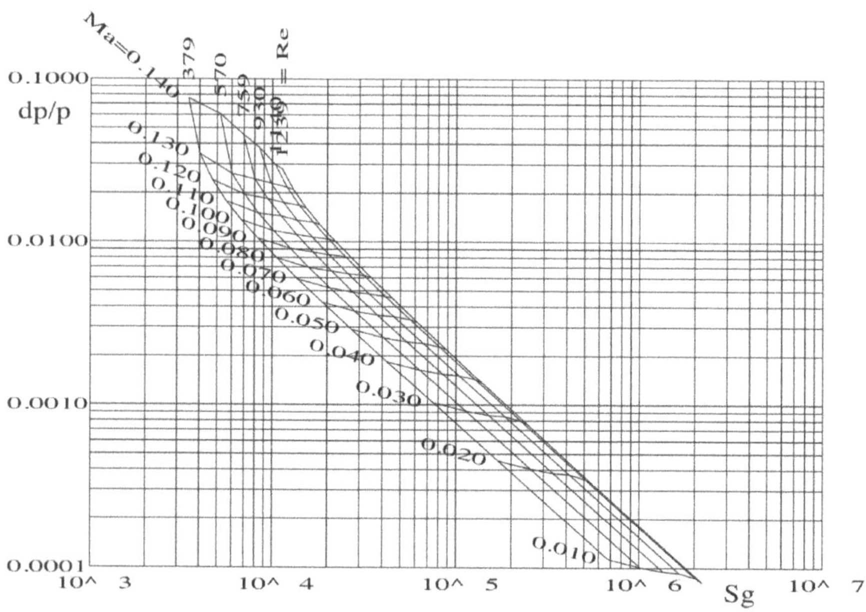

Fig. 7.8. Empirical correlation (Eq. (7.3)) plotted over the range of Su's experimental results (cf. Figs. 7.5 and 7.6).

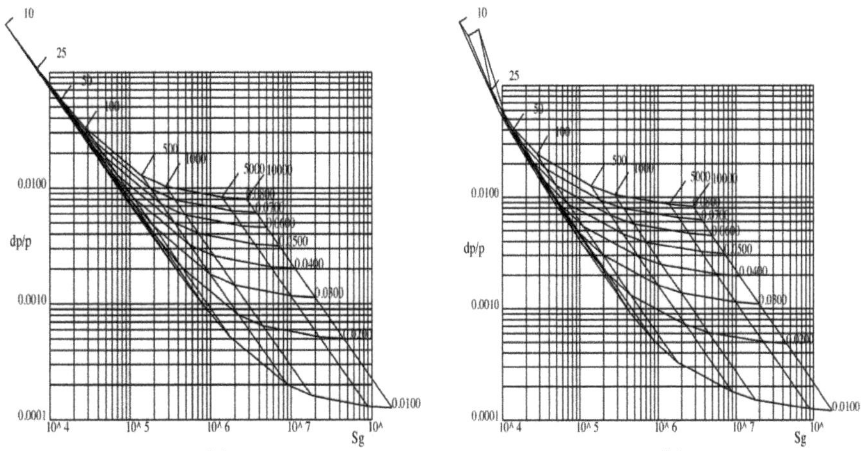

(a) (b)

Fig. 7.9. (a) K & L results at low *Re* 'reverse-engineered' to expose $\mathrm{d}p/p|_{\text{gauze}}$. (b) Steady-flow numerical model of over the same range of *Re* and *Ma*.

original. The comparison bodes well. However, it is no substitute for the extensive experimental program evidently now required — as the following section demonstrates.

7.4. Reappraisal

Suppose Kays and London had not cut corners in acquiring Fig. 7.1, i.e., had instead achieved the variation in *Re* at constant hydraulic radius r_h. The resulting correlation would still have been unusable outside the *p*, *T* and working fluid under which it was acquired: setting up the experiment to achieve a particular *Re* gives you the *Ma* that you get! (Eq. (7.1) again).

For Fig. 7.10, C_f was generated by the steady-flow simulation taking air as the medium, over a range of *Re* at back-pressure $p = 1.25$ atm. and for three values of inlet temperature *T*. The different *T* bring about different relationships between *Re* and *Ma* (Eq. (7.1)) with the predictable outcome — choking at different *Re*. Three different values of *p* at common *T* would have generated an analogous result.

A hasty conclusion is that a lighter gas — say, he — tested at low back-pressure would offer a more favourable characteristic. Figure 7.11 shows that this is not the case, with compressibility effects taking over as before, although at higher *Re*. Equation (7.1) dispels any

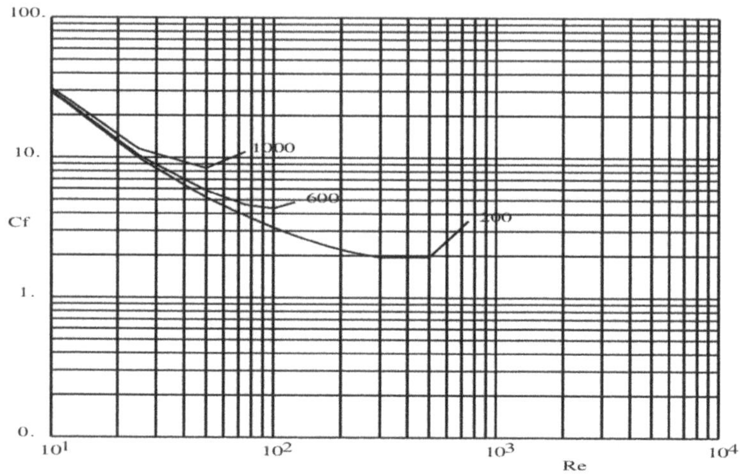

Fig. 7.10. How a single K & L correlation from Fig. 7.1 would have appeared had the variation in *Re* been achieved by varying ρu rather than r_h. Parameter values are abs. temperature. Upstream $p = 1.25$ atm. $¶_v = 0.689$.

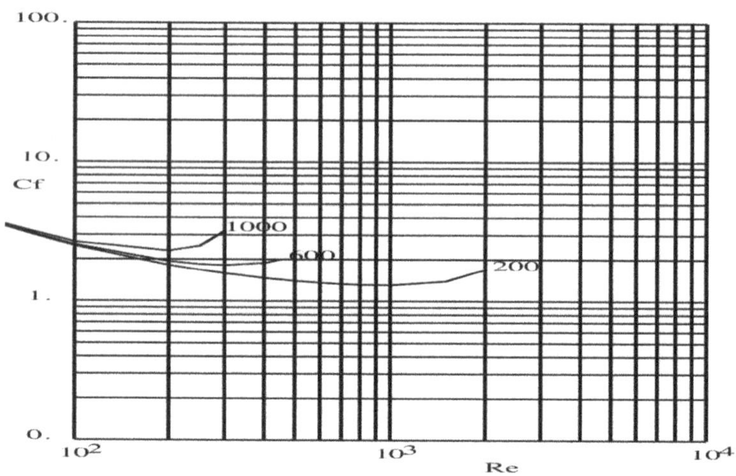

Fig. 7.11. Simulated tests of Fig. 7.10 repeated using He with $p = 15$ atm.

mystery: all other things being equal, the larger the gas constant *R* the larger is *Ma* for given *Re* (see Fig. 7.12).

Figure 7.13 is consistent with the reputation of H2 as the preferred gas for high-pressure, high-rpm operation.

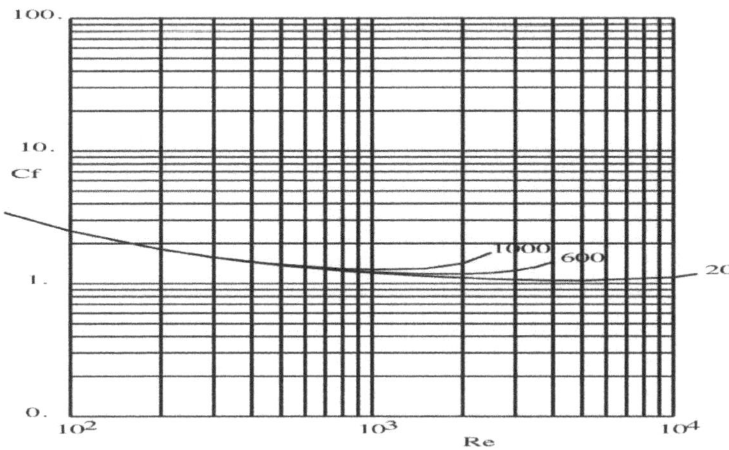

Fig. 7.12. Helium 'tested' at 110 atm. Onset of compressibility effects shifted to higher *Re*. $\P_v = 0.689$.

7.5. Interim Deductions

Friction factor is a perverse concept in the circumstances. To see why, note that $\rho u^2 = \gamma p Ma^2$ — pure algebra, and nothing *per se* to do with compressibility. C_f is acquired from experimentally-measured $\mathrm{d}p/p$ by dividing by $\gamma Ma^2 \mathrm{d}x/r_h$. The original flow may or may not have been in the compressible regime. Reconstituting to acquire $\mathrm{d}p$ involves multiplying C_f by Ma^2 acquired from *different* u, γ, $\mathrm{d}x$, r_h — again with no distinction as to context — compressible or incompressible.

Correlations uncritically acquired by Kays and London combine viscous and compressibility effects in arbitrary proportions. Such a format prevents the respective influences being retrieved retrospectively.

Such correlations remain potentially useful over the range of *Re* within which viscous effects dominate: (the curves of Figs. 7.10–7.13 are identical over that range). On the other hand, different working fluids, different back-pressures and different entry temperatures give rise to different ranges. Pressure swings by a factor of 2 over a cycle. From one end of the regenerator to the other, temperature ranges through a factor of 3.

Reducing test results to the format of Figs. 7.6–7.9 avoids the problem because, as the good Professor Kays might have said: '*A Reynolds number*

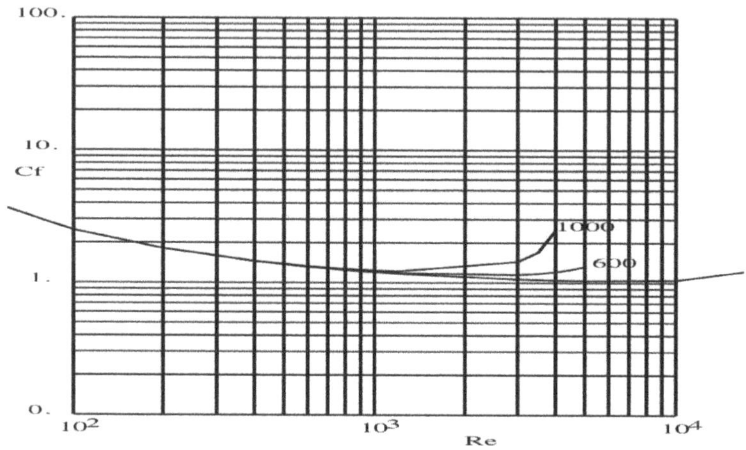

Fig. 7.13. H2 'tested' at 110 atm. $¶_v = 0.689$.

is a Reynolds number — provided it is recognized that a Mach number is a Mach number'.

On the other hand, old preferences do not give way easily. Is there a way of holding on to the $C_f - Re$ tradition without the need for an infinity of correlation curves?

Equation (7.1) to the rescue again! The term $4pr_h\sqrt{\gamma}/\mu(T)\sqrt{RT}$ is, by definition, dimensionless, enabling the relationship between *Re* and *Ma* to be reduced to an expression of the form:

$$Re = Ge.Ma, \tag{7.1'}$$

$$Ge = \frac{4pr_h\sqrt{\gamma}}{\mu(T)\sqrt{RT}}.$$

The function form of C_f allowing for compressibility becomes:

$$C_f = C_f(Re, Ge, ¶_v). \tag{7.4}$$

It thus becomes possible to envisage a manageable number of correlations covering all combinations of p, r_h, γ, R and T — and to a choice of formats:

(a) C_f vs *Re* at a fixed value of *Ge* with volume porosity $¶_v$ as parameter.
(b) C_f vs *Re* at a given $¶_v$ with *Ge* as parameter.

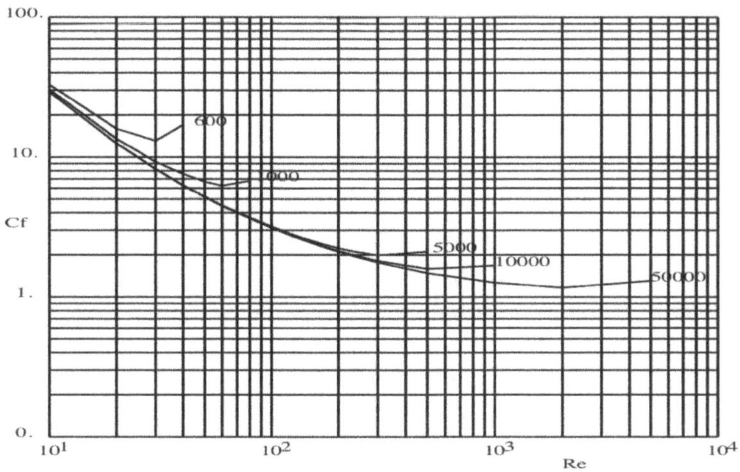

Fig. 7.14. 'Universal' curve of friction factor vs *Re* with parameter *Ge* absorbing any combination of *p*, *T* and *R*. Volume porosity $\P_v = 0.689$.

Figure 7.14 is based on the latter option, and applies to volume porosity $\P_v = 0.689$.

A numerical example will advise suggestions for updating atmospheric air-charged engines by inclusion of a fine-wire gauze regenerator. In the context, a swept volume of 600 cc would be modest, and one might speculate on a peak-load rpm of 1500. Staying with $r_h = 0.0254$ mm of Figs. 7.4–7.11 will allow cross-checking of the outcome.

Various methods are to hand for estimating peak *Re* at the expansion end of the matrix — e.g., that at p. 88 of the author's 2014 text. Any one of these yields $Re = 300$ or so, the specific figure depending particularly on the ratio of regenerator pore volume V_{dr} to swept volume V_{sw}, viz on the value of δ_r. From Eq. (7.1) *Ge* is about 660.

Figure 7.10 applies directly to this case — air at ≈ 1 atm. and r_h = 0.0254 mm. A vertical line run from $Re = 300$ cuts the left-most parameter curve $T_E = 1000\text{K} \approx 700°\text{C}$ at a value of C_f evidently strongly influenced by compressibility.

Applying a straight-edge to Fig. 7.6, joining the proposed values of *p* and r_h and reading from the left-hand graduations of the centre scale gives the maximum *Re* allowed by the cautious criterion derived earlier from Fig. 7.2, namely, $Re \leq 60$.

Use of the consolidated figure (7.14) means imagining a curve for $Ge = 660$ slightly to the right of that labelled 600 — i.e., into the region affected by compressibility. The conclusion is the same as that drawn from Fig. 7.10.

7.6. A Broader Issue

Research interest in the flow of air through wire screens can be traced back to the immediate post-war period, i.e., to the early days of jet propulsion. Ingestion of foreign objects was a concern — as was the performance penalty when using a wire screen as protection.

The main focus was on the compressible regime — but with due recognition of the rôle of Reynolds number. Regardless of flow domain, experimental results were graphed in terms of pressure coefficient $\Delta p/1/2\rho u^2$ (rather than friction factor) with aperture ratio α (or solidity $s = 1 - \alpha$) as parameter. Figure 7.15 is replotted from the account by Cornell (1958). It covers a range of *Re* with *Ma* sufficiently low to avoid the effects of compressibility and embodies the experimental results of independent studies. The eventual rise in pressure coefficient with increasing *Re* conflicts with the trend which Stirling engine studies take

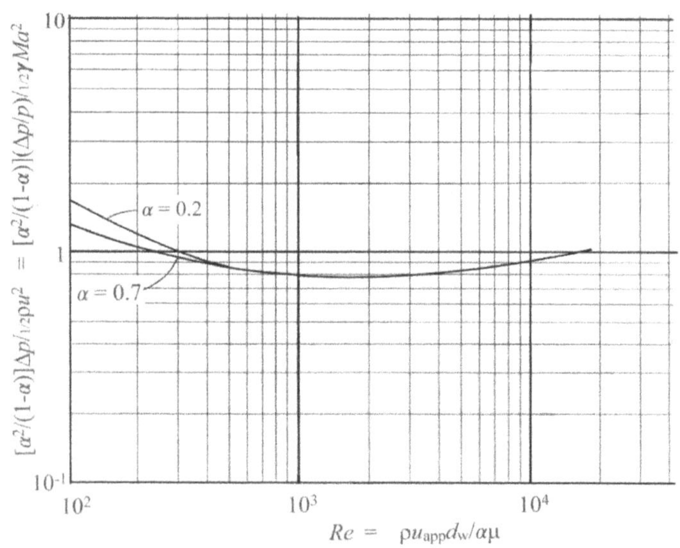

Fig. 7.15. Reproduced from Cornell (1958). At higher end of incompressible regime gradient of pressure coefficient vs *Re* reverses.

for granted — but is consistent with the C_d – *Re* characteristic of the isolated cylinder. The picture is endorsed by Annand (1953).

The science of the Stirling engine has evolved to this point while overlooking — or dismissing — *at least three* defining features of steady flow through gauzes in the presence of a temperature gradient. It would be interesting to know why.

The explanation which follows has been offered previously. However, it is of such fundamental importance that there are no apologies for a repeat — which is in terms of ultra-low specific cycle work which characterises the *genre*. General Motors' GPU-3 returned Z_{pmin} at rated operating point of 0.274 (Table 2.1 earlier). Now consider the Schmidt 'engine'. Its regenerator is *perfect*: zero flow loss and 100% thermal recovery. Fed with the specification and operating conditions of the GPU-3, the lossless Schmidt 'engine' returns Z_{pmin} = 0.439 [–]. Allowing 85% mechanical efficiency reduces this to 0.373 [–].

But from the foregoing figures, the percentage improvement relative to Z of the specimen internal combustion engine (Table 2.1) is a mere 100 × (0.373 – 0.274)/5.47 = 1.81%! (And any such hypothetical improvement is due in part to additional idealisations of the Schmidt 'engine'.)

There is only one conclusion: To date (i.e., in current embodiments) regenerator design has not been critical! It can be argued in support that, if it were otherwise, then there would be instances of 'grading' of the matrix — end-to-end variation of hydraulic radius and/or porosity — to accommodate the considerable differences in flow conditions at the extremities.

It may now be clear why there has previously been no compelling reason to take a second look at reliance on incompressible, steady-flow data such as those of Kays and London.

This is a single aspect of the overriding challenge: to make specific cycle work more competitive. This is a circular problem:

- the greater the improvement achieved, the greater the load on the regenerator;
- the operating environment becomes more demanding, with greatly-increased pressure-swing and flush ratio;

- target temperature recovery ratio and flow loss become harder to design for, calling for flow data relevant to the unique operating conditions of the Stirling engine.

And the immediate problem? Data appropriate to the task are non-existent — either for wire-mesh or for any other candidate material. It is not only pressure drop correlations which are at stake: a cycle simulation needs experimentally-acquired heat transfer data. It can be confidently assumed that these are equally vulnerable to compressibility and thus equally in need of re-acquisition with due regard to Dynamic Similarity principles.

8

A Wrong Turn

8.1. No Apologies

A senior engineer (Forschungsleiter) at Ricardo Deutschland in Stuttgart once confided that, where dynamometer testing was involved, verifying the redesign of component of a Volkswagen engine (in this instance the exhaust manifold) costs upwards of €3 million.

There is no Stirling engine in mass production, so comparison is treacherous ground. However, in terms of sophistication there is something in common with the submarine unit (earlier Fig. 1.1). The regenerator is one of the most costly single components, so it is safe to assume that development will be conservative with a view to minimising unforeseen costs. Suggestions for a radical shake-up of regenerator design are unlikely to receive a warm welcome.

At the opposite end of the scale of sophistication is a pipe-dream of this writer: a portable, air-charged, unpressurised (and thus potentially affordable) Stirling engine of about 50 cc swept volume and 100–150 W output. In the context of the legendary Honda 50 cc four-stroke engine (1800 W at 5700 rpm) the target is hardly over-ambitious. Conversely, it involves improving by a factor of 6 on the current norm of about 0.3 W/(atm.cm^3).

Half of the target would be achieved if specific cycle work N_B could be raised by the modest factor of three. Current values are depressing (Chapter 2): It can hardly be beyond the wit of man to treble such a puny fraction of potential. The outstanding factor of two is available in principle by doubling the rpm at which the improved specific work is delivered. This suggests doubling from about 1500 rpm to 3000.

Neither target is going to be achieved by side-lining crucial and obviously pertinent insights into the regenerator in favour of data from tests using gross apparatus having no relevance whatsoever to the scale of the flow events of interest.

On this basis, there are no apologies for revisiting — and belabouring — the pioneer investigations of Pinker and Herbert (1967).

8.2. From the Horse's Mouth

The authors open their synopsis thus:

The body of previous literature relating to the pressure-loss characteristics of woven wire gauzes fails to provide adequate information on the separate effects of Reynolds number and Mach number in the range of usual interest for duct flow experiments.

To appropriate the parlance of the scientific community: *You can say that again*! They continue:

For this reason, tests have been carried out on a selection of eight single gauzes placed normal to the axis of a duct, of sizes (wire diameter and mesh) such as might be commonly used to give values of porosity[1] *between 0·3 and 0·7. Pressure loss measurements were carefully made, and covered independently duct Mach numbers up to choking and a 50-fold range of Reynolds number. The test data are presented so as to provide a single relation between the incompressible loss factor and a form of Reynolds number, plus a compressibility correction which depends on gauze porosity.*

Figure 8.1 replots the original experimental readings.

The work was carried out at NGTE Pyestock, the UK's prestigious National Gas Turbine Establishment. For over 50 years NGTE led the world in gas-turbine research, testing and development. It was decommissioned in 2000, but facilities at the time were unique and comprehensive, as a visit to one of several websites[2] will confirm.

---

[1]The authors use *porosity* throughout, where the present account will use *aperture ratio.*

[2]Google Daily Mail online: *Goodbye Pyestock – haunting images,* etc.

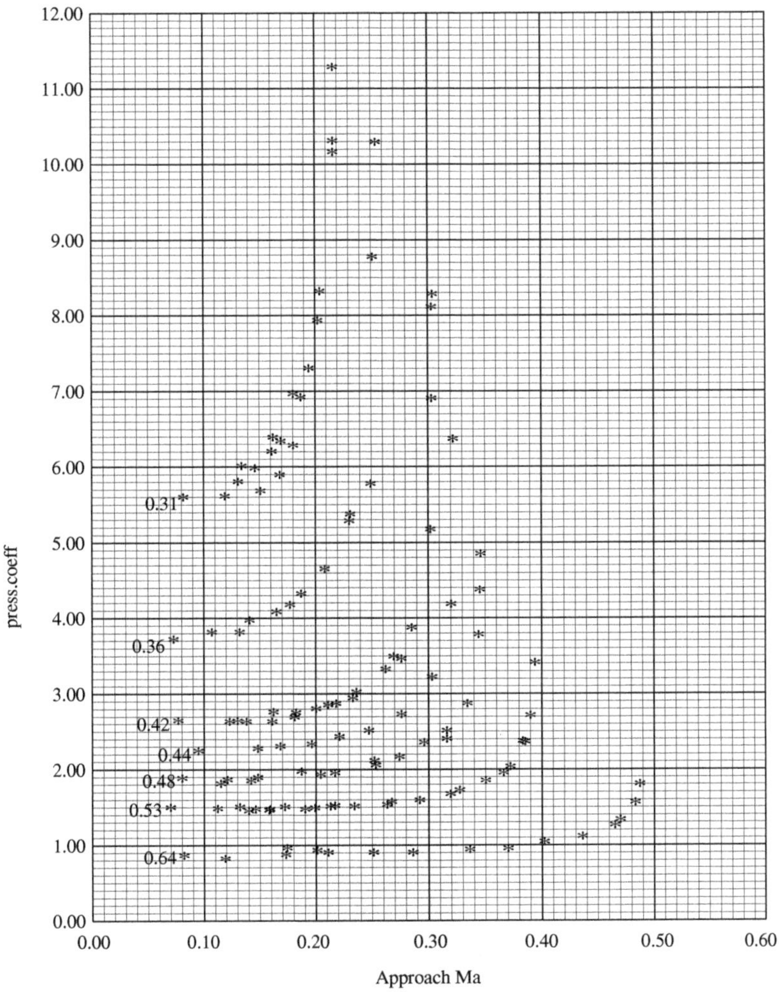

Fig. 8.1. Original experimental readings of Pinker and Herbert. For some reason, the parameter is orthogonal aperture ratio (their 'orthogonal porosity') rather than the 'sinusoidal' value they themselves promote.

A line of enquiry, already under way in 1963, had been the effect on performance of wire screens installed upstream of turbo-jet engine inlets to protect against debris ingestion. It goes without saying that experimental facilities were second-to-none. Researchers Pinker and Herbert had access to a precision-bored wind-tunnel test section of 90 mm internal diameter and variable-velocity flows at static pressures ranging from minus 0.73 atm. and (plus) 5.8 atm. (evidently *gauge* values). It would be

Table 8.1. Gauzes tested by Pinker and Herbert tabulated in order of increasing aperture ratio (i.e., with published sequence reversed). Dimensions were reported in imperial units. *Re* in right-hand column back-calculated from common 'flow-factor'.

d_w (in.)	m_w (1/inch)	$(1-d_w m_w)^2$ (–)	s'soidal ap. rat.	r_h (mm)	*Re* (–)
0.022	20	0.313	0.387	0.632	1992
0.022	18	0.364	0.457	0.669	2213
0.022	16	0.419	0.523	0.705	2490
0.048	7	0.440	0.547	0.719	5692
0.022	14	0.478	0.587	0.742	2846
0.039	7	0.528	0.636	0.771	5692
0.020	10	0.640	0.734	0.832	3984

no exaggeration to say that the work at Pyestock dwarfs the programme of Kays and London — and, indeed, all subsequent investigations.

The superficial similarity between Su's results (earlier Fig. 4.2) and those under consideration here is misleading: All of the former's experimental points were acquired at common respective values of *Re* and of volume porosity $¶_v$. Despite Pinker's and Herbert's ambitious overture, it is not *Re* which they maintain constant as parameter but *aperture ratio* (their 'porosity'). The situation is complicated by the prevalence of three definitions:

$$\alpha = 1 - d_w m_w, \tag{8.1a}$$

$$\alpha = (1 - d_w m_w)^2. \tag{8.2b}$$

The authors' *sinusoidal porosity* α' is here renamed *sinusoidal aperture ratio*. With reference to Fig. 8.2:

$$\alpha' = 1 - 2/\pi^2 \left\{ 1 + 2\left(\frac{1}{2}\pi\, m_w d_w\right)^2 \left[\frac{1}{2}\pi - \cos^{-1}\left(\frac{1}{2}\pi\, m_w d_w\right)\right] \ldots \right.$$

$$\left. 3\left(\frac{1}{2}\pi\, m_w d_w\right)\sqrt{1 - \left(\frac{1}{2}\pi\, m_w d_w\right)^2} \right\}. \tag{8.3}$$

Figure 8.3 and Table 8.1 illustrate the considerable numerical difference between the three expressions as parameter $m_w d_w$ is varied.

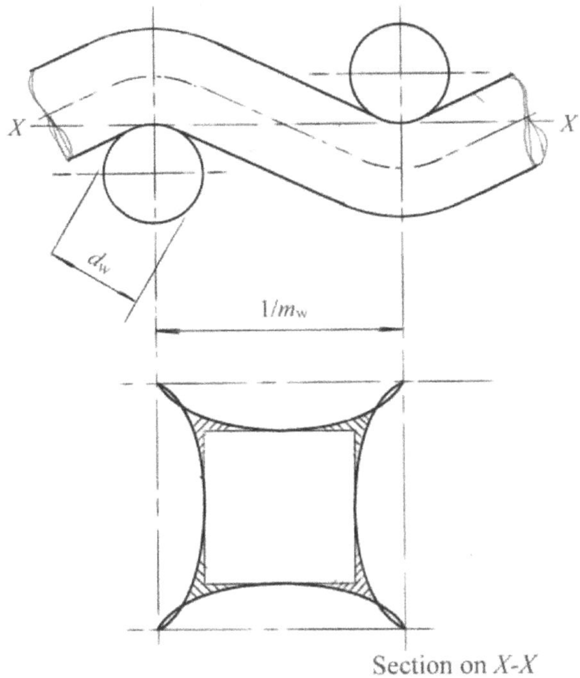

Fig. 8.2. Notation for Pinker and Herbert's 'sinusoidal porosity'.

The true parameter in Fig. 8.1 is aperture ratio α. (For some reason, numerical values of the parameter label are calculated from the compromise Eq. (8.2b) rather from Pinker's and Herbert's own, recommended definition — Eq. (8.3).)

On the other hand, both they and Su compute volume porosity $¶_v$ from the same expression:

$$¶_v = 1 - 1/4\pi d_w m_w. \tag{8.4}$$

Obviously, the three expressions for α' give three different results (Fig. 8.3). However, what matters here is that, for the square-weave wire gauze, $¶_v$ together with all three expressions for α are functions of product $d_w m_w$ *only* — which, being dimensionless, takes the same numerical value regardless of units (inch and wire/inch, mm and wire/mm). Rectangular-weave gauzes sharing the same value of $d_w m_w$ are essentially geometrically similar, and share the same respective values of $¶_v$ and α (any definition).

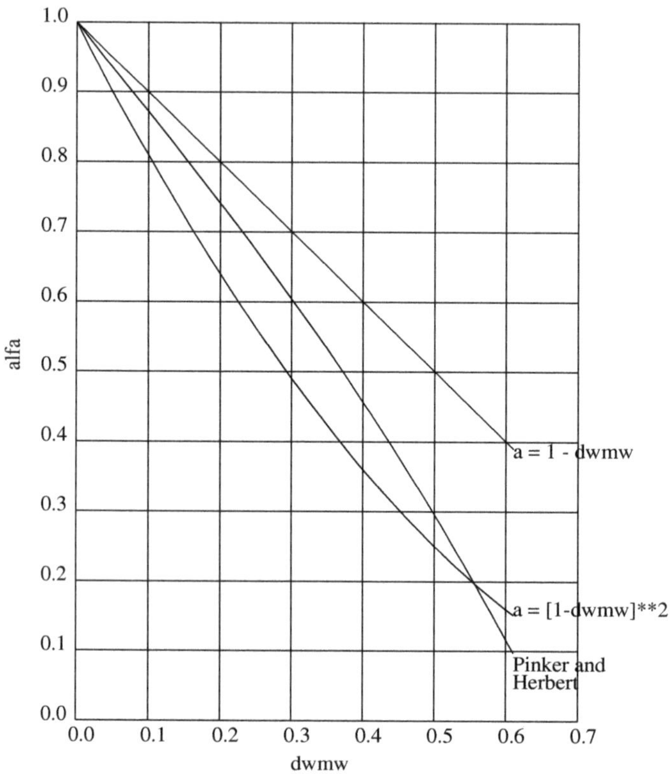

Fig. 8.3. Variation with $d_w m_w$ of the three candidate expressions for aperture ratio.

At many levels, the Pinker and Herbert paper is a model of critical review, experimental technique and exposition. For some reason, however, they chose a parameter with dimensions — their *flow parameter* $\underline{V} = pu_{app}/\mu[m^{-1}]$ — to correlate measurements. $\underline{V}$ is evidently a Reynolds number per unit length. The justification on offer is '… *uncertainty as to the correct characteristic dimension required to make it* (i.e. to make $\underline{V}$) *dimensionless*.' (Since time immemorial the characteristic dimension has been hydraulic radius r_h!)

Given the declared objective, this must be regarded as a missed opportunity — particularly in view of the fact that it is a straightforward matter to vary any one of *Re*, *Ma* or α separately while the other two are held constant, as follows:

In the present context (wire-mesh regenerator) Reynolds number $Re = 4\rho|u|r_h/\mu$ is customarily based on mean interstitial speed $u_{app}/¶_v$. Substituting the ideal gas equation:

$$Re = \frac{4pu_{app}r_h}{RT\mu ¶_v}.$$

Hydraulic radius r_h can be expressed in terms of wire diameter d_w and $¶_v$

$$r_h = 1/4\ d_w ¶_v/(1 - ¶_v).$$

Substituting and cancelling:

$$Re = \frac{pu_{app}d_w}{RT\mu(1-¶_v)}. \quad (8.5)$$

Volume porosity is a function only of the product $d_w m_w$ — as is aperture ratio (any definition), so the latter can be kept constant by maintaining $¶_v$ constant. *Ma* is constant if $u_{app}/\sqrt{RT}$ is constant. This leaves scope for independent variation in aperture ratio and *Ma* while *Re* is kept constant by keeping a constant value of the group $pu_{app}\, d_w/(1 - ¶_v)$.

For the moment is has to be accepted that Pinker and Herbert preferred to hold $\underline{V}$ constant by varying upstream velocity and static pressure. The numerical value settled on was 4.3×10^6/ft. (or 1.31×10^7/m). When back-calculated, the *Re* covered exceed by an order of magnitude values anticipated in Stirling engine operation. This reflects the order-of-magnitude difference in wire diameter d_w (contrast the *three* orders difference used by Kays and London to achieve their *Re* of 10^5).

8.3. Pinker and Herbert in Context

Figure 8.4 replots Fig. 8.1 to the format used in Chapter 4 for displaying Su's results, viz as $dp/p|_{gauze}$ vs Stirling number $Sg = pr_h/\mu u$ with *Re* back-calculated as parameter. The plot violates a golden rule — but this is justified by an insight which will be delivered by Fig. 8.5.

There is an impression of encouraging family resemblance with earlier sets of curves to the same format. This, however, is misleading, as is

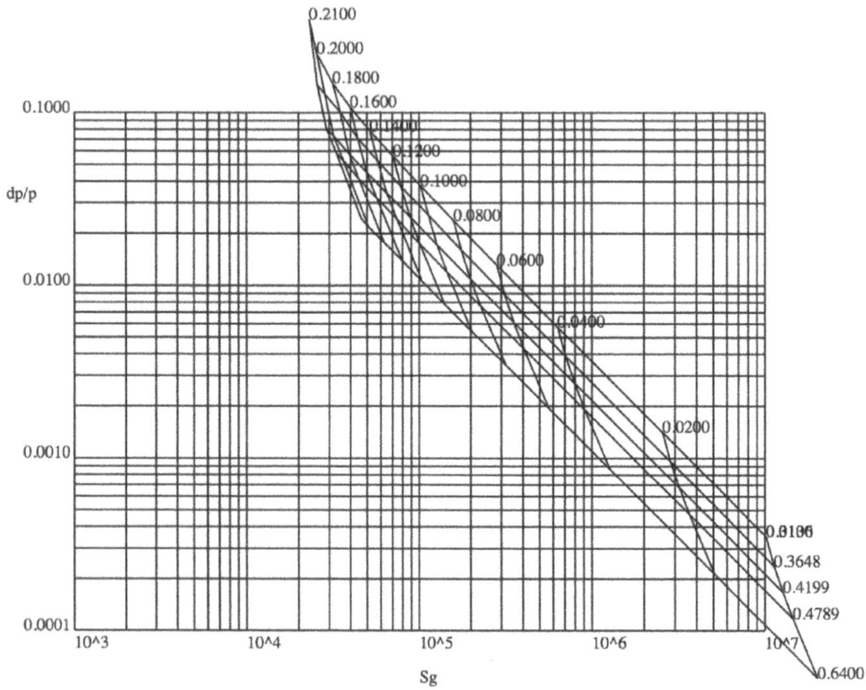

Fig. 8.4. Results of Pinker and Herbert re-plotted to dp/p vs Sg format. Parameter labels at lower right are aperture ratio α.

seen on replotting with *Re* held constant over the whole range. (An arbitrary value serves.) The result at Fig. 8.5 shows that a decrease in aperture ratio is accompanied by a large increase in d$p/p|_{\text{gauze}}$ at any given *Ma*. This contrasts with earlier plots with their modest decrease in d$p/p|_{\text{gauze}}$ with increasing *Re* when volume porosity $¶_v$ (and thus aperture ratio α) is held constant throughout.

The original experimental settings have been fed to the computer simulation which generated the maps of d$p/p|_{\text{gauze}}$ – Sg featured in Chapter 5. Figure 8.6 shows the result. Exact superposition is not achieved, but the simulation evidently responds to variations in the input parameters in all the same senses as does the experiment. This, however, is not the main contribution of simulation, which will be in synthesising a plot conforming to Dynamic Similarity principles.

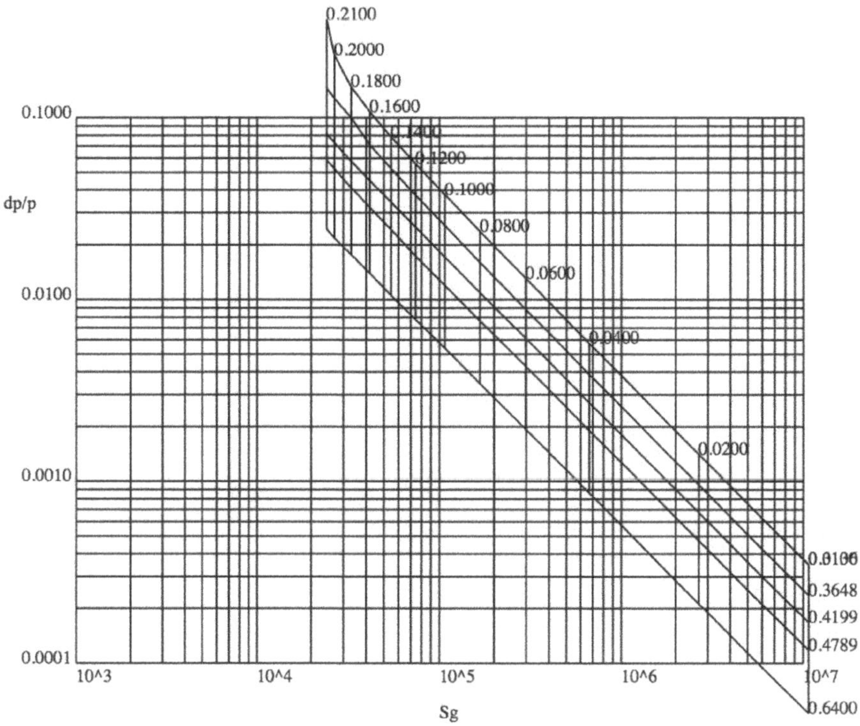

Fig. 8.5. Figure 8.4 replotted with *Re* held at common invariant value for all aperture ratios.

This is shown in Fig. 8.7. But for the fact that the simulation struggles with approach *Ma* above about 0.16, this 'legitimate' form of plot supports the earlier 'sleight-of-hand' re-processing shown in Fig. 8.5. Pressure coefficient $dp/p|_{gauze}$ shows a wider variation with aperture ratio than does the experimental counterpart — a matter which is under investigation.

8.4. A Way Forward

Pinker and Herbert pursue their campaign:

It is necessary to start by distinguishing clearly between two major effects on the pressure-loss coefficient of any particular geometry of

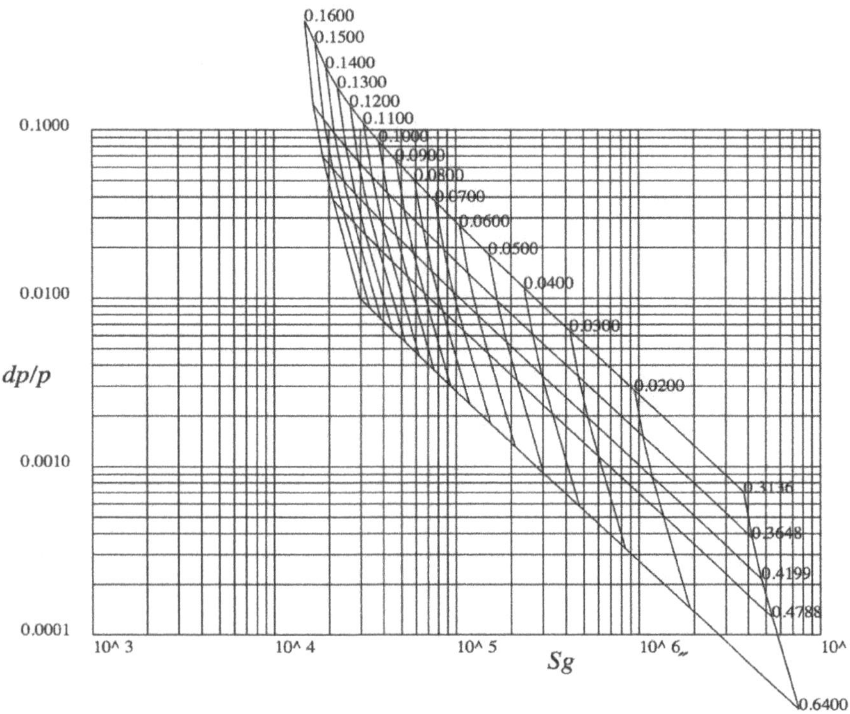

Fig. 8.6. Pressure coefficient $dp/p|_{gauze}$ calculated by steady-flow simulation over the range of values covered in Pinker's and Herbert's experiments.

gauze — namely, the effect of Re (normally correlated under incompressible flow conditions), and that of Ma at given Re. The hope may be entertained that the variation with Mach number can be expressed in a form which is effectively independent of the particular value of Re, provided that value is constant. Thus the two effects may be separable for the purpose of correlation and prediction.

Some previous authors have created confusion by presenting data which purport to show the effect of Re, but include significant variation in Ma (e.g., reference. [5]) or vice versa (e.g., reference [4]): ... ***In none of the known literature is it possible to find data for compressible conditions in which the two effects can be properly isolated (Appendix 1).*** (This author's emphasis.)

What the authors are driving home, with the aid of a certain amount of repetition, is that pressure-drop in *steady, uniform-pressure,*

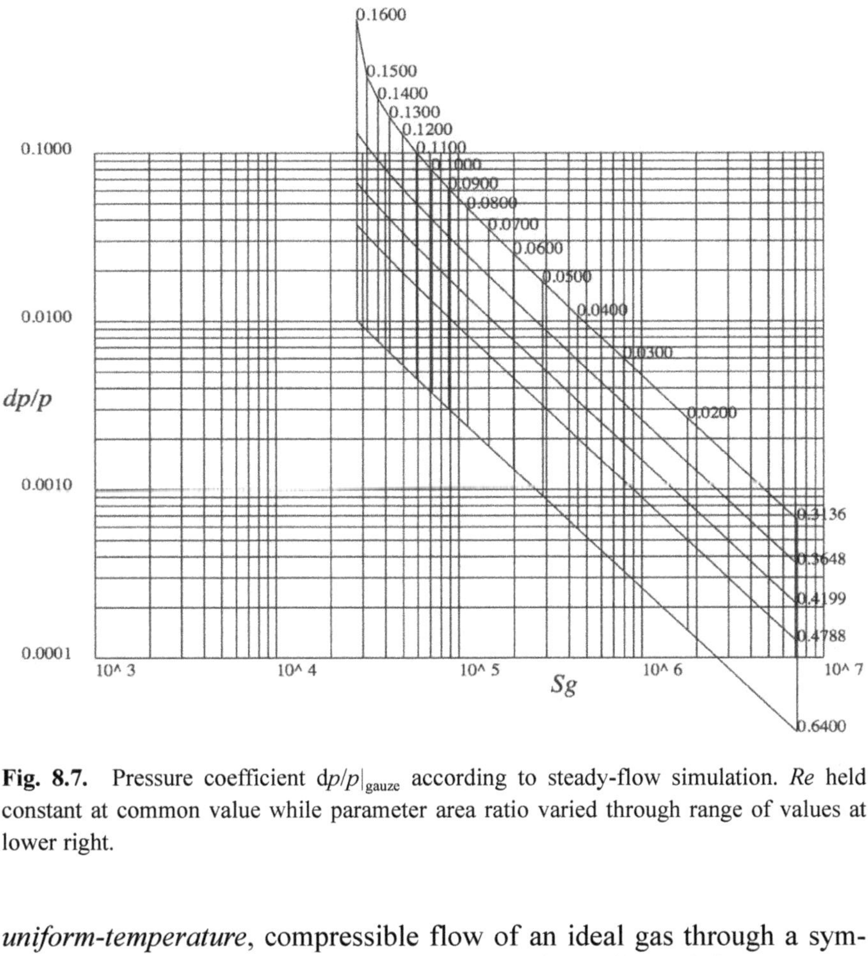

Fig. 8.7. Pressure coefficient $\mathrm{d}p/p|_{\mathrm{gauze}}$ according to steady-flow simulation. *Re* held constant at common value while parameter area ratio varied through range of values at lower right.

uniform-temperature, compressible flow of an ideal gas through a symmetrical wire gauze can be characterised symbolically as follows:

$$\mathrm{d}p/p|_{\mathrm{gauze}} = f\{Re,\ Ma,\ ¶_{\mathrm{v}} \text{ or } \alpha\}. \tag{8.6}$$

They did not have the Stirling engine in mind, so can be excused for omitting the other influences identified and illustrated in Chapter 5 — absolute temperature, T and pressure p — or, indeed, the dimensionless parameter *Ge* into which they can be combined.

Stirling engines are charged to pressures p_{ref} ranging over a factor of 100, temperatures over factor of 3 or so. All other things being equal, significant independent variation in either variable alters the relationship between *Re* and *Ma* — and thus shape of the correlation.

For as long as poor working-fluid utilisation continues to be compensated by pressurisation, the matter can be brushed under the carpet. Success in raising specific cycle work by a factor of, say, 3 through improved working fluid utilisation would mean a regenerator more heavily-loaded — and loaded differently, with pressure swing accounting for a much greater part of thermal load.

8.5. Application to Other Matrix Materials

Attention has focused on the wire gauze as the material offering the widest range of aperture ratio and hydraulic radius consistent with predictable flow-passage geometry. However, it is not the sole candidate for regenerator use which is susceptible to compressibility. Figure 8.8 is a depiction by Cornell (*op. cit.*) of the contraction downstream of flow through a sharp-edged orifice. (Sharp edges have been observed separating the pores of Retimet™ metal foam.)

In Fig. 8.9, a limited number of experimental points has been superimposed on Cornell's theoretical study, for which *Re* were assumed sufficiently high that compressibility dominated. The trend (with both aperture ratio and *Ma*) is as for the experimental curves of Su (Fig. 5.3).

8.6. Résumé

Pinker and Herbert saw flow through the wire gauze first and foremost in terms of compressibility and choking, with viscous effects as a secondary influence:

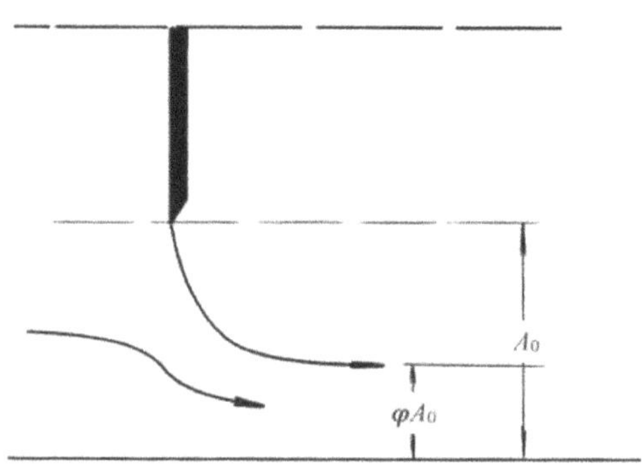

Fig. 8.8. After Cornell (1958). Contraction downstream of sharp-edged orifice.

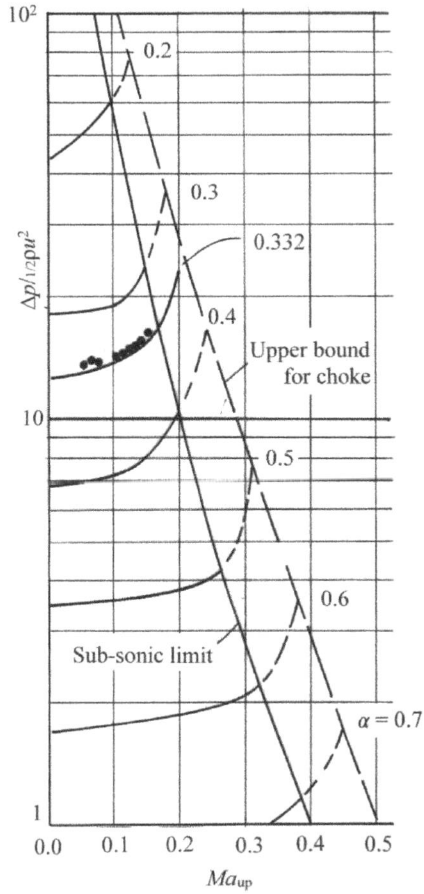

Fig. 8.9. Experimental points plotted into Cornell's theoretical study of compressible flow through a sharp-edged orifice. The parameter of the original curves was solidity *s*, relabelled here as aperture ratio 1 – *s*.

One can only infer that, in the range of flow conditions of present interest, the level of Re is sufficiently high for this particular parameter to play a comparatively small part, while the dominant effect is Mach number.

In contrast, the start-point for Kays' and London's work, first published three years previously in 1964 (but embarked upon considerably earlier), is that flow is essentially incompressible and susceptible to viscous and momentum effects only.

Clearly, both cannot be correct. It would be interesting to know how Stirling engine science came to evolve around the latter view to the total exclusion of the former. If there is to be progress, incompressible and compressible flow domains have to be treated as a continuum.

At first sight, the undertaking required to establish the form of Eq. (8.6) by experiment is daunting: 20 points each for *Re* and *Ma*, and 10 for $¶_v$ (or α) suggests 4000 experiments. On the other hand, a suitable range of specimen points might validate completion of the picture with the aid of steady-flow simulation.

The challenge remains of acquiring counterpart heat transfer data — but progress is achieved one step at a time!

9

Heat Transfer Correlations

9.1. Status Quo

The material of the previous chapter took up the approach of Pinker and Herbert (1967), whose studies of pressure loss in flow through gauzes were an attempt to separate the rôles of compressibility from those of momentum change and viscous dissipation. It now becomes impossible not to query the validity and relevance of counterpart heat transfer data, i.e., those correlating $StPr^{2/3}$ with Re.

Two difficulties threaten: (a) as far as this writer is aware, there are no precedents — neither relevant experimental work nor applicable theoretical studies and (b) the challenge of devising a competent experiment vastly exceeds that of the corresponding isothermal pressure-drop test. Moreover, even the form the experimental results might take is difficult to anticipate — apart, that is, from a couple of near-certainties:

(1) Vulnerability will be in the ranges of Re and Ma already identified as affecting $\Delta p/p$.
(2) Reynolds' analogy, which advises the form of C_f vs Re and $StPr^{3/2}$ vs Re correlations, cannot be expected to do so under the effects of compressibility.

The outcome of the present enquiry will be correlations in the spirit of those of the earlier chapter, i.e., in terms of $\Delta p/p$ vs Sg and $\Delta p/p$ vs Re. Derivation will be empirical, and with only indirect laboratory confirmation. While the resulting correlations may eventually have a role in guiding the design of experiment, they should *not* be relied upon for first-principles design — except to the extent that they caution against reliance on the established alternative.

9.2. A Fresh Start

The design of a laboratory experiment reflects the capabilities of available instrumentation. Heat transfer data which continue to be relied upon for Stirling regenerator design date from the late 1950s, when it was common to monitor temperature transients using DC equipment — thermocouples, potentiometers (Leeds and Northrup K2 if one was fortunate), mirror galvanometers, etc.

Repeating the early experiments would not be the best use of the latest in real-time data capture. Likewise, the design of a symbolic model has to be in terms of available resources — preferably the most basic to avoid biasing the outcome towards earlier conclusions.

Heat transfer data for the single wire can be expected to be reliable when used under the conditions for which they were originally acquired, namely, steady flow perpendicular to the wire. A correlation advocated by Eckert and Drake (1959, p. 242) between Nusselt number Nu_{dw} and Reynolds number based on diameter Re_{dw} is:

$$Nu_{dw} = 0.43 + 0.48\sqrt{Re_{dw}}, \quad 1 \le Re_{dw} \le 4000. \tag{9.1}$$

It derives from experiments with wires and cylinders having diameters ranging from 0.0189 mm to 150 mm (0.0007 inch to slightly less than 6 inches) — evidently spanning all d_w likely to be encountered in Stirling regenerators.

Nominal conversion from Re_{dw} to Re based on hydraulic radius r_h can be made using:

$$r_h = \frac{1}{4}\, d_w\P_v/(1 - \P_v), \tag{9.2}$$

and substituting into $Re_{dw} = \rho u d_w/\mu$. ('Nominal' because the algebraic device does not reflect a physical counterpart.)

The proposed heat transfer model requires a 'seed' value of Stanton number St. Making use of the identity $St = Nu/RePr$ allows Eq. (9.1) to be rewritten:

$$St = 0.43[Re(1 - \P_v)/\P_v]^{-1} + 0.48[Re(1 - \P_v)/\P_v]^{-1/2}. \qquad (9.3)$$

In Eq. (9.3), *Re* is the form preferred for internal flows, viz, $4\rho u r_h/\mu$.

9.3. Assumptions

Figure 9.1 indicates a 'virtual' regenerator. Gas enters from the left at steady, uniform speed u and temperature T_{gin} [K], exiting via the right-most gauze layer at temperature T_{g2} [K] depending on the prevailing value of $StPr^{2/3}$ (to be determined). Temperature gradient dT_w/dx within the core is linear between T_E and T_C and is invariant. This last condition calls for either infinite thermal capacity or steady input of heat by some mechanism yet to be devised (but see the final section of this chapter). The matrix core is a uniformly close-packed stack of gauzes of wire diameter d_w and mesh number m_w. It is assumed:

(1) That the $\Delta p/p$ analysis employed earlier, and derived under the assumption of isothermal flow, remains valid when applied across a small temperature difference, say, 1°C or thereabouts.

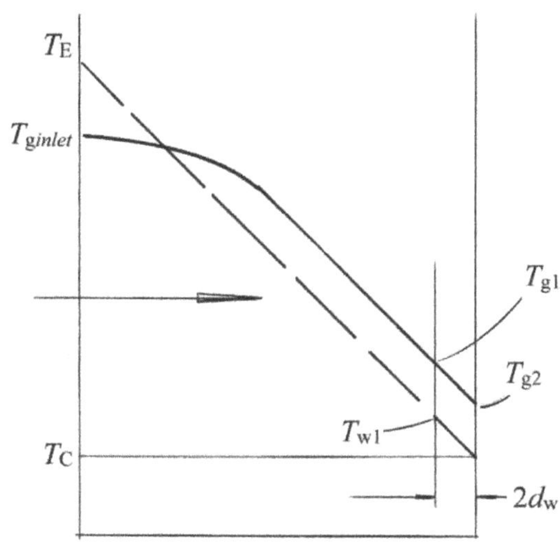

Fig. 9.1. 'Virtual' regenerator, indicating invariant linear temperature of matrix (dashed line) and gas temperature profile typical of *NTU* sufficiently high for useful function as a regenerator.

(2) That certain staples from traditional regenerator thinking apply. For example, that temperature recovery ratio η_T under (hypothetical) steady-state conditions at viable *NTU* (Number of Transfer Units = StL_{reg}/r_h) is given by Eq. (9.6) to follow.

With reference to Fig. 9.1, temperature recovery ratio η_T in the 'forward' (left-to-right) direction can be expressed as the ratio of temperature drop achieved to the maximum possible temperature drop. With $N_T = T_E/T_C$ and $\tau_g = T_g/T_C$

$$\eta_T = \frac{T_E - T_{g2}}{T_E - T_C} = \frac{N_T - \tau_{g2}}{N_T - 1}. \tag{9.4}$$

According to a derivation by the author (2014, 201 pp.) temperature difference at any location may be written in terms Number of Transfer Units, $NTU = StL/r_h$. For the entire stack of overall length L_{reg}, $NTU = St \times L_{reg}/r_h$. For the individual gauze of thickness $2d_w$ in the flow direction, $NTU = St \times 2d_w/r_h$.

$$T_g - T_w = -(1 - N_T)/NTU + [T_{gin} - T_w + (1 - N_T)/NTU](e^{-NTU} - 1).$$

For inlet temperature difference $T_{gin} - T_w$ realistically small by comparison with the term $(1 - N_T)$, exit difference $T_{g2} - T_w$ is equal to $T_{g2} - T_C$ and

$$T_{g2} - T_C \approx -(1 - N_T)/(1 + NTU). \tag{9.5}$$

Hausen (1927) may have been first to note that, when η_T is sufficiently high to be useful:

$$\eta_T = (NTU - 1)/NTU = 1 - 1/NTU \approx 1/(1 + NTU). \tag{9.6}$$

Inserting values corresponding to exit conditions and using the preceding expressions to eliminate η_T gives a first approximation to exit temperature T_{g2}:

$$T_{g2} = \frac{N_T - 1}{St \times L_{reg}/r_h} - T_C. \tag{9.4}$$

The development now requires a first approximation to the inlet temperature T_{g2} of the specimen gauze (in this case, that at exit). This should reflect the influence of pressure-drop dp/p. The steady-flow element of the energy equation for compressible flow as formulated by Shapiro (1954) is

$$q'\rho A_{\mathrm{ff}}\mathrm{d}x = \frac{\partial}{\partial} x\left[\rho u A_{\mathrm{ff}}\mathrm{d}x\left(\rho \mathrm{c_v} T + p/\rho + \frac{1}{2}\right)\rho u^2\right]\mathrm{d}x.$$

To the scale of the mesh-stack a single gauze (axial thickness $2d_w$) serves as elemental length dx. The expression is rewritten in terms of the exit gauze, approximating d(u^2) by $2u$du. The term q' is Shapiro's heat rate per unit mass. The left-hand term therefore becomes $hp_w(\underline{T}_g - T_w)\mathrm{d}x$ where $\underline{T}_g$ is gas mean temperature $1/2(T_{g2} + T_{g1})$. Noting that wire temperature T_w at exit is T_C:

$$hA_w(T_{g2} - T_C)\mathrm{d}x \approx \rho u A_{\mathrm{ff}}\mathrm{d}x[\rho \mathrm{c_v}(T_{g2} - T_{g1}) + (p_2 - p_1)/\rho + \rho u_1(u_2 - u_1)].$$

The term $(p_2 - p_1)/\rho$ is reexpressed via the ideal gas equation as $R\underline{T}_g\Delta p/p$ and the group $\rho u A_{\mathrm{ff}}\mathrm{d}x$ transferred to become the denominator of the left-hand side:

Recalling the definition $c_p/c_v = \gamma$, noting that $h/\rho|u|\mathrm{c_p}$ is Stanton number and making use of the ideal gas equation as appropriate gives:

$$\gamma St(2d_w/r_h)(T_{g2} - T_C)/T_C \approx (T_{g2} - T_{g1})/T_C + \cdots$$
$$\Delta p/p(\gamma - 1)\underline{T}/T_C + (\gamma - 1)u_1(u_2 - u_1)/R\underline{T}. \quad (9.5)$$

In Eq. (9.5), $\underline{T}$ is mean temperature $1/2(T_{g2} + T_{g1})$. Its use is a formality: the eventual solution is sensitive to *differences* $u_2 - u_1$, etc. Whether the arithmetic proceeds in terms of $\underline{T}$ or, e.g., T_{g1} has no significant influence on the outcome.

For given values of overall temperature ratio N_T and mesh geometry, the latter expressed in terms of volume porosity $¶_v$, the parameters of the formulation are boundary conditions *Re* and *Ma*. The 'seed' value of *St* is calculated by using the former in Eq. (9.3). Numerical values are available for all variables except T_{g1}: $\Delta p/p$ is generated by the compressible flow model (Chapter 5). This comes across as an absolute value, and needs to be given the appropriate sign (negative).

Equating input values of *Re* and *Ma* and cancelling gives $u_1 = 4p_1r_h\gamma/\mu$, where base pressure (atmospheric) serves for p_1. The term du is eliminated by use of the continuity equation $\rho u A_{ff}$ = constant in conjunction with the ideal gas equation, leading to

$$du/u = dT/T - dp/p. \tag{9.6}$$

A value of T_{g2} comes from Eq. (9.4), allowing the value of T_{g1} to be extracted explicitly. This, however, is tied to the 'seed' value of *St* provided by a Eq. (9.7) which is a function of *Re* only. How is this going to suggest a correlation reflecting the influence of compressibility?

One option is to calculate a value of heat transfer coefficient *h* as was done to evaluate *St* in Kays' and London's original correlations (1964) — i.e., by using values of T_{g2} and T_{g1} measured in the counterpart laboratory experiment. Convection coefficient *h* [W/m²K] is defined as $q'/A_w(T_g - T_w)$, where q' is heat rate from gas to gauze.

Equating q' to experimentally-measured enthalpy rate $\rho u A_{ff} c_p(T_{g2} - T_{g1})$ gives

$$St(Re,\ Ma,\ \P_v, \ldots) = \frac{(T_{g2} - T_{g1})}{(T_g - T_w)2d_w/r_h}. \tag{9.7}$$

The result may now be substituted for the 'seed' value of Eq. (9.5) and the calculation iterated to convergence.

9.4. Simulated Correlations

Figure 9.2 shows $StPr^{2/3}$ plotted against *Re* with curves of constant Ma_1 and of constant *Sg* superimposed for comparison with the corresponding plots of $\Delta p/p$ vs *Re* (Chapter 5). The initial impression of similarity with the $StPr^{2/3}$ vs *Re* of Kays and London is an illusion: in the latter case the parameter was volume porosity $\P_v$. In the present case it is *Ma,* the figure being constructed for a gauze with $d_w = 0.04$ mm, $m_w = 9$ wires/mm and corresponding $\P_v$ (from Eq. (8.4)) = 0.717. With this duly noted, a curve from Fig. 9.2 which coincides remarkably well with Kays' and London's curve for $\P_v = 0.725$ is that for lowest *Ma* — as should be the case.

Nothing in the algebra nor in the FORTRAN code has been 'tweaked' to achieve this outcome.

The symbolic model is only as good as the calculation of $\Delta p/p$. It will be recalled that this struggled where low *Re* coincided with the high end

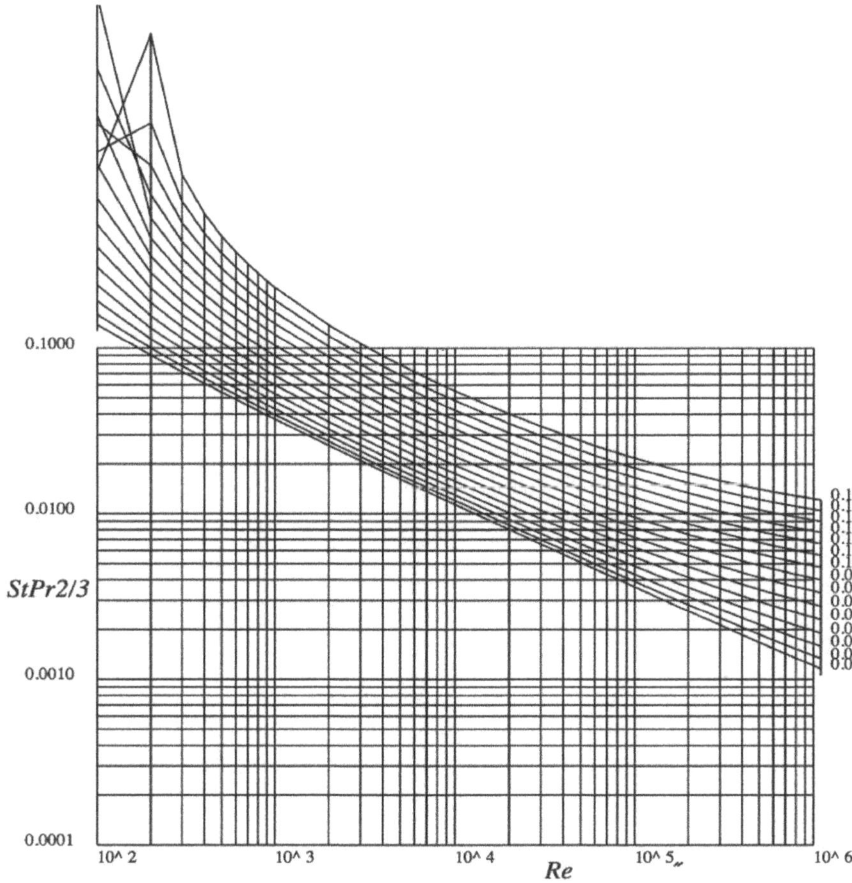

Fig. 9.2. Computed $StPr^{2/3}$ as a function of *Re* with *Ma* as parameter. Volume porosity ¶$_v$ = 0.717 throughout. Lowermost curve (lowest *Ma*) coincides closely with the corresponding Kays and London curve at their ¶$_v$ = 0.749. (***Not to be relied upon for design!***)

of the *Ma* range. The difficulty carries over to this latest formulation, as evidenced in the disrupted traces at upper left.

Sensitivity of the formulation to ¶$_v$ is indicated by comparison with Fig. 9.3, for which m_w has been decreased by some 10% to 8 wires/mm. The corresponding increase in ¶$_v$ of a bare $4\frac{1}{2}$% from 0.717 to 0.749 increases *Re* achieved in the convergent-divergent flow channel and reduces the numerical problem seen in Fig. 9.2.

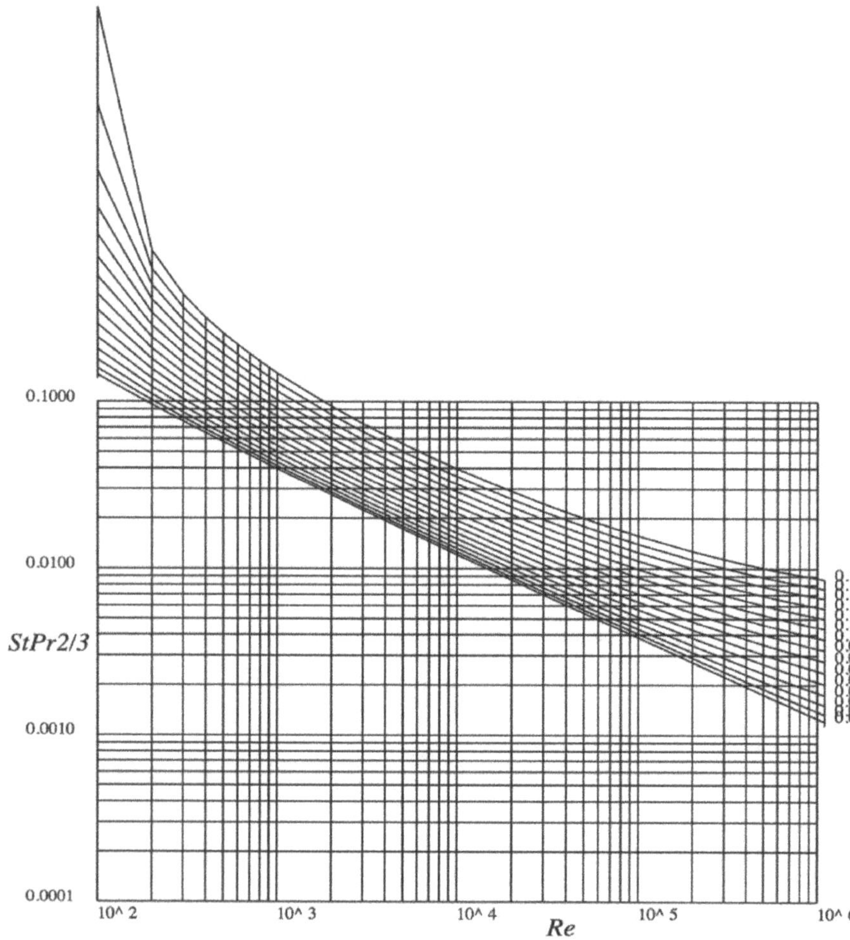

Fig. 9.3. As for Fig. 9.2, but with m_w reduced to 8 wires/mm giving $¶_v$ = 0.749. The steady-flow simulation which generates d*p*/*p* now copes with the combination of minimum *Re* and maximum *Ma*. (***Not to be relied upon for design!***)

9.5. The Case for a Return to the Laboratory

Kays and London acquired results at high *Re* in a tunnel of circular cross-section using crossed cylindrical rods to simulate woven mesh. Testing of wire screens was '... *in an entirely different test apparatus* ...' (The 1964 text, p. 133). The latter was a small, cylindrical test-section tunnel of 2.93-inch diameter.

The 'matrix' in the larger-scale tests consisted of 3/8-inch (9-mm) diameter rods. But for a single rod, all were of plastic. The exception was of copper, and contained a thermo-couple. The procedure involved heating the single rod '... *to about 30 deg. above air-flow temperature, inserting it in the desired location in the core and recording its time-temperature history on cooling*.' The contrary condition assumed in the Stirling engine is that, at any instant, each one of the hundreds of individual wires in a screen are at the same, uniform temperature.

There is no reported attempt to keep *Ma* constant while *Re* was varied, yet between lowest and highest *Re* the 'correlations' are presented as smooth, un-interrupted and without inflexion! It is safe to say that they do not reflect the last word in experimental technique.

9.6. A Candidate Experiment

Figure 9.4 represents a rectangle of gauze from which a number of wires have been stripped along all four edges. The protruding wires at opposite sides are connected to a bar of conducting material, probably by spot-welding. If a voltage — alternating or direct — is applied between the bars, the horizontal wires can be expected to share the total current equally, since there is no potential between adjacent parallel wires and thus no tendency for current to short-circuit between them.

The points of metal-to-metal contact between warp and weft are tiny cinch marks, consistent with a poor thermal conduction path. To this extent, the system of Fig. 9.4 does not guarantee uniform heating of all four sides of the individual flow passage as required by the experiment. The situation is unlikely to benefit from applying the voltage simultaneously across both orthogonal sets of wires (right-hand figure), since the current will then take the path of least resistance, i.e., the shortest. On the other hand, use of micro-processor and power-diodes in suitable combination could allow rapid switching of the electrical supply between warp and weft directions, thereby eliminating electrical shorting. Instantaneous power input could be captured and integrated in real time. A matter to be explored is whether the resistivity of the parent material would be sufficiently temperature-sensitive to provide a read-out of wire temperature. If the answer is no, a calibrated radiation pyrometer would be equally non-invasive. If yes, then a complementary check is never wasted!

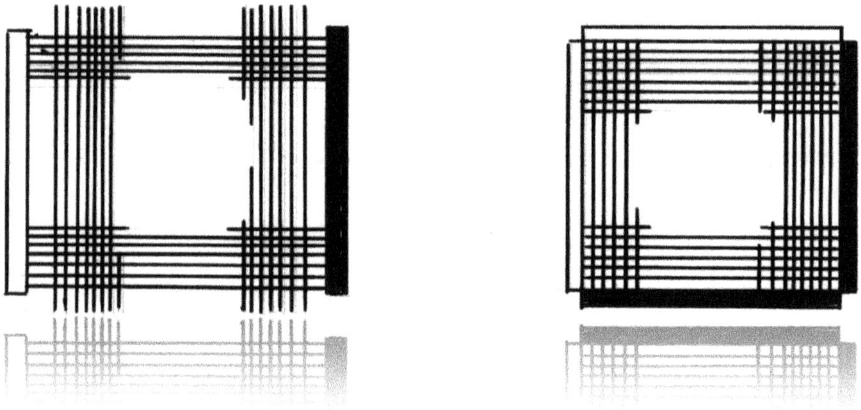

Fig. 9.4. Scheme for achieving steady, uniform electric heating of gauze. Applied voltage switched at high frequency between warp and weft directions.

9.7. Conclusion/Discussion

Robert Stirling saw his engine as a specimen application of the core invention — the regenerator. For as long as design of the key component hangs on seven-decades-old data of questionable relevance, Stirling engine studies cannot be called a science.

With little by way of prior art,[1] scope for elaboration is limited. There will doubtless be objections to the 'home-baked' formulation. The counter argument is that the priority at this stage is not to achieve the definitive treatment, but to address the more urgent matter of getting the ball rolling.

The results of the present study show encouraging overlap with earlier experimental results under precisely that flow condition where overlap might be expected, namely the incompressible regime. This could justify use of the symbolic model to explore basing $St(Re, Ma, ¶_v, \ldots)$, i.e., Eq. (9.7), on *static* pressure distribution in the flow channel, available as a by-product of the $\Delta p/p$ simulation.

[1] To borrow from the vocabulary of patent drafting.

10

Practical Regenerator Design

10.1. Academic Responsibility

Few individuals outside industry and research laboratories have contributed more to the debate on Stirling engines than has Geoff Vaizey. It had long been believed (e.g., Finkelstein, 1960a, b) that dead-space — or unswept volume — was unconditionally a performance penalty. Experiments with Vaizey's handsome rhombic drive engine (illustrated at Fig. 16.1, author's 2007 text) served to refocus attention on the true culprit — uncompensated adiabatic temperature swing.

His V-twin thermal-lag engine ran at 2000 rpm almost vibration-free, and demonstrated how far this intuition-defying concept can be taken with the aid of exemplary machining.

Vaizey has since rejoined the ranks of steam-power modellers, to their considerable gain — and at a loss to the Stirling community. He moved on with a characteristically thoughtful parting shot:

An engine whose functioning relies on pumping working fluid back-and-forth through a flow resistance starts with a crippling disadvantage. (This writer's paraphrase.)

It would not be wide-of-the-mark to say that Vaizey has been let down by the academics. Collectively we are guilty of a wide range of sins: pursuit of mirages (Carnot efficiency); perpetuating out-of-date concepts; failure to verify sources; arbitrary choice of numerical methods for partial-differential equations and this writer's speciality: good, old-fashioned carelessness. Most reprehensibly in the present context, we have failed, after two centuries of pondering, to put together a viable science of regenerator design.

Walker may have been the first to draw attention to the regenerator as more than a challenge in unsteady heat exchange. Under heading *Regenerator design — a practical guide* at p. 150 of his 1980 book he sees design as an exercise in reconciling conflicting priorities:

for maximum heat capacity — a large, solid matrix
for minimum flow losses — a small, highly porous matrix
for minimum dead-space — a small, dense matrix
for maximum heat transfer — a large, finely-divided matrix
for minimum contamination[1] *— a matrix with no obstruction* (*sic*!)

Things have since moved on, particularly in the case of the wire-gauze option. A matter which has come to prominence during exchanges with Dutch engineer Camille van Rijn is that r_h and $¶_v$ *can be varied independently*. Reynolds number *Re*, C_f and *NTU* and are functions of hydraulic radius r_h. Susceptibility to compressibility effects is strongly influenced by volume porosity $¶_v$ (or, equally, of area ratio α). These features means that, within limits, one can design *against* the onset of compressibility effects whilst simultaneously designing *for* required heat transfer. The two parameters become inter-dependent only when mesh number m_w is finalised.

The task clarifies further on addressing a specific performance requirement. If the un-pressurised, air-charged engine is to serve a useful purpose there has to be a dramatic increase in specific power. Two areas for improvement have already been identified. One is specific cycle work, or Beale number, N_B. The other is the rotational speed at which rated N_B is achieved. A three-fold improvement in each suggests a net gain of 9 — almost an order of magnitude. The present chapter focuses on the search for higher rpm.

[1]Contamination might be addressed as problem in own right rather than being allowed to compromise regenerator design.

10.2. Designing for *rpm*

The power curve of the 1-atm. air-engine typically falls off above 500 – 600 rpm. It is possible that this reflects the onset of choking in the regenerator, but lack of data gets in the way of further enquiry. Conversely, the desirability of deferring onset of choking to higher rpm has already been justified: from Fig. 8.1 and Table 8.1 of the earlier chapter the condition is the largest possible area ratio α (or highest porosity $\P_v$) consistent with required hydraulic radius r_h.

The arithmetic starts from the expression for volume porosity in terms of wire diameter d_w and mesh number m_w:

$$\P_v = 1 - \tfrac{1}{4}\pi d_w m_w, \tag{10.1}$$

(and not as regrettably mis-printed at p. 163 of the author's 2007 text). Extracting d_w:

$$d_w = 4(1 - \P_v)/\pi m_w. \tag{10.2}$$

A relationship already established between r_h and d_w is

$$r_h/d_w = 1/4\ \P_v/(1 - \P_v). \tag{10.3}$$

Eliminating d_w between Eqs. (10.2) and (10.3):

$$m_w = \P_v/\pi r_h. \tag{10.4}$$

With volume porosity $\P_v$ selected in terms of compressibility criteria, feeding Eq. (10.4) with the value of r_h indicated by heat transfer and/or flow friction calculations gives required mesh number in the units of r_h — e.g., wires/mm if r_h is in mm. Corresponding wire diameter [mm] can then be back-calculated via Eq. (10.1).

The combination of d_w and m_w resulting from a first iteration may well be outside the range of manufacturers' stock sizes, where aperture ratio tends to hover around the low value of 0.31. A call to the UK branch of the Swiss company G Bopp and Co. confirms that the company will weave to m_w and d_w specified by the customer (although to a minimum order of 150 metre length).

An expensive exercise threatens! But one is either serious about Stirling engines or one is not. If the former, there is a strong incentive

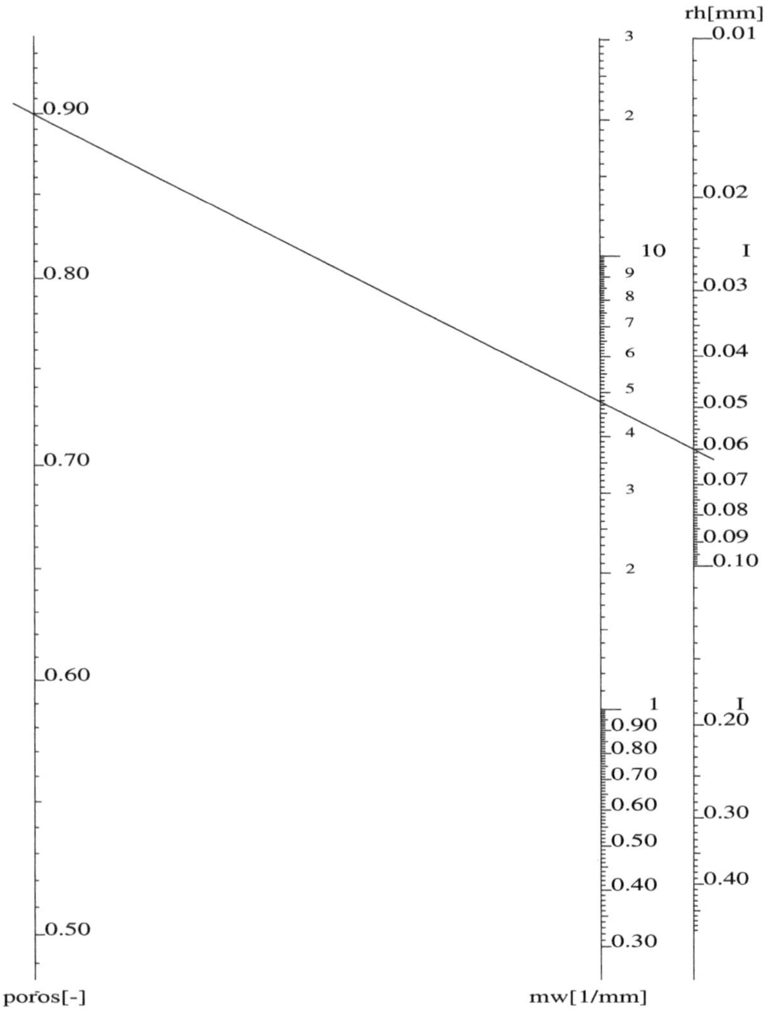

Fig. 10.1. Graphical display of Eq. (10.4). The oblique line intersects the m_w scale at the value of m_w corresponding to $¶_v = 90\%$ and $r_h = 0.06$ mm. Right-hand graduations of the m_w (middle) scale gives wires/mm.

to get the design right first time. This in turn suggests the likelihood of several iterations. Figures 10.1 and 10.2 may help. The former gives m_w [wires/mm] in terms of $¶_v$ and r_h [mm]. The latter back substitutes to give d_w.

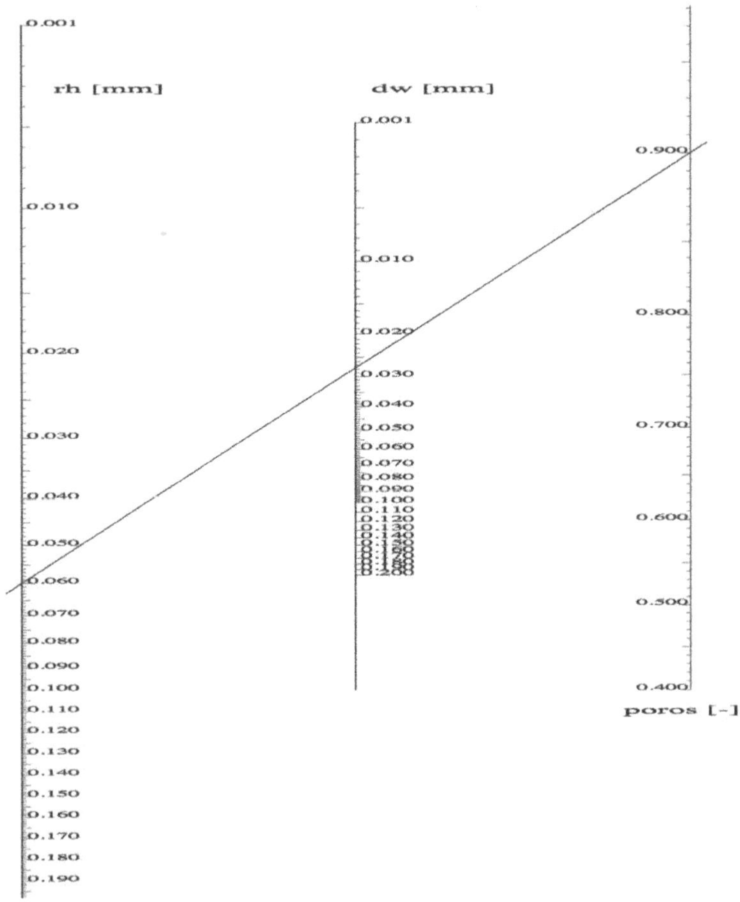

Fig. 10.2. Graphical display of Eq. (10.1) with Eq. (10.3). The oblique line intersects the d_w scale at the value of Eq. (10.1) corresponding to $\P_v = 90\%$ and $r_h = 0.06$ mm.

The oblique line in both cases is the 'solution' corresponding to the specimen choice of $\P_v = 90\%$ and $r_h = 0.06$ mm.

10.3. Unexplored Territory

Under an earlier heading it was deduced that specific cycle work (Beale number, N_B) of the reciprocating, fossil-fuelled heat engine has an upper limit around 7, with values of 5 or so demonstrated in practice. By contrast, the Stirling engine typically achieves some 2% of the latter figure.

The comparison will draw different interpretation from different observers. To this writer it suggests grossly under-unexploited potential. Useful *power* per unit charge pressure and per unit swept volume is in proportion to the product of N_B with *rpm*. On showing to date, the ultimate *rpm* limit of the air-charged Stirling engine is lower than that of the internal combustion engine of comparable cubic displacement. On the other hand, nobody is in a position to say what that apparent limit owes to the depressingly low N_B.

Which leads to the point of this section: exploring the potential (if any) while minimising outlay requires thermodynamic data appropriate to the territory. Loosely speaking, this is the range between the modest *rpm* currently demonstrated and unprecedentedly high values.

For flow speeds in the range $Ma < 1.0$, the relevant transport phenomena are a continuum. Available data, by contrast are anything but! Indeed, refugees from other disciplines might consider them to be in disarray. In the flow domain unaffected by compressibility (more accurately, *assumed to be* unaffected) the dependent variable friction factor C_f is expressed in terms of Reynolds number *Re* with volume porosity $¶_v$ as a mildly-influential parameter. In the region where compressibility begins to take over, pressure coefficient $\Delta p/1/2\rho u^2$ is a function of *Ma* with parameter area ratio α having a dominant influence.

Attempts to unify the two approaches are thwarted by inconsistencies. Crossed rods used to 'simulate' screens do not actually do so; they offer area ratio α given accurately by $1 - m_w d_w$ per layer of thickness d_w (half the thickness of the gauze of the same m_w and d_w), and $¶_v$ given *precisely* by Eq. (10.1). For the woven screen, α is given only *approximately* by $(1 - m_w d_w)^2$ and $¶_v$ is only *approximated* by Eq. (10.1). Taking $m_w d_w = 0.4$ as an instance, α is either 0.6 or 0.36 depending upon choice of definition — a difference of 40% or 66% depending on whether it is looked at as an over- or under-estimate.

The level of consistency and accuracy necessary to design a Stirling engine would be a matter for debate[2]. However, it is unlikely that performance will ever approach potential without a raising of standards — or even that design will become competitive.

[2] Scientists find it necessary to define the metre (for example) as the length of the path travelled by light in a vacuum in 1/(299792458) of a second, i.e., to an accuracy of some +/− 0.0000003%. On this basis, study of the Stirling engine has a way to go before graduating to the status of a science.

With increasing speed of flow (more appropriately, with increasing *Ma*), the transition from incompressible to compressible behaviour is smooth and continuous. Looking for this reality in existing design correlations calls for interchangeability of $¶_v$ and α, and, in turn, for mutually-consistent definitions. The section which follows attempts the necessary rationalisation.

10.4. Inter-Conversion of $¶_v$ and α for Rectangular Screens

The process will be in three steps, the first providing an approximation for kick-starting computation of later more ambitious expressions — and to give confidence in the result.

An approximation to aperture ratio of the woven screen is:

$$\alpha \approx (1 - d_w m_w)^2. \tag{10.5}$$

Inverting Eq. (10.1) gives $d_w m_w = 4(1 - ¶_v)/\pi$. Substituting into Eq. (10.5):

$$\alpha \approx [1 - 4(1 - ¶_v)/\pi]^2. \tag{10.6}$$

Curve 1 of Fig. 10.3 shows how α depends on $¶_v$ according to Eq. (10.6). A provisional deduction is that, at low values of $¶_v$, use of the numerical value of the one gives a misleading guide to the value of the other. Of more pressing concern is whether either represents reality to the extent necessary for the purposes of first-principles design.

An expression for $¶_v$ which should be an improvement over Eq. (10.1) acknowledges the increased length of the individual wire element beyond the notional value of $1/m_w$:

$$¶_v \approx 1 - 1/4\pi\, d_w m_w \sqrt{1 + (d_w m_w)^2}. \tag{10.7}$$

Expanding the earlier Eq. (10.5) yields a quadratic equation in $d_w m_w$, the standard solution of which reduces to

$$d_w m_w \approx 1 \pm \sqrt{\alpha}. \tag{10.8}$$

The only possibility is the negative sign. Substituting for $d_w m_w$: into Eq. (10.7)

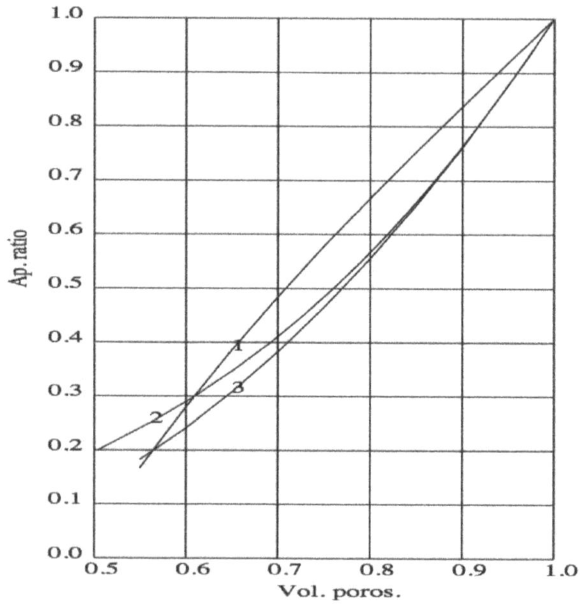

Fig. 10.3. Aperture ratio plotted against volume porosity according to 3 representations of flow-passage geometry. Curves 1, 2 and 3 in order of assumed realism.

$$¶_v \approx 1 - \frac{1}{4}\pi(1-\sqrt{\alpha})\sqrt{1+(1-\sqrt{\alpha})^2}. \tag{10.9}$$

The relationship of Eq. (10.9) is superimposed on Fig. 10.3 as curve 2.

The most convincing expression for α — here denoted α' – is due to Pinker and Herbert (1967). With reference to the earlier Fig. 6.2, effective aperture ratio α' (Eq. (6.3)) was

$$\alpha' = 1 - 2/\pi^2 \left\{1 + 2\left(\frac{1}{2}\pi m_w d_w\right)^2 \left[\frac{1}{2}\pi - \cos^{-1}\left(\frac{1}{2}\pi m_w d_w\right)\right] \ldots \right.$$

$$\left. 3\left(\frac{1}{2}\pi m_w d_w\right)\sqrt{\left(1 - \left(\frac{1}{2}\pi m_w d_w\right)^2\right)}\right\}. \tag{10.10}$$

An improved estimate of $¶_v$ is available on approximating the likely path of the wire element by a sin curve:

$$y=\frac{1}{2}d_w \sin\theta, \quad -\pi/2<\theta<\pi/2. \tag{10.11}$$

The standard expression for length ds of a line following curve y = sin$x$ is $\mathrm{d}s=\mathrm{d}x\sqrt{1+(\mathrm{d}y/\mathrm{d}x)^2}$. The relevant x in this case has the value $-\frac{1}{2}m_w$ for $\theta = -\pi/2$. Intermediate values of θ are therefore given by $\theta = \pi m_w x$. Using the chain rule, dy/dx = (dy/dθ)(dθ/dx) and

$$\mathrm{d}s = m_w^{-1}\sqrt{\left[1+\left(\frac{1}{2}\pi\cos x\right)^2\right]}\mathrm{d}x. \tag{10.12a}$$

Equation (10.12a) appears not to be a standard integral, but is readily dealt with numerically by summing over the range $-\pi/2m_w < x < \pi/2m_w$ in successive increments of $\Delta x = 1/(n.m_w)$:

$$s \approx m_w^{-1}\sum_{j=n/2}^{n/2}\sqrt{\left\{1+\left[\frac{1}{2}\pi\cos(j\Delta x)\right]^2\right\}}\Delta x. \tag{10.12b}$$

An individual ‘cell’ of mesh has envelope volume $2d_w/m_w^2$. The volume of solid material which intrudes is now $2 \times 1/4\pi d_w^2 s$. The corresponding estimate of volume porosity is

$$¶_v \approx 1-1/4\pi d_w m_w \times m_w s. \tag{10.13}$$

Equation (10.13) is used to eliminate $\pi d_w m_w$ from Eq. (10.9):

$$\pi d_w m_w \approx (1-¶_v)/1/4m_w s. \tag{10.14}$$

This is Eq. (10.7) with straightline wire-element path-length $m_w^{-1}\sqrt{1+(d_w m_w)^2}$ replaced by the revised estimate, viz by $m_w^{-1}s$. Substituting into Eq. (10.10) for α' (‘effective’ area ratio according to Pinker and Herbert) gives curve 3 of Fig. 10.3. Comparison with curve 1 shows the scope for improvement, with a maximum of ~20%.

The relationships derived in this section may serve as an interim measure while results are awaited from tests properly isolating the influence of *Ma* from that of *Re*.

10.5. Common Ground

At *Ma* below unity, incompressible merges into compressible forming a continuum. First-principles design would benefit by having experimental data acquired and displayed accordingly.

The function form of direct use for Stirling regenerator design is in terms of fractional pressure drop per gauze, $\Delta p/p$ (reasons having been given elsewhere):

$$\Delta p/p = \Delta p/p[Sg \text{ (or } Re), Ma + \text{a parameter to define geometry}].$$

It has not been possible to locate published data, incompressible or compressible, which conform. Moreover, accounts are in insufficient detail to allow retrospective conversion over the relevant ranges of *Re* and *Ma*. It is, however, possible to display incompressible and compressible results in terms *either* of pressure coefficient $f = \Delta p/1/2\rho u^2$ *or* of friction factor $C_f = 2fd_w/r_h$. The preference here is f on the basis (a) that it has been the unanimous choice for reporting tests in compressible flow (b) that the number of independent sources outnumbers the competition (incompressible) and (c) that the hijacking of C_f by Kays and London from the loosely-related context of pipe flow is highly questionable.

The $2d_w/r_h$ of the proposed conversion can be retrieved from Eq. (10.3).

$$2d_w/r_h = 8(1 - \P_v)/\P_v. \tag{10.15}$$

An essential step is to settle on a common geometric parameter. At first sight, the alternatives are porosity $\P_v$ and aperture ratio α. Three definitions of $\P_v$ are in current use, and three for α, so that a gauze (unambiguously) defined in terms of d_w and m_w can be reported in terms of no fewer than nine different pairs of numerical values!

A way forward is now obvious: all six competing expressions are functions of the parameter uniquely defining the flow-passage geometry of the woven wire screen, i.e., of $d_w m_w$. Moreover, the numerical value $d_w m_w$ can be retrieved unambiguously in every case for which the choice of definition — $\P_v$ or α — is declared in the source. Retrieval is explicit as per Eq. (10.8), or by elementary inversion as in the case of Eq. (10.1), or with the aid of a graph such as Fig. 10.3 or, in the case of more sophisticated expressions which may eventually prove desirable, by iteration. Any of the less accurate expressions can provide a seed value, so the iteration option can hardly fail.

Figure 10.4 is a hybrid: For the left-hand image, pressure coefficient f converted from the classic Kays and London correlation (their Fig. 7.9) is replotted over the range of Re covered by their experiments with gauzes (i.e., stopping short of results achieving high Re by use of unrepresentative crossed-bars). Parameter values have been converted back from $\P_v$ to $d_w m_w$ via the equation declared by Kays and London, i.e., Eq. (10.1) above.

The right-hand image is a reworking from Annand (1953) whose steady-flow experiments went some way to separating Ma from Re by testing at high values of the latter where its influence (according to Annand — but see later) is minimal. The parameter of the original curves was solidity s, i.e., $1 - \alpha$, here converted in two steps to the equivalent $d_w m_w$.

Before the two sets of curves are compared it is necessary to return to the left-hand set. In the original from which they have been

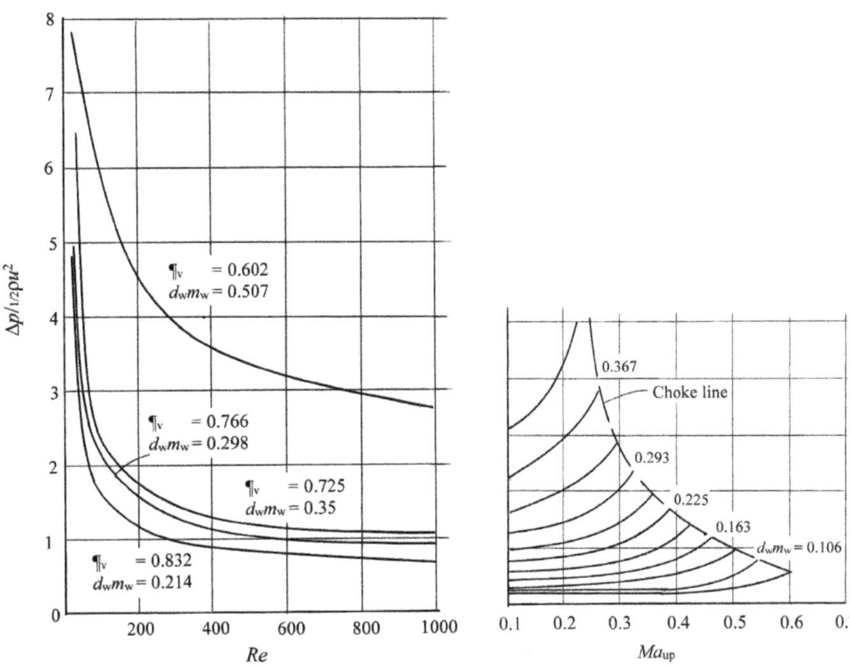

Fig. 10.4. Left: Kays' and London's *Cf-Re* correlation (region acquired using wire screens) replotted in terms of pressure coefficient and universal geometric parameter $d_w m_w$. Right: Cornell's correlation (compressible flow) *f–Ma* correlation with parameter curves replotted and relabelled in terms of universal geometric parameter $d_w m_w$.

converted, the curve for $¶_v = 0.602$ twice crosses that for $¶_v = 0.832$. This conflicts with intuition — and with the fact that the curves of the companion correlation ($StPr^{2/3}$ vs *Re* — Fig. 7.8 of the 1964 original) do not cross. Moreover, whereas the latter show monotonic increase in $StPr^{2/3}$ with increasing/decreasing $¶_v$, the counterpart $C_f - Re$ curves show no consistent trend.

So it is perhaps remarkable that, when converted to the form of Fig. 10.4 (left) (f vs *Re* with $d_w m_w$ as parameter), the curves no longer cross, and the trend of f with $d_w m_w$ is monotonic in the same sense as for the (original) $StPr^{2/3}$ vs *Re* curves. It would be gratifying to be able to claim this in vindication of the choice of $d_w m_w$ for parameter. That, however, cannot be justified until the matter is understood. In the meantime, Figs. 10.4 (left) and 10.4 (right) have to be compared as they stand.

Having different abscissae (*Re* vs *Ma*), the plots *per se* do not illustrate a continuum. Nevertheless, the expectation is for continuity — or at least tangency — between incompressible and compressible curves of comparable $d_w m_w$. This might be claimed in the case of $d_w m_w = 0.214$ and $d_w m_w = 0.298$ of the left-hand figure, but some stretch of the imagination would be required to reconcile the higher values. In any case, the expected tangency implies that the extremities of curves either side of the divide should be horizontal.

Either the assumption of continuity between incompressible and compressible regimes is flawed, or the data replotted here have been achieved under invalid assumptions. The latter interpretation can be probed using an empirical fit commonly recommended for the original Kays and London curves, namely $C_f = c/Re + d$. Different c and d are required to fit different $¶_v$, but the variation either side of $c = 60$ and $d = 0.25$ is not great. Recalling that $\frac{1}{2}\rho u^2 = \frac{1}{2}\gamma pMa^2$, fractional pressure drop in incompressible flow is fitted approximately by

$$\Delta p/p \approx \frac{1}{2}\gamma Ma^2(60/\mathrm{Re} + 0.25). \tag{10.16}$$

As Professor Kays would doubtless insist, a Mach number is a Mach number. Indeed, the *Ma* of the present treatment is the same *Ma* whether referred to incompressible flow or compressible. The caution is that Eq. (10.16) does not describe compressible flow.

This is the point at which to return to the assertion by Annand and by Cornell that the high *Re* of their respective curves of f vs *Ma* were independent of *Re*. The highest *Re* achieved by Kays and London in tests on woven screens has been inferred earlier as 600 or so. Assuming a similar figure for Annand *et al.*, the viscous term of Eq. (10.1) takes the value 60/600 = 0.1 — suggesting strong dependence of $\Delta p/p$ on *Re*.

Far from validating the exercise, the outcome reinforces the need for laboratory tests appropriate to the task — that of separating the influence of *Re* from that of *Ma*.

10.6. Application to Reality

Under an earlier heading it was proposed to explore the possibility of increasing by a factor of three the rated rpm of the 1-atm. air engine. This should evidently be pursued in terms of the values of *Ma* and *Re* likely to arise.

Under the assumption of near-simple-harmonic motion, an estimate of peak particle speed is available in terms of peak piston speed, ωr; where r is crank throw — or $S/2$, where S is stroke. Approximating ω by $rpm/10$, denoting piston area as A_p and noting that SA_p equals swept volume V_{sw} gives peak particle speed at the piston face $\hat{u}_p$,

$$\hat{u}_p = \frac{1}{2} rpm \times V_{sw}/10A_p.$$

Estimating *Ma* at approach to the regenerator involves pro-rating $\hat{u}_p$ by the ratio of A_p/A_{fr} where A_{fr} is matrix frontal area. The latter in turn is equal to $A_{ffr}/¶_v$ where A_{ffr} is matrix free-flow area. In terms of earlier definitions of L_{ref} as $V_{sw}^{1/3}$ and α_{ffr} as $A_{ffr}/V_{sw}^{2/3}$ the estimate of peak approach speed $\hat{u}$ is

$$\hat{u} = \frac{1}{2} rpm \times L_{ref}/(10 \times ¶_v \times \alpha_{ffr}). \tag{10.17}$$

Using values for Philips' MP1002CA air engine with rated *rpm* increased by the proposed factor of three (i.e., to 4500) suggests $\hat{u}$ would be 17.76 m/s, or slightly over *Ma* = 0.05. The origin of Fig. 10.4 (right) is $Ma_{app} = 0.1$, but it is evident that compressibility effects arise at lower values, and that they do so increasingly as $d_w m_w$ increases (as $¶_v$ decreases).

And this is far from the complete picture: The data for Fig. 10.4 (left) were acquired at *Re* claimed to be sufficiently high that *Re* had no

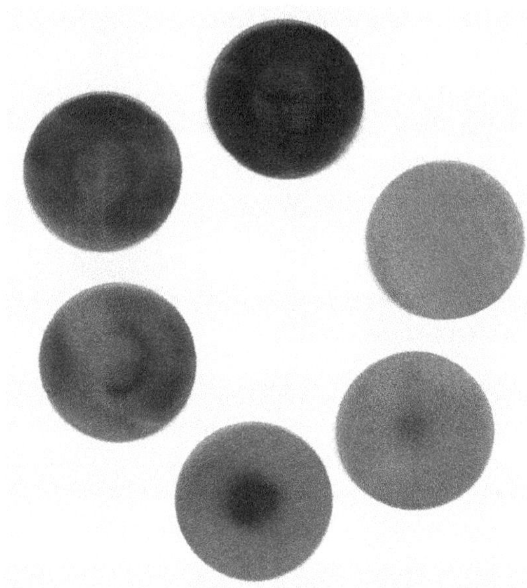

Fig. 10.5. Evidence of 'jetting' from a central manifold penetrating successive gauze layers.

influence. For air at 1 atm. of the proposed specification ($\rho = 1.2$) $Re = 4 \times 1.2 \times \hat{u} \times r_h/0.017 \times 10^{-3}$. Hydraulic radius r_h of the MP1002CA has been estimated at 0.056 mm, giving peak $Re = 351$. This is hardly a high value — and if it was the peak of the experimental range, imagine the influence on the picture of Fig. 10.4 of intermediate values!

10.7. Flow Non-Uniformity

It has been an implied assumption to this point that flow normal to the plane of the gauze (both in wind-tunnel test and *in situ*) is uniform. In fact, there is ample evidence of 'jetting', involving non-uniformity not only in flow but in temperature. Figure 10.5, which speaks for itself, shows specimen gauzes from the hot end of a regenerator fed from a central manifold. Representative approach *Ma* evidently needs to be calculated in terms of area ratio at exit from the exchanger (or manifold). The relevant area ratio in the case of the MP1002CA is α_{ffxe} with a value of 0.0910. This raises the estimate of $\hat{u}$ in the jet to 87.4 m/s and corresponding *Ma* to 0.25.

11

Mechanical Power with No Recurring CO_2 Penalty

11.1. Background

The Stirling renaissance initiated by Philips spawned a number of ambitious claims. Prominent among these was low pollution.

That, however, was when mention of pollution raised concerns about photo-chemical smog. It has since been appreciated that the real enemy — the elephant in the room — is carbon dioxide (CO_2). Hydrocarbon-fuelled engines are all equally guilty. Stirling engines in general are less efficient than the latest diesels. By today's criteria this makes them *more* polluting than the competition.

Reducing CO_2 emissions without cutting back on motive power means burning less fossil fuel. It is difficult to see how this can be achieved while continuing to rely on the internal combustion engine.

The Stirling engine, by contrast, can function from the heat of concentrated sunlight. Moreover, the internally-focussing variant (Finkelstein, 1961) offers potential for raising expansion temperature T_E above the usual metallurgical limit of 650°C.

Any such engine would be in direct competition with the *p–v* panel — a contest which the Stirling would be unlikely to win outright. On the other hand:

(a) a detailed embodied-energy calculation (Chapter 20) will compare *p–v* panel and internally-focusing Stirling engine, both operating from one m^2 absorber area, on the basis of number of hours' operation before generating 'carbon-neutral'. Surprisingly, the arithmetic strongly favours the Stirling;

(b) day-to-day observation suggests that it is *availability* which drives demand rather than the other way around.[1]

Nothing, of course, will happen until a simple, affordable, reliable Stirling engine is available! The first step might be an outline feasibility study.

11.2. Skeleton Design Procedure

Design of a conventionally-fuelled Stirling would probably proceed by choosing a combination of swept volume V_{sw}, charge pressure p_{ref} and speed *rpm* promising target power, by providing for heat input sufficient to maintain rated T_E and by keeping fingers crossed for respectable efficiency.

In the context of thermal radiation, heat supply, collector area A_C, incident power, absorber area A_R and absorber temperature T_R cannot be chosen independently. Among other considerations, T_R cannot be arbitrary — and certainly cannot be allowed to exceed the metallurgical limit should the engine stop while still receiving radiation. With the aid of a few idealisations, however, the interrelationship is predictable:

Peak temperature T_p of radiation focused by a perfectly reflecting mirror, or by a non-absorbing lens, is in terms of the surface temperature of the sun, viz, 5780 K or thereabouts. With A_C for collector area (lens or mirror) and A_R for internal receiver area

$$T_p = 5780\{(A_C/A_R)/46200\}^{1/4}. \tag{11.1}$$

Let target engine power be P [W]. Assuming 15% overall thermal efficiency and incident radiation to be 1 kW/m^2, required collector area $A_C = P[\mathrm{W}]/\{1[\mathrm{W/m^2}] \times 0.15\ [\mathrm{W/W}]\} = 1\ \mathrm{m^2}$. Anticipating a circular internal absorber of diameter d and substituting into Eq. (11.1) gives a relationship between d, power P and temperature T_p:

---

[1]Mobile phones whose storage and processing speed surpass by orders of magnitude those of the computers which assisted the first moon landing; motorcycles whose engine capacity (1,650 cc) exceeds that of many cars.

$$T_p = 5780\{(4P/0.15\pi d^2)/46200\}^{1/4}. \qquad (11.2)$$

Figure 11.1 is the nomogram equivalent of Eq. (11.2). Known values for any pair of variables joined by a straight line yield the unknown value at the intersection with the third axis. The oblique line is a specimen 'solution': It suggests that the absorber of a 150 W prototype made of refractory material and functioning at 1873 K (1600°C) requires $d = 50$ mm.

The skeleton design might take into account that the bore/stroke ratio D/S of a large number of post-renaissance Stirling engines is 'over-square'. Opting for a value of 1.5:1 suggests $S = D/1.5$, or 33.3 mm, which provisionally yields swept volume V_{sw} = 64.8 cc.

Further progress requires an arbitrary assumption. This in turn calls for a brief digression.

The coaxial, internally-focusing configuration is uniquely adapted to benefit from efforts (Organ, 2019) to improve the lamentable working-fluid utilisation of the genre — and the equally unimpressive specific cycle work N_B:

$$N_B = 60P/(V_{sw} p_{ref} \text{ rpm}). \qquad (11.3)$$

The increase in allowable temperature ratio promises to compound any improvement — which will show up in an increase in N_B. Provisionally (and optimistically) presupposing N_B increased by a factor of 4 (to 0.6) permits p_{ref} to be expressed in terms of rpm via Eq. (11.3). Inserting numerical values already settled:

$$p_{ref} = 2314/\text{rpm}. \qquad (11.4)$$

At the 1500 rpm tentatively assumed, p_{ref} = 1.5 bar. This is an unexpectedly modest value. At first sight it invites raising to, say, to 2 bar, thereby increasing the chance of achieving design power of 150 W. This, however, fails to take into account a basic reality: that with receiver size defined, input heat rate Q' is fixed (at 1kW in this case). Output power P is the product of Q' with thermal efficiency η. Unless η can be changed nothing is gained by altering p_{ref}.

There is a stronger case for embarking on the thermodynamic design phase on the basis of p_{ref} = 1 bar, i.e., avoidance of pressurisation. Retaining earlier assumptions, the Beale number arithmetic reduces to:

$$1\ [\text{bar}] \times V_{sw}\ [\text{cm}^3] \times \text{rpm} = 10^3 \times Pwr. \qquad (11.3')$$

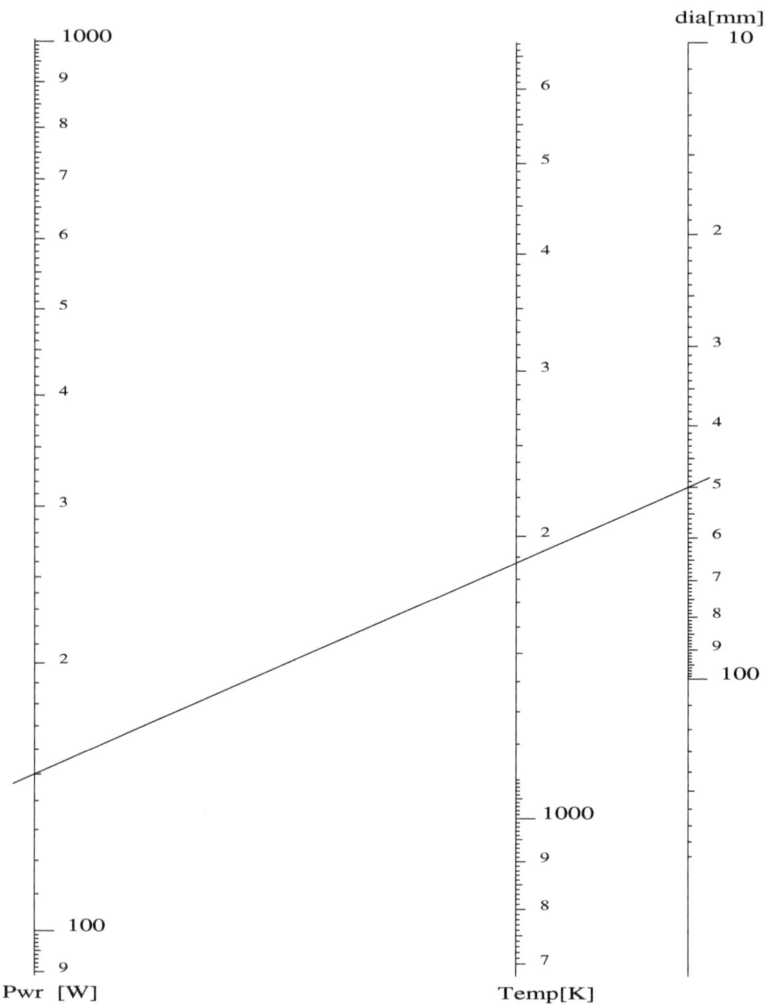

Fig. 11.1. Nomogram display of Eq. (11.2). A straight line joining any two known values (as shown) crosses the third scale at the numerical value of the third. In the example (specimen oblique line), an output power requirement of 150 W at peak absorber temperature of 1873 K (1600°C) calls for absorber diameter *d* of 50 mm.

With a conversion factor of 10^3 from cm^3 to mm^3 Eq. (11.3′) can be combined with Eq. (11.2):

$$T[\mathrm{K}] = 101.8 \times \mathrm{rpm}^{1/4} \times d^{1/4}. \tag{11.4}$$

In Eq. (11.4), *d* is in mm. Figure 11.2 is the equivalent nomogram. It offers to speed repetitious arithmetic at the early stages.

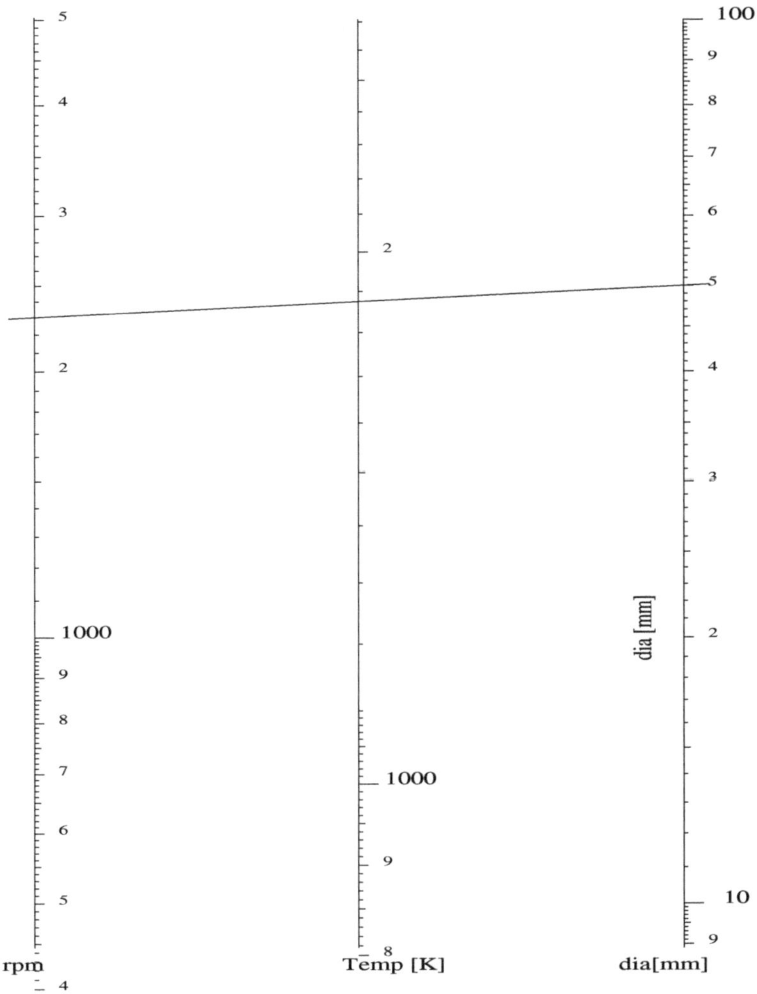

Fig. 11.2. Equation (11.4) in nomogram form. Internal absorber diameter (mm) in terms of allowable absorber temperature and target rpm. Constructed for $p_{ref} = 1$ atm. and 1 kW/m^2 radiation incident on a collector of 1 m^2.

11.3. Precedents

The inspiring instance of lateral thinking by Finkelstein represents the first and only radical reworking of Stirling concept since the invention of 1816. Figure 11.3(a) is the schematic cross-section which accompanied his original disclosure. Figure 11.3(b) shows the engine running (possibly powering its own cooling fan, although the matter is not clarified).

Remarkably, follow-up studies have been few, demonstrations being limited to table-top models with air at atmospheric pressure as working fluid. Features unique to the internally-focusing embodiment include:

- Heat to the cycle is communicated to the working gas without the losses associated with conduction through containing walls.
- The hottest part of the engine (the receiver/absorber) is not required to contain the working pressure of the cycle. It can therefore run at higher temperature than the customary 600°C or so.
- Use of a suitable refractory material for the absorber might permit operation at the T_E of the gasoline engine, i.e., an increase in temperature ratio N_T from about 3 to about 7. This promises increased efficiency.
- Reduction of expansion exchanger dead-space to near zero assists achievement of improved working fluid distribution.
- Compression ratio may be increased without sacrificing the benefits to the adverse effects of adiabatic heating.

The window material employed by Finkelstein was clear fused quartz (Table 11.1). The properties making it a candidate are ultra-low coefficient of thermal expansion, high optical transmissivity and low thermal conductivity. The combination suggests the ability to contain working pressure of around 1 bar at up to 1650°C without problematic thermal stress or conduction loss under severe temperature gradient. (These latter properties raise the possibility of designing for absorber temperature *above* that of the window itself — i.e., for $T_E > 1650$°C, maintained by cooling the outside surface of the window by a current of air.) Additionally, attention can now turn from use of conducting materials to

Table 11.1. Selected properties of clear fused quartz.

Tensile strength: 48.3 MPa
Compressive strength: >1.1 GPa
Coefficient of thermal expansion: 5.5×10^{-7}/°C (average from 20°C to 320°C)
Thermal conductivity: 1.3 W/(m·K)
Specific heat capacity: 45.3 J/(mol·K)
Softening point: ≈1665°C

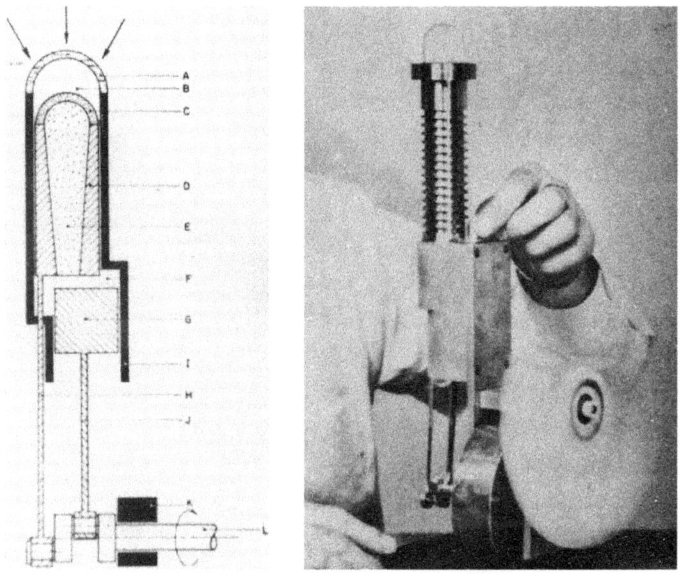

(a) (b)

Fig. 11.3. (a) Finkelstein's schematic cross-section through his internally-focusing engine. (b) Exterior view (presumably running, although radiation source not declared). Reproduced (a) with permission of the American Society of Mechanical Engineers and (b) with the permission of the Battelle Memorial Institute.

their insulating counterpart, considerably reducing thermal short penalty.

11.4. A Promising Experiment

An exception to the earlier claim (shortage of follow-up studies) is the unashamed imitation at Fig. 11.4, taken from the author's 1962 dissertation. Of 7.12 cc displacement (converted from imperial dimensions) it is shown running — unloaded — at about 400 rpm on radiation focused by a glass lens 8 inches in diameter ($\approx$0.03 m^2). At 1kW/m^2, incident radiation would have been 33 W.

Evidently some of the light focused at the window made it through the aperture (5/8-inch diameter $\approx$15.8 mm). There would have been absorption in both lens and window and re-reflection from the target. These factors suggest a net heat input rate closer to 25 W.

Fig. 11.4. The author's imitation of the Finkelstein concept running at about 400 rpm (unloaded) on radiation concentrated by an 8-inch diameter glass lens.

The figure can be considered an endorsement of the internally-focusing principle on the following basis: from a structural point of view the engine was grossly over-designed (by the writer's current standards it was not 'designed' at all). Crank pins were of 6 mm (1/4-inch) diameter, as was the displacer drive rod. The main journal ran on a pair of SKF 602 deep-groove ball-races (static load-carrying capacity 806 N vs the 10 N or so actual load). Both features added up to many-times the unavoidable friction.

Over a one-hour period, a 'night-light' recently tested consumed 2 gm of paraffin-wax. Assuming 45 MJ/kg for the heating value of the wax suggests 25 W. There is no way under the sun (an appropriate pun, maybe) that the puny flame would have warmed the bulky expansion cylinder by more than a degree or so. There is clearly something to be said for

Fig. 11.5. Running on Ontario winter sunshine concentrated by a 10-inch diameter plastic-film Fresnel lens and observed by personnel of the Department of Mechanical Engineering, Toronto University.

releasing the heat inside the engine — to which the internal combustion engine amply testifies. If the obvious advantage scales, then the internally-focussing engine is off to a head-start.

The engine subsequently became much-travelled, being transported in the author's cabin trunk aboard the steam-ship Homeric on a transatlantic crossing to Montreal, and from there via a stretch of the trans-Canada railway to Toronto. The photograph of Fig. 11.5 was taken on the roof of the Mechanical Engineering building at the University of Toronto[2]. On the

[2]Followers of aviation history may be interested in the identity of the observer nearest the camera. Dr Charlie Miller had been one of the small group of elite engineers at Orenda Engines who developed the *Iroquois* gas turbine for the legendary Avro Arrow. Over many coffee-breaks this author had the privilege of hearing at first-hand accounts subsequently immortalised in documentary/dramatisations such as *The Arrow* (1997).

temperature difference between the intense sunlight and biting cold of an Ontario December the engine ran as never before — although this is not the point of including the photograph. The collector was a 10-inch diameter (≈250 mm) plastic Fresnel lens purchased new for a couple of dollars and which had arrived by mail rolled-up in a tube. Engine components do not come much lighter, or more easily transported — or cheaper!

Eventually back in the UK the engine was later integrated with a small DC generator. Using the armature as crankshaft main journal kept total number of bearings the same, and friction sufficiently low to allow indoor operation. The source was a mercury-discharge lamp, part of a high-speed camera set-up (Fastax — 10,000 frames/sec). The collector was a 20-inch Mangin (rear-silvered) mirror, and in this configuration the engine featured in a BBC *Tomorrow's World* programme with Raymond Baxter breaking the previous record for length of unbroken commentary.[3] (Over the following few days, audience response to the programme taxed the University of Birmingham switchboard to the limit!)

11.5. Internally-Focusing Solar Stirling Engine Mk II

The foregoing material will guide design of a follow-up 'engine'. The quotes emphasise that the intended function is first and foremost as a test bed. A novel crank mechanism affording bespoke volume variations will allow an earlier claim to be put to practical test. The claim cited poor working fluid utilisation. The proposed remedy is based on intuition. It can now be checked in the only way that counts — by practical experiment. At the same time, Finkelstein's elegant scheme, so far demonstrated only at table-top scale, can be implemented closer to power-producing size.

The chapters to follow will work incrementally towards a full set of dimensioned, tolerance engineering drawings, calling on relevant aspects of Stirling engine science as available and appropriate.

[3] TV-quality video-tape existed but the means to edit it apparently did not. Individual items therefore had to be self-contained and the commentary spoken without hesitation. Raymond Baxter's day-long rehearsal was an object lesson in professionalism.

12

In Search of the Efficient, Affordable, Environment-Friendly 150 W Air Engine

12.1. Goal

The overall strategy remains unchanged from that of earlier chapters: to achieve by improved working fluid utilisation what is currently achieved by pressurisation. (Recall, the internal combustion engine does not require turbo-charging to 110 atm.!) The best prospects appear to lie with the coaxial 'beta' configuration, and with piston and displacer kinematically-coupled rather than by exploiting the dynamics of the free-piston type.

It is the drive mechanism which is responsible for the cyclic variations in volume. A step in the right direction should be a mechanism which causes the maximum possible fraction of the working fluid charge to be purged from the expansion space for the compression phase, and likewise the maximum to be purged from the compression space for the expansion process.

Increased pressure swing is inevitable. All things being equal, higher mechanical loads and higher friction forces will have to be accommodated. This suggests steering clear of cams and sliders. For the same reason, rolling-element bearings (e.g., needle rollers) should be understood where the accompanying illustrations suggest plain bushes.

A planar mechanism having promising kinematic flexibility has been described in an article for *Stirling News* (Organ, 2017). The account was, however, at fault in failing to clarify that the proportions of the schematic illustration did not yield the motion diagram which accompanied it.

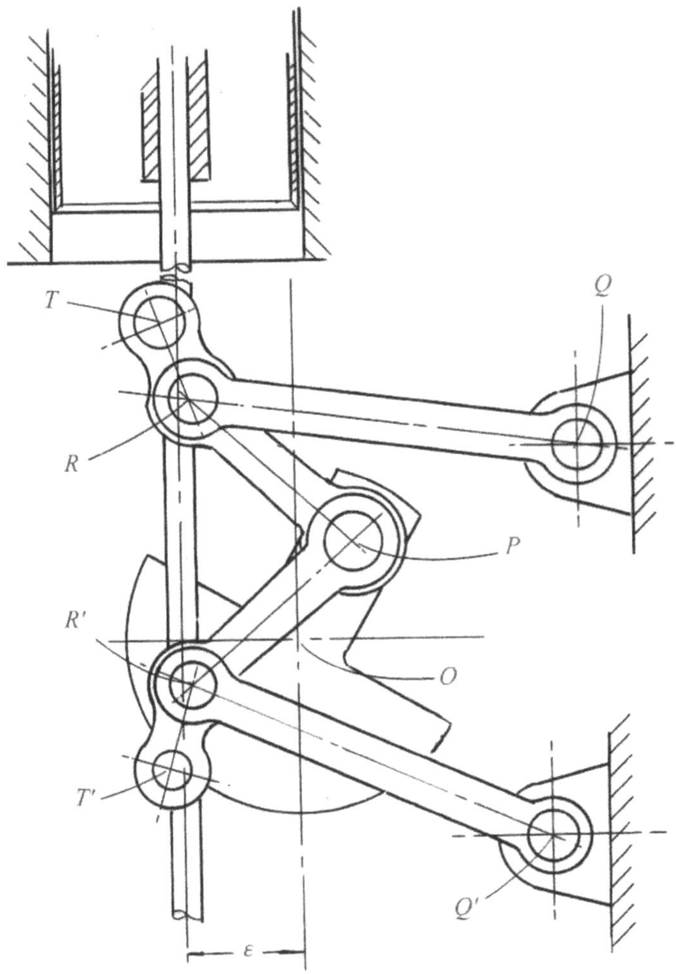

Fig. 12.1. Core components of 'pseudo-half-rhombic' mechanism.

The matter is now academic, because a mechanism has since been devised which offers vastly greater potential.

Figure 12.1 is adapted from the drawing which accompanied the patent application (Organ, 2019). It shows that the underlying angular phase displacement is achieved exactly as in the celebrated rhombic drive, i.e., by offsetting vertical axes of crankshaft and cylinder by a lateral amount ε comparable to the semi-throw of the crank, and by driving both connecting rods from a common crank-pin. To that extent, this is one-half of the rhombic drive.

However, instead of horizontal forces being contained within the mirror-image arrangement of the rhombic drive, they are reacted by binary links *R–Q* and *R'–Q'*. If the latter are provisionally imagined to be infinitely long, then the motions of points *R* and *R'* are purely vertical, and respectively identical to those of a rhombic drive of the same crank radius r_p, the same connecting-rod centre-distances and the same lateral offset ε.

But the links are of finite length. Moreover, attachment points *Q* and *Q'* are open to choice. The effect is to afford a wide range of 'higher harmonics' over and above the fundamental angular displacement between the vertical motions of points *R* and *R'*. Wholesale re-siting of points *Q* and *Q'* opens up further scope. Figure 12.2 shows the principal options.

Yet further kinematic flexibility results from replacing conventional 'binary' connecting-rods *P–R* and *P–R'* by 'ternary' links or 'couplers'

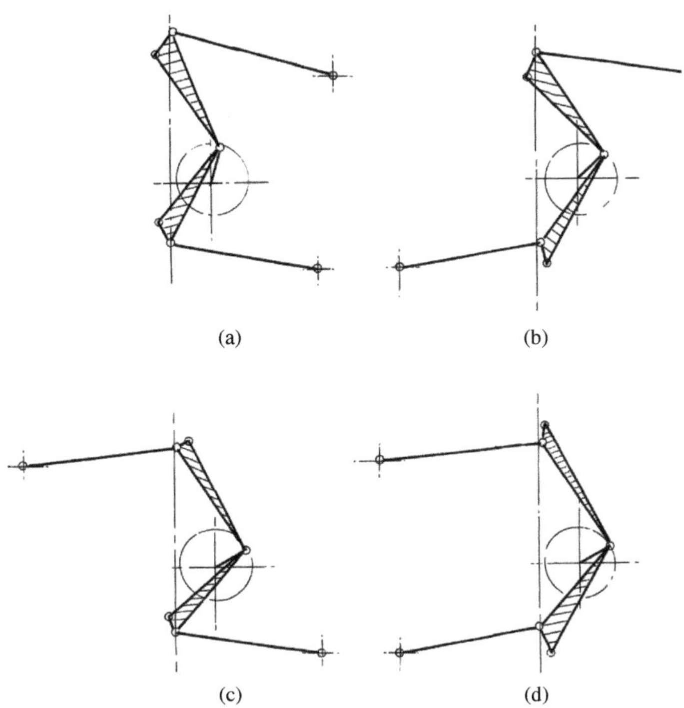

Fig. 12.2. Alternative configurations based on a common component inventory.

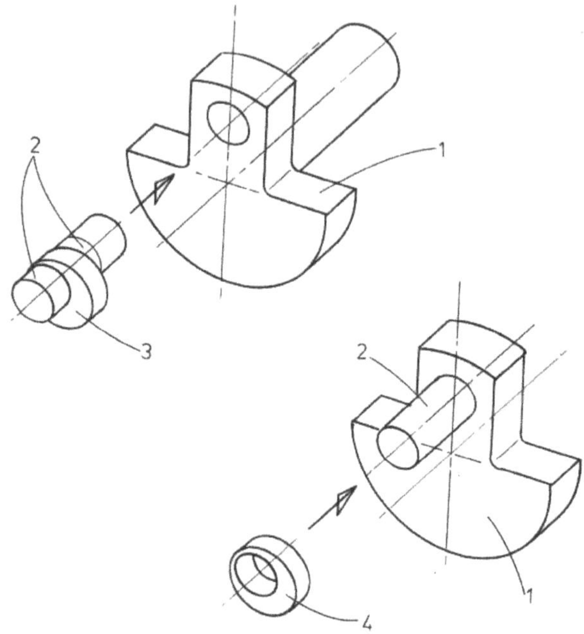

Fig. 12.3. Compound crank-pin for achieving the effect of a two-throw crank-shaft.

P–R–T and *P–R′–T′* and by driving piston and displacer (by means not indicated) from *T* and *T′* respectively.

Finally, effective phase-shift and relative amplitude can be further 'fine-tuned' by introducing a two-throw crankshaft. This might be achieved by incorporating an eccentric sheave into the crank-pin as suggested in Fig. 12.3. Either piston or displacer can be driven off the sheave, but it would make sense for the more heavily-loaded piston coupler rod *P–R–T* to be in the form of two, parallel plates symmetric either side of the sheave, sharing the load directly off the crank-pin.

The compound-pin arrangement can also be used at coupler points *R* and *T* and/or *R′* and *T′* when centre-to-centre distances are otherwise inconveniently close.

There are too many options to allow experiment or intuition to predict the combination of linear and angular dimensions giving best working fluid distribution. The specimen motion diagrams which follow have been arrived at by first coding the trigonometry of the kinematics for computer (Section 12.4 of this chapter). This becomes the basis for calculating the variations in compression-space and expansion-space volumes over a cycle. Feeding these to an 'isothermal' or 'adiabatic' analysis gives an estimate of the mass-content of the working spaces at, say, one-degree intervals over a representative cycle. Summing the elemental changes to compression-space mass during overall volume decrease (compression phase), and doing likewise for the expansion space while overall volume is increasing (expansion phase) give a numerical criterion, or 'figure-of-merit', for the extent to which the design objective (appropriate flushing) is being achieved.

The code provides for systematic variation ('perturbation') of all defining linear and angular dimensions and keeping track of respective effects — improvement or deterioration. This does not mean that the mechanism is a good candidate for mechanised optimisation. Far from it: the numerical algorithm is distracted by the large number of 'local optima'.

A combination of human intervention and mechanised search is nevertheless yielding promising results: In Fig. 12.4, the horizontal (x) coordinate is crank angle. The lower trace represents the upper face of the work piston, the upper trace corresponds to the lower face of the displacer. The figure amounts to 'work-in-progress': the piston linkage is shown 'tweaked' in three stages. The trend is to exaggerate the dwell at outer dead-centre, allowing the displacer to reach its top dead-centre before compression begins. Applying the computer-automated search to the displacer mechanism results in accelerated upwards stroke in advance of compression.

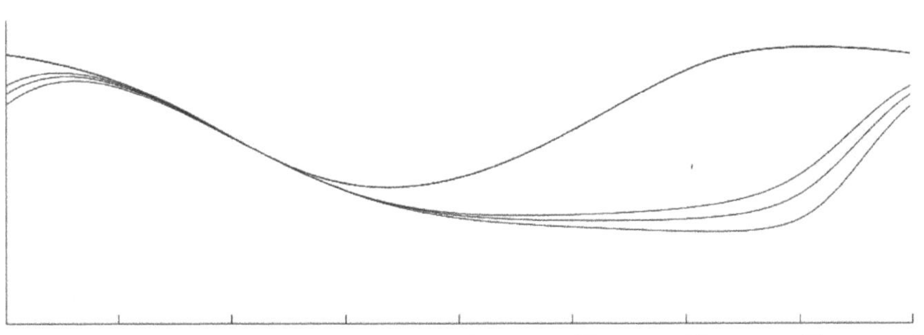

Fig. 12.4. 'Work-in-progress'. Lower traces: work piston during three stages of 'tweaking' of linkage. Upper trace: displacer linkage left unmodified during this process.

The vertical separation between traces at any crank angle is an indication of compression-space volume. Even at this early stage it is effectively zero over some 90 degrees of the expansion phase.

Figure 12.5 illustrates the successive changes in the indicator diagrams as the mechanised search proceeds.

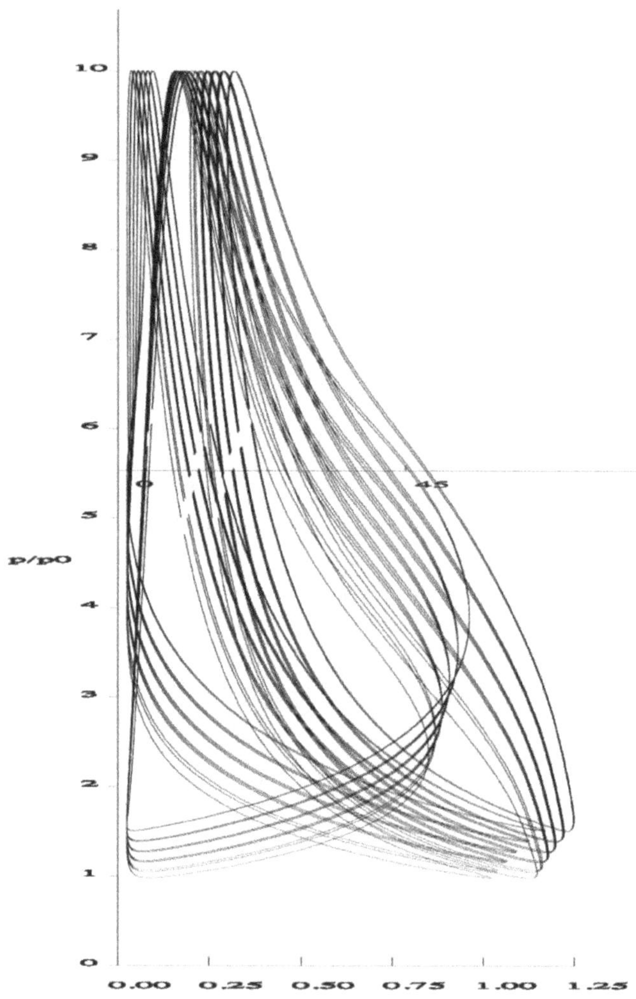

Fig. 12.5. Succession of indicator diagrams corresponding to Fig. 12.4 and calculated from an 'isothermal' model.

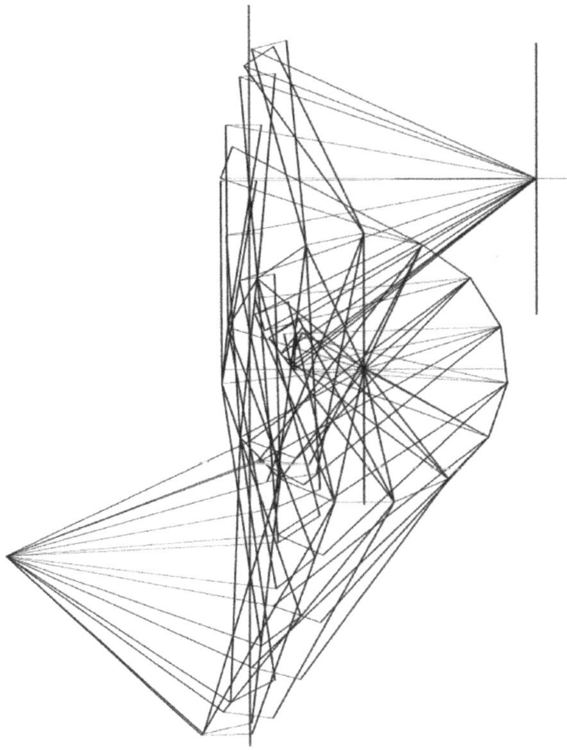

Fig. 12.6. Skeleton linkage shown in eight successive positions.

Figure 12.6 is not the work of a demented, mechanically-inclined spider! The number of angular increments of crankshaft rotation has been reduced (from the usual 360) to show the linkage in eight positions.

12.2. Discussion

This stage of the enquiry should be understood against the background of a harsh reality, namely, that *the one and only way of raising work per cycle* (of *any* heat engine, incidentally) *is by increasing the difference between heat admitted per cycle and heat rejected per cycle*. If fluid distribution is improved, heat exchanger and regenerator response must improve accordingly — either by redesign, or spontaneously by virtue of built-in over-capacity.

A telling comparison between reciprocating prime-movers — of which the Stirling engine is only one — is in terms of *specific cycle work* N_B defined as work per cycle per swept volume V_{sw} per unit charge pressure p_{ref}. Values for the Stirling engine are typically from 1% to 2% of those for an internal combustion engine of the same V_{sw}.

The main contributory factor is the manifest failure of the *genre* to make effective use of its working-fluid charge: only a small fraction — if any — is processed through a complete cycle between upper and lower temperature extremes T_E and T_C.

A subsidiary problem arises: If any Stirling regenerator has ever been critically *designed*, the account has not come to the attention of this author. There are several sides to this: In a low-pressure engine (e.g., Philips' MP1002-CA), adequate thermal capacity ratio and recovery ratio are readily achieved with a range of matrix materials, so the effect of an improvement is barely detectable. Poor working fluid utilisation means relative insensitivity to changes in thermal recovery ratio, since the latter is not the performance limitation.

Woven wire mesh gets specified as matrix material *not* because it is superior from the thermal or hydrodynamic point of view, but because it is available off-the-shelf in a wide range of porosity and hydraulic radius, is readily punched to form discs, and because the usual stock materials are heat-resistant in the working temperature range.

Similar arguments apply to the 'isothermal' exchangers.

In summary: a crank mechanism which ticks all the boxes will not, of itself, transform specific work. In the jargon of the mathematician, it is a necessary — but not sufficient — provision.

12.3. Interim Summary

To place earlier claims in proper perspective, it should be noted that implementation of the mechanism in isolation is unlikely to result in performance improvement — and may even bring about deterioration: the gas path of the Stirling engine is in open communication, and the effect of a thermodynamic event occurring in a given location is felt throughout the entire gas path — and not only at that instant but over a representative cycle.

Attending to poor working-fluid utilisation has a secondary effect: it raises cyclic pressure swing. All things being equal, this means an increase in the adiabatic component of local temperature excursions. Realising the potential benefits calls for attention to the dead-space and thermal loading of the heat exchangers. In particular, the regenerator now requires to be designed for the specific flow conditions and pressure excursions anticipated in the design under consideration. This is new territory: in established theories of regenerator operation the 'driving function' is variable flow at constant (and usually uniform) pressure. The cycle of operation now includes phases of near-zero flow during wide swings of pressure.

Evidently, a problem of dynamic balance also arises. This is looked into in a later chapter.

12.4. Algebra of Kinematics

Figure 12.7 isolates the links forming the drive train for the work piston. Crank angle ϕ is reckoned positive as shown, although operation as a prime mover requires anti-clockwise rotation as indicated by the arrow.

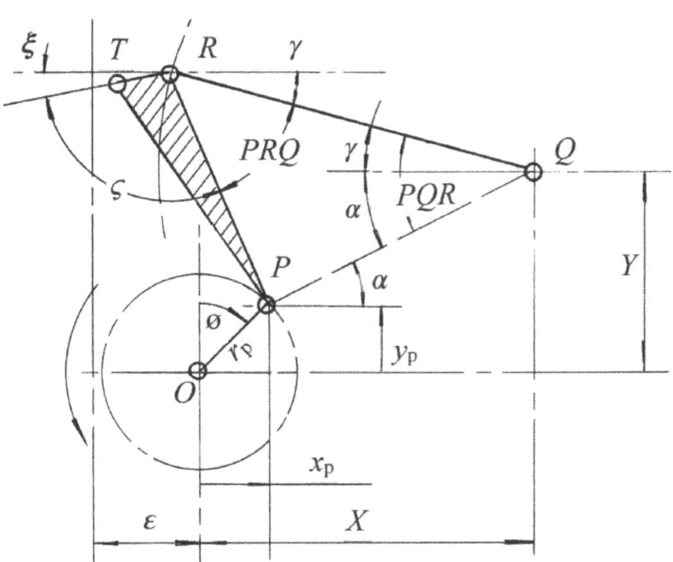

Fig. 12.7. Notation for algebra of work-piston drive components (Fig. 12.1). NB: Layout chosen for clarity rather than to represent viable proportions.

The algebra arrives at the vertical and horizontal motions of coupler point *T*, viz, of y_T and x_T as functions of crank angle ϕ. It starts with the instantaneous coordinates of crank-pin *P*:

$$x_p = r_p\sin(\phi); \qquad y_p = r_p\cos(\phi). \tag{12.1}$$

Lengths *PR*, *QR* and *PQ* form a triangle of which all internal angles and length of side *PQ* vary with rotation of the crankshaft. The process of specifying instantaneous values in function of ϕ starts by determining length *PQ*. Pythagoras' theorem comes to the rescue:

$$PQ = \sqrt{(X - x_p)^2 + (Y - y_p)^2}. \tag{12.2}$$

Angle *PQR* is available from the standard trigonometric relationship:

$$\cos(PQR) = \frac{PQ^2 + QR^2 - PR^2}{2PQ \times QR}. \tag{12.3}$$

Angle α is readily obtained from the definition of the tangent function:

$$\tan(\alpha) = (Y - y_p)/(X - x_p). \tag{12.4a}$$

If *X* less than zero (as in Fig. 12.2(c)):

$$\tan(\alpha) = (Y - y_p)/(-X + x_p). \tag{12.4b}$$

Angle γ follows as angle *PQR* minus α.
Reusing the earlier trigonometric formula:

$$\cos(PQR) = \frac{PR^2 + QR^2 - PQ^2}{2PR \times QR}. \tag{12.5}$$

Relative to origin *O*, horizontal and vertical components of the instantaneous location of point *R* are as follows:

$$x_R = X - QR\cos(\gamma), \qquad y_R = Y + QR\sin(\gamma).$$

If X less than zero (as in Fig. 12.2(c)):

$$x_R = X + QR\sin(\gamma).$$

It remains to determine angle ξ. Angle ς is to be specified by the designer. Angles ς, γ, PRQ and unknown ξ sum to a total of π radians:

$$\xi = \pi - \varsigma - \gamma - PRQ. \tag{12.6}$$

y_T and x_T can now be calculated in relation to coordinates x_R and y_R of point R:

$$x_T = x_R - RT\cos(\xi), \qquad y_T = y_R - RT\sin(\xi). \tag{12.7}$$

With slight and obvious modification, the algebra applies to the remaining three configurations of Fig. 12.2.

If the crank-pin is equipped with a second throw (Fig. 12.3), the algebra for the drive off the sheave of effective throw r_p' and angular advance β starts from crank-pin coordinates:

$$x_p' = r_p' \sin(\phi - \beta), \qquad y_p' = r_p'\cos(\phi - \beta), \tag{12.1'}$$

and is otherwise unchanged.

13

Sizing the Wire-Mesh Regenerator

13.1. Outline

Temperature recovery ratio η_T and pressure-drop ratio $\Delta p/p$ are both functions of Reynolds number *Re*. The problem of designing to specific respective values of η_T and $\Delta p/p$ simplifies dramatically on taking an algebraic step which eliminates *Re* between the equations while leaving it intrinsic to the development.

Further algebraic manipulation leads to a quasi-explicit expression for L_r/r_h — ratio of stack length to hydraulic radius giving the combination of matrix temperature swing ΔT_{sw}, $\Delta p/p$ and η_T specified by the designer. Length/diameter ratio L_r/D emerges as a crucial parameter and reveals the overriding design target: lowest practicable L_r/D (consistent with containing thermal short).

Specimen ΔT_{sw}, $\Delta p/p$ and η_T are then back-calculated from benchmark engines (GPU-3, V-160, etc.) to provide guidelines for *ab initio* design.

A candidate design intended for operation on air at atmospheric pressure is specified in terms of operating conditions (rpm, V_{sw}, etc.). This permits construction of a chart from which L_r, wire diameter d_w [mm] and mesh number m_w [mm^{-1} or wire/inch] are calculated in terms of required ΔT_{sw}, $\Delta p/p$ and η_T.

The study sheds light on the phenomenon whereby unpressurised air engines have generally been more successful without formal regenerator provision than with.

13.2. The Problem

The regenerator is the core component of machines operating on the Stirling cycle. Indeed, the original patent (Stirling, 1816) was for the regenerator, the engine being cited as one of several practical applications.

A substantial literature dating from the 1920s addresses the analytical challenge of predicting transient temperature response — sometimes called the '*conjugate heat exchange problem*'. The masterful exposition by Jakob (1957) — although dated — has yet to be improved upon. On the other hand, dealing with thermal response is only one aspect of matching the regenerator function to the requirements of a practical application, where maximising temperature recovery conflicts with minimising pumping loss. Moreover, there is no evidence that classic treatments have usefully advised the design of any 'benchmark' Stirling engine — GPU-3, Philips' MP1002CA, V-160, etc.

A 'first-principles' approach would tackle the partial-differential equations governing the conjugate heat exchange problem. On the other hand, the criteria for best performance are self-evident: (a) minimum flow loss $\Delta p/p$ consistent with (b) maximum temperature recovery ratio η_T together with (c) minimum practicable matrix temperature swing ΔT_{sw} (matrix thermal capacity matched to *b*). Realising *a*, *b* and *c* in isolation would present little challenge: the problem is achieving them simultaneously.

This chapter gains substantial simplification by eliminating *Re* between two of the defining equations. The remaining terms form a quasi-explicit expression for L_r/r_h. The expression achieves the combination of $\Delta p/p$ and η_T specified by the designer.

Despite its obvious relevance, length/diameter ratio L_r/D has not previously made an appearance as a parameter of Stirling regenerator design. Formulating the algebra with L_r/D in mind leads to an explicit criterion[1] for best thermal recovery ratio η_T consistent with lowest possible pressure drop ratio $\Delta p/p$. By specifying thermal capacity ratio N_{TC}, the designer is

[1]The criterion is not to be confused with the 'obvious' but misguided 'figure of merit' C_f/St proposed by Ibrahim *et al.* (2012). While the product of *St* with L_r/r_h does, indeed, determine temperature recovery ratio, and while CfL_r/r_h does, indeed, feature in the expression for $\Delta p/p$, the latter is not determined until CfL_r/r_h is multiplied by a value of variable *Ma* — see Eq. (13.7) and paragraph preceding.

now able, with the aid of three charts, to arrive at a full specification of a regenerator to perform to chosen η_T and $\Delta p/p$ — wire diameter d_w, mesh number m_w, stack length L_r and equivalent housing diameter D_{eq}.

13.3. Assumptions

The matrix is a cylindrical stack of square-weave wire gauze of individual wire diameter d_w [m] and mesh number m_w [m^{-1}].

On a macroscopic scale the stack is of uniform volume porosity $¶_v$:

> Flow is uniform at all cross-sections.
>
> Thermal conduction through the matrix in the flow direction may be calculated separately.
>
> Flow behaves in incompressible fashion at all locations and at all instants. (Flow perpendicular to a wire screen or screen stack is subject to compressibility effects at values of approach Mach number Ma_{app} *an order of magnitude below* the criterion of 0.3 for steady pipe flow, threshold value being dependent on Reynolds number *Re* (Pinker and Herbert, 1967; Su, 1986; Organ, 2014).

The stack is a close fit in a cylindrical housing of internal diameter D_o and axial length L_r. Pore (or 'dead') volume V_{dr} is the product of housing (envelope) volume with volume porosity, viz, $¶_v(\frac{1}{4}\pi D^2 L_r)$. (Results to be discussed extend to the parallel, annular regenerator of outside and inside diameters D_o and D_i respectively by evaluating an equivalent diameter D_{eq} as $\sqrt{(D_o{}^2 - D_i{}^2)}$. This account proceeds in terms of D, which means D_{eq} where appropriate.)

Steady-flow heat transfer and flow friction characteristics of the stack are of the form $StPr^{2/3} = a/Re^b$ and $C_f = c/Re + d$, respectively. The method extends with negligible modification to other matrix materials for which similar correlations are available.

Pressure drop and temperature recovery ratio are characterised by $\Delta p/p$ and η_T calculated for the peak flow condition.

For the purposes of specific examples, properties of the individual wire are those of stainless-steel to AISI 316, viz, $7.87 \leq \rho_w \leq 8.87 \times 10^6$ kg/m^3; $490 \leq c_w \leq 530$ J/kgK.

13.4. How Much Thermal Mass?

Overall temperature swing of the gas at exit is ultimately determined by the thermal storage capacity of the matrix. For parent material of given density ρ_w and specific heat c_w, this defines the housing volume which has to be provided. Dead volume V_{dr} follows from matrix porosity $¶_v$. Target matrix temperature swing ΔT_{sw} thus becomes the start point for design.

Assuming ΔT_{sw} is sufficiently small for viable function as a regenerator, design can be approached by working in terms of maximum possible ΔT_{sw}.

Under linear variation of temperature between T_E and T_C, gas mass content m_r of a stack of envelope volume V_{env} at uniform pressure p is

$$m_r = ¶_v(pV_{env}/RT_C)\log_e N_T/(N_T - 1).$$

The mass of working fluid nominally processed per cycle by the regenerator is the product of m_r with flush ratio, N_{FL}. For engines of conventional configuration and with kinematic actuation, flush ratio is approximately equal to the inverse of fractional regenerator dead-space δ_r.

Total mass of wire matrix material of density ρ_w is $V_{env}\rho_w(1 - ¶_v)$. Using c_w and c_p to denote specific heats of wire material and gas respectively, allowing (constant) charge pressure p_{ref} to represent variable p and recalling the identity $R/c_p = (\gamma - 1)/\gamma$, the ratio of thermal capacities N_{TC} solid/interacting gas is:

$$N_{TC} = \delta_r \frac{\rho_w c_w T_C}{p_{ref}} \frac{1 - ¶_v}{¶_v} \frac{\gamma - 1}{\gamma} \frac{(N_T - 1)}{\log_e N_T}. \tag{13.1}$$

Nominal temperature excursion of the gas is $\pm (T_E - T_C)$. Corresponding matrix swing ΔT_{sw} follows in terms of Eq. (13.1):

$$\Delta T_{sw} = \pm (T_E - T_C)/N_{TC} = \pm T_C(N_T - 1)/N_{TC}. \tag{13.2}$$

The N_{TC} value evidently sets an upper limit on ΔT_{sw} and a corresponding limit to *net* temperature recovery ratio η_T.

The sizing algorithm starts with a check of dead space ratio δ_r required to hold ΔT_{sw} to a value prescribed by the designer. Figure 13.1 facilitates by combining Eqs. (13.1) and (13.2) into nomogram format. The ΔT_{sw} of 1°C in the example (oblique line) is realistic for an atmospheric engine (p_{ref} = 1 atm. — see *Exercise in Sizing* which follows) but is otherwise over-optimistic.

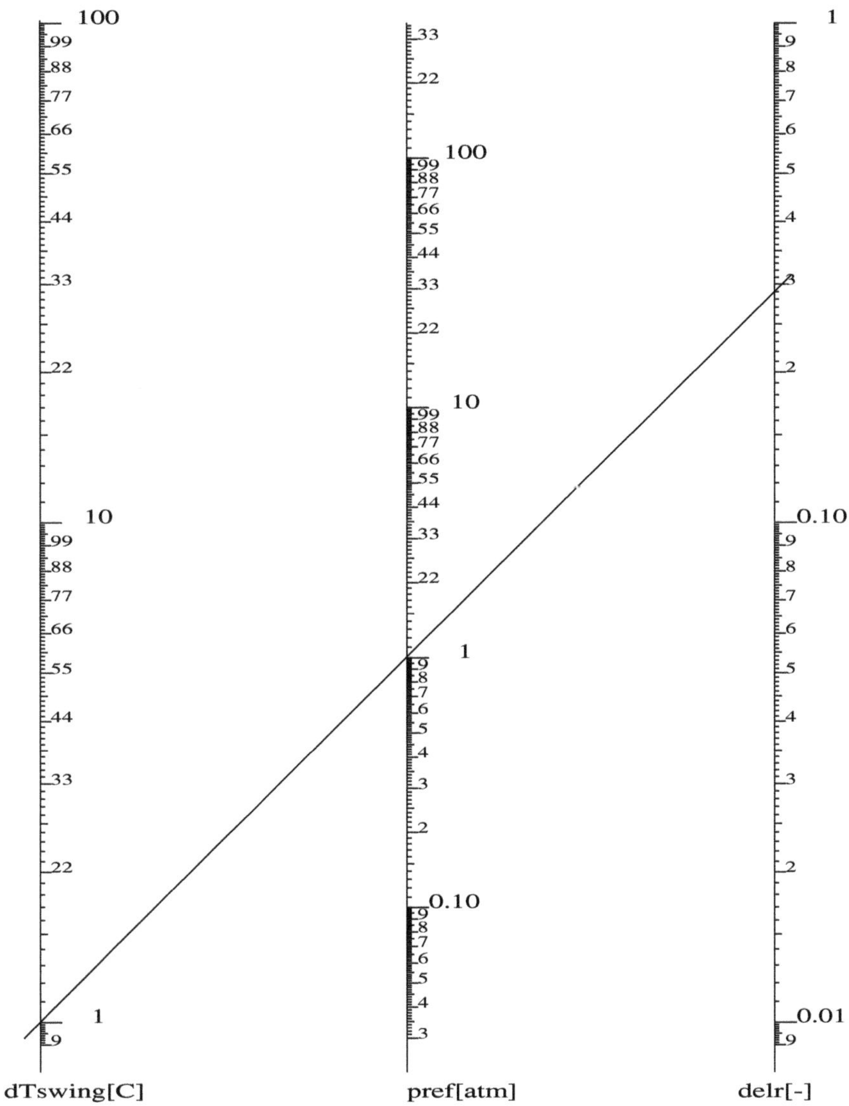

Fig. 13.1. Dead-space requirement δ_r in terms of target ΔT_{sw} and proposed charge pressure p_{ref}. Valid for diatomic gases and for matrices of wire gauze square woven from stainless steel (AISI 316) to porosity $¶_v$ of Table 13.1.

13.5. Temperature Recovery Ratio η_T

Temperature recovery ratio η_T can be expressed in terms of Number of Transfer Units NTU, where $NTU = StL_r/r_h$ and where St is Stanton number $h/(\rho|u|c_p)$. For NTU sufficiently high to be of relevance to the functioning Stirling engine (the author's 2014 text):

$$\eta_T \approx (NTU - 1)/NTU. \tag{13.3}$$

Equation (13.3) is inverted to give the NTU required to achieve specified η_T:

$$NTU = 1/(1 - \eta_T). \tag{13.4}$$

Stanton number St in incompressible flow through wire screens is related to Reynolds number Re through $StPr^{2/3} \approx a/Re^b$. (Prandtl number Pr is assumed constant throughout this development.) Substituting into Eq. (13.4) and inverting returns the Re predicted to give specified η_T:

$$Re = \{aN_{pr}{}^{-2/3}L_r/r_h(1 - \eta_T)\}^{1/b}. \tag{13.5}$$

Appendix A goes further, re-expressing the usual definition ($Re = 4\rho \underline{u} r_h/\mu$) in terms of the dimensionless variables and parameters of the Stirling cycle. There are now two equations for Re — and two things which can be done with them: (a) evaluate Re or (b) eliminate it. The latter course will yield useful simplification. Dividing Eq. (13.2) by Eq. (13.5):

$$\{aL_r/r_h N_{pr}^{-2/3}(1-\eta_T)\}^{1/b} = \frac{4\sigma' N_{RE}(L_r/L_{ref})^2}{\psi \delta_r L_r/r_h}.$$

Rearrangement enables L_r/r_h to be extracted:

$$\{aN_{pr}^{-2/3}(1-\eta_T)\}^{1/b}(L_r/r_h)^{(1+b)/b} = (L_r/L_{ref})^2 \frac{4\sigma' N_{RE}}{\psi \delta_r f(\mu)}. \tag{13.6}$$

13.6. Pressure Drop Ratio $\Delta p/p$

Pressure drop Δp in steady flow may be written $\Delta p =$ (minus) $\frac{1}{2}\rho u^2 CfL_r/r_h$. It appears to be not generally recognised that, when the fluid is an ideal gas, $\frac{1}{2}\rho u^2$ may be replaced by $\frac{1}{2}p\gamma Ma^2$. (Appendix A re-expresses Ma in terms of the dimensionless variables and parameters of Stirling machine operation.) Noting that, for steady incompressible flow through stacks of wire screens $Cf \approx c/Re + d$, an expression emerges for Δp as a fraction of local, instantaneous pressure p:

$$\Delta p/p = (-)\tfrac{1}{2}(L_r/L_{ref})^2\left\{\frac{\sigma' N_{MA}}{\psi\delta_r}\right\}^2(c/\text{Eq.}(13.5)+d)L_r/r_h. \tag{13.7}$$

Equations (13.6) and (13.7) are of the form:

$$a_{11}(L_r/r_h)^{(1+b)/b} = a_{12}(L_r/L_{ref})^2, \qquad a_{21} = a_{22}(L_r/L_{ref})^2 L_r/r_h.$$

(Coefficient a_{22} is not a pure constant because Eq. (13.6) features L_r/r_h). The term L_r/L_{ref} may be eliminated by division, leaving.

$$\Delta p/p = \{(a_{11}\times a_{22})(L_r/r_h)^{b/(1+2b)}\}/a_{12}. \tag{13.8}$$

Substituting leads to two occurrences of $N_{MA}{}^2$ — which cancel:

$$\Delta p/p = \frac{\sigma'}{16N_{SG}\psi\delta_r}(L_r/r_h)^2[c+d(L_r/r_h)^{1/b}\{(aN_{pr}{}^{-2/3}(1-\eta_T)\}^{1/b}]. \tag{13.9}$$

Seldom does an equation speak so clearly — or such volumes. The closer η_T comes to unity (100% recovery ratio) the smaller is the second of the terms between square brackets, and the closer does Eq. (13.9) approach the simple quadratic form:

$$\Delta p/p \approx \frac{\sigma'}{16N_{SG}\psi\delta_r}c(L_r/r_h)^2. \tag{13.10}$$

13.7. General Design Guidelines

Design guidelines of any generality have been thin on the ground to date. However, Eq. (13.10) yields the following two results.

13.7.1. Length/diameter ratio L_r/D

But for the fact that regenerator studies in general manage without mention of length/diameter ratio L_r/D, it might be thought remarkable that so ubiquitous a parameter has not made an appearance: All other things being

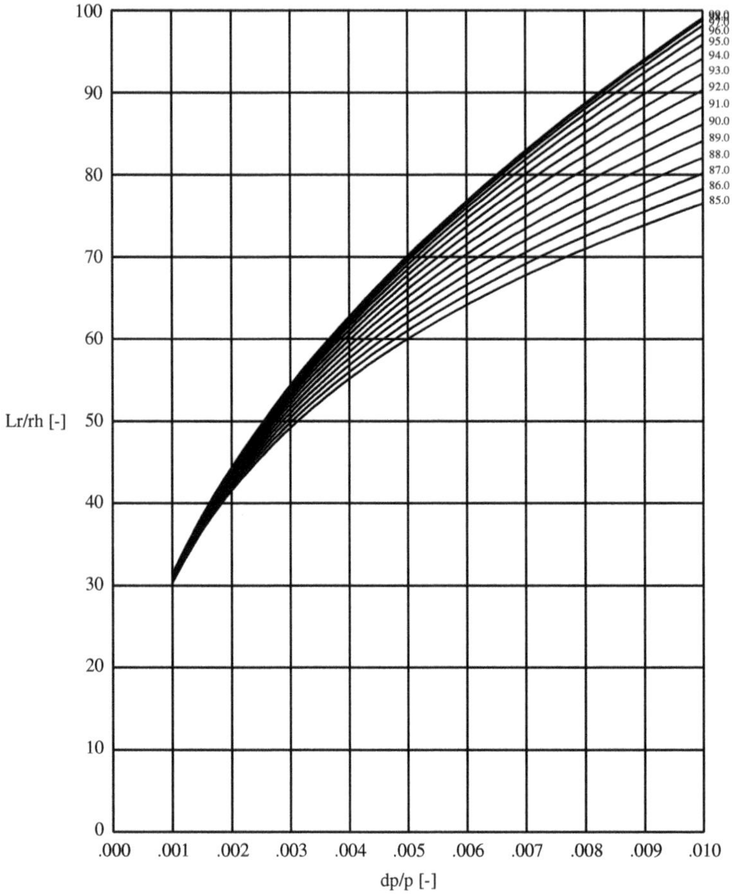

Fig. 13.2. Graphical equivalent of Eq. (13.9). Parameter is percentage temperature recovery ratio η_T [%].

equal, low L_r/D connotes low thermal recovery ratio, low pressure drop, non-uniformity of flow over the cross-section (not overlooking high axial heat leak!!). High L_r/D suggests the opposite: high recovery ratio, high pressure drop, but improved flow uniformity (and low axial heat leak).

L_r/D has not emerged spontaneously from the algebra to this stage, but is intrinsic to the definition of dead-space ratio $\delta_r = V_{dr}/V_{sw}$. For the cylindrical stack this may be rendered:

$$\delta_r = \tfrac{1}{4}\pi ¶_v D^2 L_r/V_{sw}.$$

L_{ref} is $V_{sw}^{1/3}$ by definition, and so:

$$\delta_r = \tfrac{1}{4}\pi\P_v(L_r/D)^{-2}(L_r/L_{ref})^3.$$

Inverting:

$$L_r/L_{ref} = \{4\delta_r(L_r/D)^2/\pi\P_v\}^{1/3}. \tag{13.11}$$

Combining Eqs. (13.9) and (13.11) and plotting (Fig. 13.3) gives, at-a-glance, the condition under which the ideal combination of L_r and r_{hr} will be found, viz, where $\Delta p/p$ is minimum and η_T maximum. All other things being equal, this is where the numerical value of aspect ratio L_r/D is a minimum.

13.7.2. Hydraulic radius r_h in relation to charge pressure p_{ref}

Stirling parameter N_{SG} is defined $p_{ref}/\omega\mu$. Substituting into Eq. (13.10):

$$\Delta p/p = \frac{\omega\mu\sigma'}{16p_{ref}\psi\delta_r}c(L_r/r_h)^2. \tag{13.10'}$$

According to Eq. (13.10′), $\Delta p/p$ increases as charge pressure p_{ref} decreases and as μ increases. The fine wire cloth which serves at $p_{ref} = 110$ atm. is unlikely to be suitable for an engine charged to a single atm. with air or N2, where a much coarser grade is indicated.

13.8. Exercise in Sizing

Coolers operating on the Stirling cycle now play a multi-million-dollar rôle in the temperature control of stored biological specimens. The presence in prime-mover mode is far smaller numerically, although Stirling-based auxiliary power generation in submarines is likewise a multi-million business.

Neither role is particularly public, leaving the type virtually unheard-of in applications dominated by the small, single-cylinder internal combustion engine. In an era of low-power electronics and LED illumination,

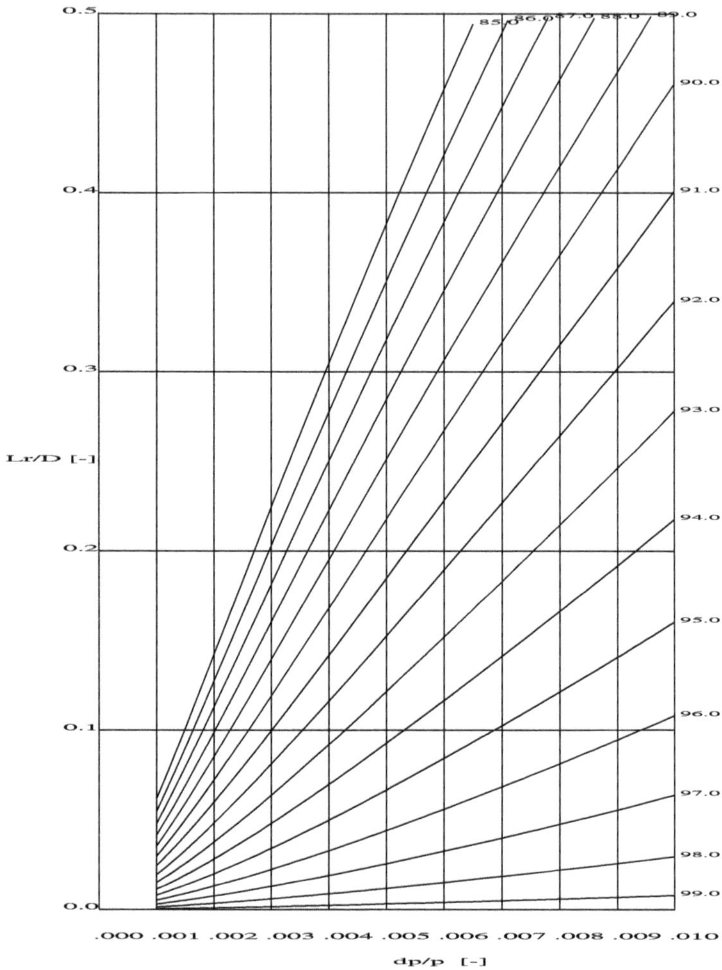

Fig. 13.3. Graphical representation of Eq. (13.9) after embodying Eq. (13.11): best recovery ratio and best $\Delta p/p$ occur at lowest L_r/D.

however, this need not remain the case: 50 W generated unobtrusively by an engine to the outline specification of Table 13.1 could well satisfy a niche.

From this point the design process reflects a crucial principle: An engine with a regenerator having temperature recovery ratio η_T *below*

Table 13.1. Outline specification of 100 cc air/N2-charged engine intended for 50 W net output.

Swept volume V_{sw}	100 cm^3
Bore, stroke	65 mm, 30 mm
Charge pressure p_{ref}	10^5 Pa (1 atm.)
rpm	2000 min^{-1}
Working fluid	air/N2
Target output, *Pwr*	50W
Beale number $Pwr/(p_{ref}V_{sw}\text{rpm}/60)$ =	0.15
Regenerator square-weave wire mesh	
Material	AISI 316 stainless steel
Specific heat c_w	0.5 kJ/kgK
Density ρ_w of parent wire mat'l	7.9×10^3 kg/m^3
Volume porosity $¶_v$	0.754
Target ΔT_{sw}	1.0°C(±0.5°C)

target may well function. An engine whose regenerator causes excessive pressure-drop ratio $\Delta p/p$ (as can be the result of aiming for too high η_T) will not function.

Two constraints arise: (a) fractional pressure drop $\Delta p/p$ must be held at or below the value at which engines of the genre are known to function and (b) length L_r and outside diameter D must fit the space allocated. As regards (a), reverse-engineering the $\Delta p/p$ for a number of "serious" engines reveals that peak cycle $\Delta p/p$ lie in the range $0.005 \leq \Delta p/p \leq 0.007$ (or $0.5\% \leq dp/p \leq 0.7\%$). (The reverse-engineering process is not 'exact': it can be only as faithful to reality as the analysis on which it is based. However, a forward design process based on the same analysis offers a fighting chance of reproducing the same $\Delta p/p$.)

Accepting this same $\Delta p/p$ (0.007) allows the analytical work of design to be embodied in a single chart, use of which will yield a temperature recovery ratio η_T equal to at least 90%.

With δ_r already settled in terms of proposed ΔT_{sw} (Section 13.3 above), design proceeds with calculation of Stirling number N_{SG}, defined as $60 \times p_{ref}/(2\pi \times \text{rpm} \times \mu)$, in which rpm is revolutions per minute (2000 in this case), p_{ref} is nominal charge pressure (abs. in Pa) which, for this example, is 10^5 [Pa]. μ [Pas] is coefficient of dynamic viscosity at reference temperature (ambient) which, for air, is 0.017×10^{-3} Pas. Substituting numbers for symbols: $N_{SG} = 60 \times 10^5/(2\pi \times 2000 \times 0.017 \times 10^{-3}) = 0.2809 \times 10^8$. This point is located on the left-hand scale of Fig. 13.4 and joined with a straight line to the right-hand scale at the value of δ_r

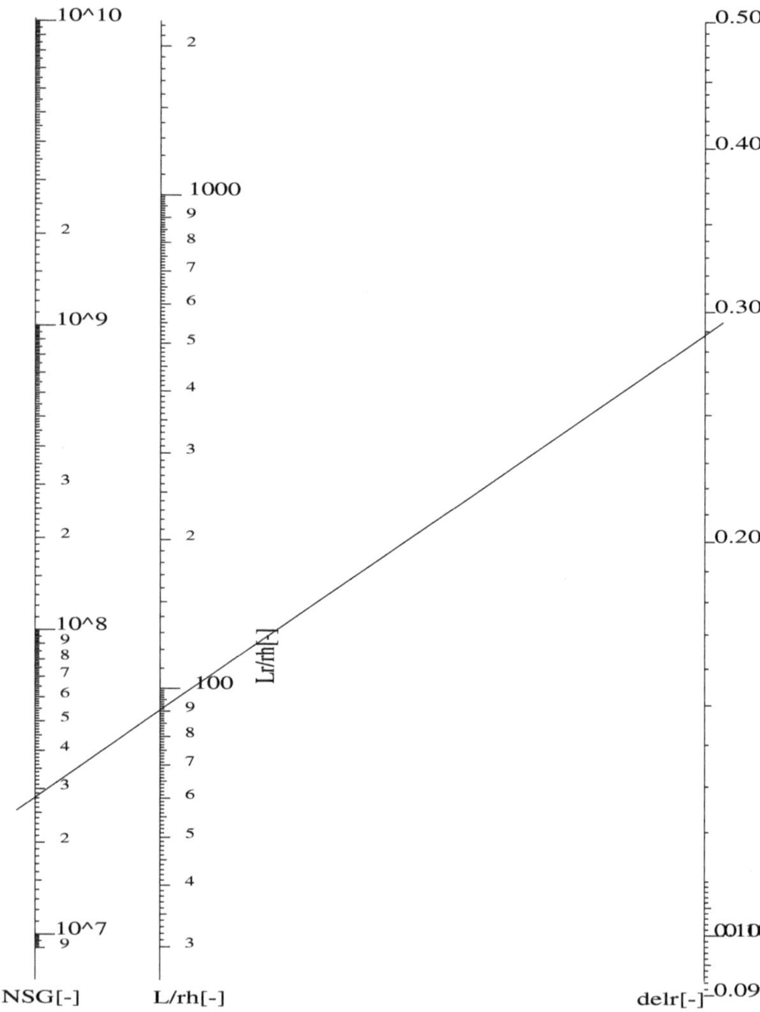

Fig. 13.4. Accepting $\Delta p/p = 0.007$ allows L_r/r_{hr} [–] to be displayed directly in terms of δ_r [–] (calculated earlier) and N_{SG} [–].

calculated earlier. The line intersects the centre scale at the value of L_r/r_{hr} which is going to define the regenerator — the ratio of stack length L_r [mm] to hydraulic radius r_{hr} [mm]. In the present example this is read off as 90.45 [–].

The design specification outlined in Table 13.1 is under evaluation by this writer. It has the regenerator contained within the body of the displacer and moving with it (cf. the inspired concept of the internally-focusing solar engine (Finkelstein, 1958). An overriding design requirement is therefore that the stack should fit within the proposed bore of 65 mm. The design arithmetic accordingly proceeds with a tentative enclosure diameter D of 50.0 mm. This determines stack length L_r via the expression for dead volume, viz, $V_{dr} = \frac{1}{4}\pi D^2 L_r \times ¶_v$, where $¶_v$ is porosity. Inverting this expression to make L_r explicit and substituting known numerical values:

$$L_r[\text{mm}] = 4V_{dr}/(\pi D^2 \times ¶_v) = \frac{4 \times [28.8 \times 10^3]\text{mm}^3}{\pi \times 50.0^2 \times 0.754} = 19.45 \text{ mm.}$$

Inverting the earlier expression for L_r/r_{hr} = 90.45 [–] yields r_{hr} = L_r/90.45 = 19.45[mm]/90.45 = 0.215 [mm].

For the square-weave wire gauze, wire diameter d_w relates via porosity $¶_v$ to hydraulic radius as $r_{hr} = \frac{1}{4} d_w ¶_v/(1 - ¶_v)$. Inverting this: $d_w = 4r_{hr}(1 - ¶_v)/¶_v$. Substituting numerical values:

$$d_w \text{ [mm]} = 4 \times 0.215 \text{ [mm]} \times (1 - 0.754)/0.754 = 0.281 \text{ [mm]}.$$

Mesh number m_w comes by application of the standard formula for m_w in terms of $¶_v$ and d_w, viz, $m_w = 213.4 \times (1 - ¶_v)/(\frac{1}{4}\pi d_w)$. The arithmetic yields 28.34 wire/inch. However, only integral values are meaningful, so m_w = 29 wire/inch.

Nominal number of gauzes n_g is stack height divided by gauze thickness, i.e., by twice wire diameter:

$$n_g = L_r/2d_w = 19.45[\text{mm}]/(2 \times 0.215[\text{mm}])$$
$$= (\text{say})\ 35\ \text{gauzes.}$$

The regenerator is now completely specified.

Against the standards of pressurised engines the mesh is coarse. This aptly reflects the vastly different working environment. There are accompanying benefits:

- *handling is facilitated — particularly cutting/punching;*
- *the stack comprises a small number of gauzes vs the hundreds required in the high-pressure context;*
- *the unit benefits from a degree of self-support (not for nothing are the finer gauzes referred to as wire 'cloth');*
- *bypass leakage is reduced because diametral inaccuracy of the punching process can be a smaller fraction of hydraulic radius.*

Little has been achieved unless wire gauze to the required d_w and m_w is available 'off-the-shelf'. The German manufacturer Haver and Boecker Anon (1998) lists a weave with $d_w = 0.28$ mm at 29 imperial mesh as part of their 'industrial' range. (The grade of stainless steel is not specified.) Weaves having $d_w = 0.28$ mm at both 28 mesh and 30 mesh are available from the 'standard' range of weaves manufactured by G Bopp and Co. AG and can be supplied in stainless steel to AISI 304 or 316 (Manufacturer's catalogue 2002).

13.9. Authentication

There are no known experimental measurements of ΔT_{sw} of η_T or of $\Delta p/p$ having relevance to the present study. Dimensional and performance data are, however, available for a small number of 'benchmark' engines — General Motors' GPU-3, the United Stirling (USS) V-160 unit, the Alison PD-46, for Philips' 400 horse-power/cylinder design (evaluated by means of the 'part-cylinder' concept) and for the USS P-40.

The foregoing design sequence can be run in reverse to see whether there are combinations of η_T and yielding the published figures for L_r, d_w and mesh number of respective engine specifications. (m_w figures are conversions to Imperial units, i.e., wire/inch.) Table 13.2 shows results. Actual (published) values are in bold. At first sight, the generally high η_T are consistent with claims attributed to Philips in the early stages of their Stirling programme. On the other hand, the underlying back calculation presupposes infinite heat capacity of the matrix. Respective matrix temperature swings ΔT_{sw} (fourth row down) give an over-optimistic idea of likely in-service η_T.

Table 13.2. Matrix temperature swing ΔT_{sw} back-calculated via Eq. (13.10). ΔT_{sw}, η_T, $\Delta p/p$ and W_p/W_{ideal} from published operating conditions and respective L_r, d_w and mesh number of the table.

Equipment	400 hp/cyl	GPU-3	V-160	PD-46	P-40
L_r [mm]	**75.0**	**22.2**	**30.0**	**20.32**	**44.0**
d_w [mm]	**0.140**	**0.04**	**0.049**	**0.040**	**0.05**
mesh #	**[1/in] 96.5**	**200**	**213**	**250**	**200**
ΔT_{sw} [°C]	34.7	42.8	73.8	20.9	32.6
η_T [%]	98.98	98.65	98.58	98.89	98.85
$\Delta p/p$ [–]	0.0057	0.0064	0.0069	0.0036	0.0067
W_p/W_{ideal} [–]	0.0064	0.0089	0.0075	0.0054	0.0137

13.10. Corroboration

It would be reassuring (though not conclusive) to know that the method correctly predicts the L_r/r_{hr} of an existing regenerator.

$\Delta p/p$ back-calculated for the USS V-160 is 0.0069, i.e., almost precisely the 0.007 built into the present method. δ_r for the V-160 is 0.305 [–], while N_{SG} at rated output is 11.62×10^9 [–]. Entering Fig. 13.4 with those values gives L_r/r_{hr} = 1300 [–]. One published specification of the V–160 has $L_r = 30.0$ mm, $r_{hr} = 0.0279$, or $L_r/r_{hr} = 1075$.

A discrepancy of 21% is hardly negligible. On the other hand, this writer knows of no other method offering explicit guidance on sizing (Table 13.2).

13.11. Sizing vs Scaling

The author has previously advocated gas-path design by *scaling* — largely for lack of a proven alternative. The full title is *Scaling of gas-path superficial geometry*, i.e., of heat-transfer surface geometry. Regenerator thermal capacity ratio does not — except by coincidence — scale in sympathy with surface geometry, so neither does matrix temperature swing. The two methods can be expected to give differing results. Sizing appears to give the preferred result when, as here, prior choice of thermal capacity ratio constrains the scaling outcome.

Allowing judgement to enter may not be welcomed by those to whom Dynamic Similarity is a rigid discipline affording a numerical outcome unique to several decimal places. This may reflect a prevalence of examples in the syllabus which favour that outcome: what may not be widely appreciated is the extent to which the academic course content is driven by the need for examination questions which can be answered in twenty minutes and which can be marked in two.

Appendix I applies Dynamic Similarity to a 'real-world' problem, which may be the more convincing for its context: no connection whatsoever with Stirling engines — Elon Musk's Virgin/HyperLoop concept!

13.12. Discussion

The inconvenience of evaluating Reynolds number *Re* has been circumvented by eliminating its explicit appearance.

An important parameter of regenerator design not previously exploited is aspect ratio L_r/D: by aiming for lowest practicable L_r/D, highest recovery ratio η_T is achieved at lowest pressure drop ratio $\Delta p/p$. (Towards the limit, marginal gains will be overwhelmed by losses due to increasing conduction. However, there is still a balance point, and locating it promises a relatively straightforward analytical challenge.)

An engine can function despite η_T below the norm and/or large matrix temperature swing ΔT_{sw} well above. An engine with excessive $\Delta p/p$ will not function.

Mimicking the regenerator of an existing machine other than by formal scaling is almost guaranteed to have an unsatisfactory outcome.

It is highly unlikely that the common expedient of creating a regenerator by stuffing a space of arbitrary volume with wire wool will achieve performance potential.

14

Mechanism Forces

14.1. Computational Strategy

This chapter returns to the mechanical design of the internally-focussing solar concept.

Net bearing loads at any a given crank-angle are the sum of instantaneous gas loads and inertia forces.

Even for fixed mechanism proportions, the routine approach calls for dedicated recalculation corresponding to every candidate data set — piston mass m_p and cross-sectional area A_p, with similar recalculation for changes to m_d and A_d of the displacer.

Conversely, if it can be assumed that m_p and m_d make the dominant contribution to inertia loads, minor approximations will allow a single calculation run to be scaled to any combination of rpm, charge pressure, respective areas and masses.

Figure 14.1(a) shows the power-piston drive with the crankshaft in arbitrary angular location ϕ, and with load F_T at point T reapplied to joint R and adjusted slightly so as to act purely vertically.

Loads in members PR and QP at any instant are in direct proportion to F_T, allowing the algebra to be formatted with F_T set to unity.

Force analysis for any given crank-pin position can be by application of the celebrated *Bow's notation* of the undergraduate course in structural engineering. In the assumed absence of significant bearing friction, forces in individual members are aligned with respective members, and can therefore be represented as lines to scale. For example, a load of 1 N in member QP can be shown as a line of unit length 10 cm aligned with QP.

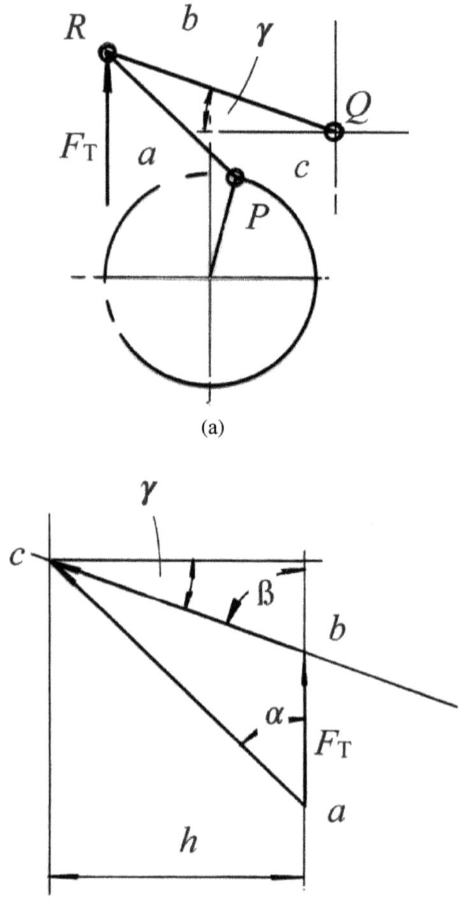

Fig. 14.1. (a) 'Space' diagram with piston drive isolated showing Bow's convention for labelling spaces between lines of action of forces. (b) Corresponding force diagram.

The 'spaces' between lines of action of individual forces (including F_T) are labelled, e.g., *a*, *b*, and *c*. With F_T positive in the *y*-direction, proceeding clockwise from space *a* across F_T to space *b* yields vertical line $a - b$ in the force diagram (Fig. 14.1(b)).

The scaled magnitude $b - c$ of the force in *RQ* is not yet known, but point *c* must lie along a line parallel to member *QR* of the space diagram.

Crossing member *PR* from space *c* to space *a* must close the force polygon by ending at *a*. This leaves only one possibility for point *c*, as shown.

On the drawing-board, length *c* – *a* in the space diagram of Fig. 14.1 measured 67 mm. With F_T drawn as 30 mm, this makes force *PR* $67/30 \times F_T = 2.13 \times F_T$, with similar elementary arithmetic for force *QR*.

Application of standard trigonometry converts the force diagram to algebra, which can be rendered as computer code and thence as graphical output. Principal steps are as follows.

Recalling the earlier Fig. 10.1 and combining the notation with that of Fig. 14.1, the angle θ made by *RQ* with the horizontal is

$$\theta = \mathrm{a}\tan\frac{y_R - y_{pin}}{x_{pin} - x_R}.$$

Other angles follow immediately:

$$\alpha = \pi/2 - \theta; \qquad \beta = \pi/2 - \gamma.$$

The length of side *h* is obtained from the standard formula for sides and angles of a plane triangle (Hodgman, 1954):

$$h = F_T \frac{\sin\alpha \sin\beta}{\sin(\beta - \alpha)}. \tag{14.1}$$

Length *h* is evidently the *x*-component of force in *PR* per unit F_T. There are no other horizontal components, so this is equal and opposite to the *x*-component in *PQ*. The *y*-component in *PR* is $F_{PR}\sin\theta$. The *y*-component in *QR* is that in *PR* after subtraction of F_T.

Possibly because the direction of action of net force varies over a cycle, the polar diagram is the usual choice for depicting loads on rotary bearings. Load is displayed in terms of magnitude (radial extent in Fig. 14.2), of direction of action, and of angles marked at intervals around the loop. The latter give the crank position at which the respective value arises.

If the polar plot is generated *per unit input load*, i.e., for F_T = unity, the result is a 'generalised' solution from which 'real' loads (inertia and gas pressure) can be scaled without the need for reformulation from scratch: the radial extent of the force vector as read from one of the scales

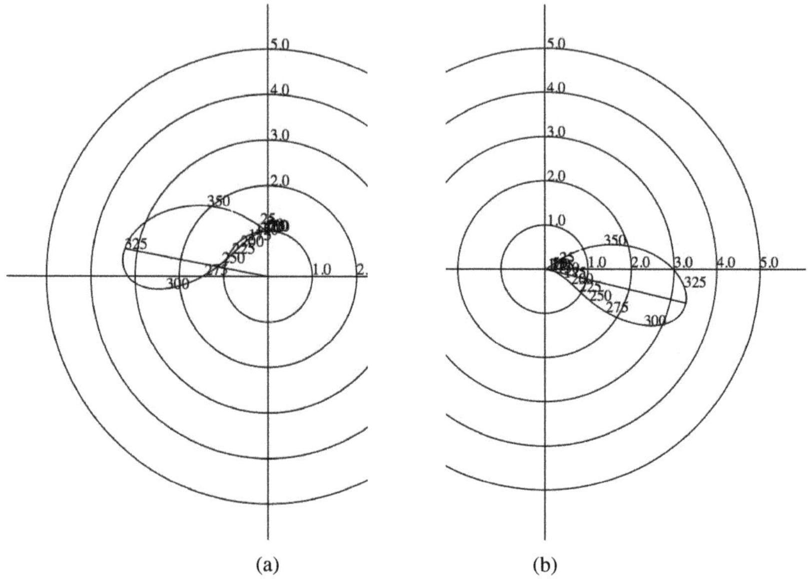

Fig. 14.2. Polar diagrams for load in *PQ* (a) and *PR* (b) generated for unit input load F_T at *R*.

(horizontal or vertical) is the numerical factor by which the load at *T* (whether gas-pressure, inertia or a combination) is amplified when crank position ϕ is that of the label on the curve. From the polar on the left, the force in *PQ* (and therefore the load on the crank-pin at *P*) is approximately 3.4 times the load at T when ϕ is at 325°. (Recall: ϕ is measured *anti-clockwise* from the vertical.)

The fact that the region around the origin is too cluttered to resolve is not a problem: At a first iteration, bearing selection is in terms of *peak* load. The relevant region of the curve is — by definition — far from the origin and remains legible.

14.2. Actual Loads (in Newton, N)

The 'input' forces which the linkage has to contain are inertia and gas loads aligned with F_T, i.e., are those due to respective vertical accelerations of piston and displacer, and to the effect of differences in the cyclically-varying gas pressures across each. In the case of the displacer (only), it will be assumed — and justified later — that pressure drop is small and that inertia accordingly dominates.

Kinematic diagrams referenced to this point (e.g., Fig. 14.1(a)) are dimensionless: they are scaled to crank radius r_p, and the independent variable is crank angle ϕ rather than 'real' time t. Vertical accelerations of displacer attachment point y_T calculated by direct numerical differentiation are thus dimensionless, i.e., are of the form $\partial^2(y_R/r)/\partial\,\phi^2$. Multiplying F_T (unity) by $\partial^2(y_R/r)/\partial\phi^2$ calculated over the interval $0 < \phi < 2\pi$ yields polar diagrams of force per unit m_d, per unit crank radius r_d. and per unit of rpm^2.

The polar plots of Figs. 14.3(a) and 14.3(b) remain dimensionless — and to this extent 'universal': they represent displacer loads in links PQ' and $Q'R'$ in an engine of any cubic displacement, rpm and displacer mass m_d whose mechanism has the proportions of Fig. 14.1.

Conversion to 'real' forces is achieved on multiplying up by the product m_d[kg] $\times$ r_p[m] $\times$ ω^2[rad/sec]^2. The units of the result are Newton, N. If more convenient, multiplication can be by m_d[kg] $\times$ r_p[mm] $\times$ rpm^2, the outcome then requiring division by 10^5 mm/m $\times$ (rpm/(rad/sec))^2.

A numerical example will illustrate: mass m_d of displacer, including drive-rod and all components rigidly attached is estimated at 250 g, or 0.25 kg. Crank radius is 12 mm, and target operating speed 2000 rpm. The type of bearing in mind for the crank-pin end of link PR' is a needle-roller.

According to Fig. 14.2(b), peak (dimensionless) load occurs close to $\phi = 240$ and at a value of about 8.5 [–].

Actual peak load F_{max} is computed as

$$F_{max} = \frac{8.5\,[-] \times 0.25\,[\text{kg}] \times 12\,[\text{mm}] \times 2000^2\,[\text{rpm}^2]}{10^5\left[\text{mm/m} \times (\text{rpm/(rad/sec)})^2\right]} = 1020\text{ N.}$$

A look ahead to Appendix B on bearing selection will reveal that *all needle-roller bearings*, from the smallest (3 mm i.d.) upwards, have basic dynamic load ratings C [N] in excess of the foregoing 1020 [N]. Although the life calculation remains to be carried out, design can proceed in the confidence that it will be possible to specify a needle-roller bearing from the standard range which will be up to the job.

Appearance of the term $m_d \times r_p \times \omega^2$ offers insight: it is the magnitude of the centrifugal force needed to restrain mass m_d [kg] at radius r [m] rotating at ω [rad/sec]. Its vertical component is the force experienced by a crank-pin rotating at ω rad/sec ($\approx$ rpm/10) and driving a scotch-yoke and displacer of combined mass m_d at radius r, i.e., in simple-harmonic motion (SHM). The polar diagram therefore displays directly the factor by which

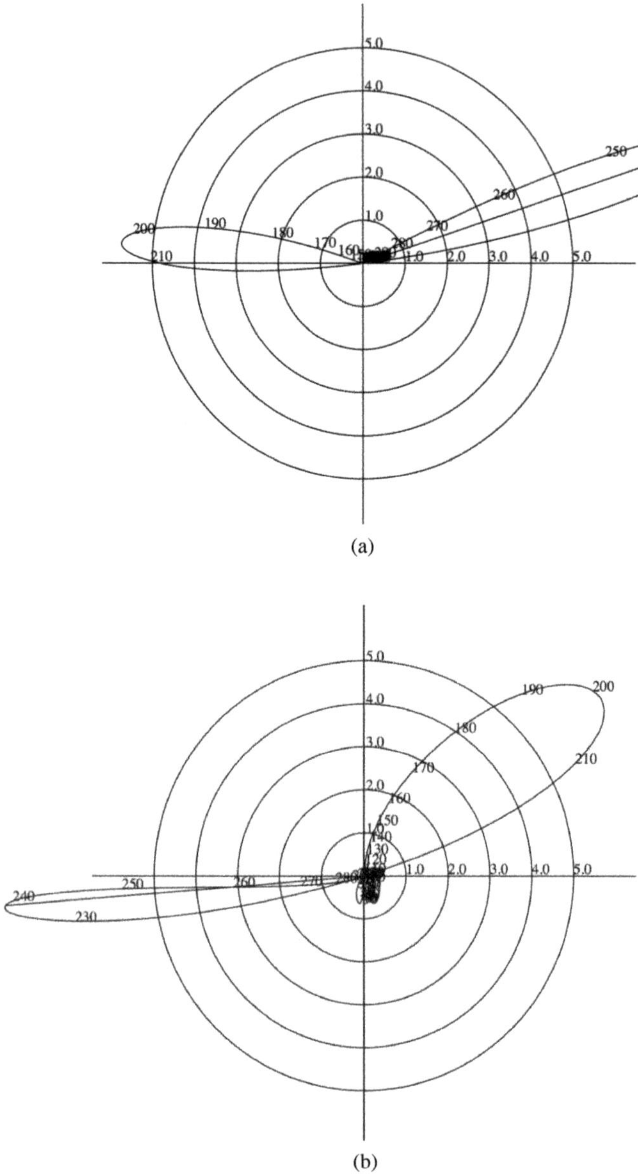

Fig. 14.3. Polar plots of 'dimensionless acceleration' or 'dimensionless load'. (a) Link Q′R′. (b) Link PQ′.

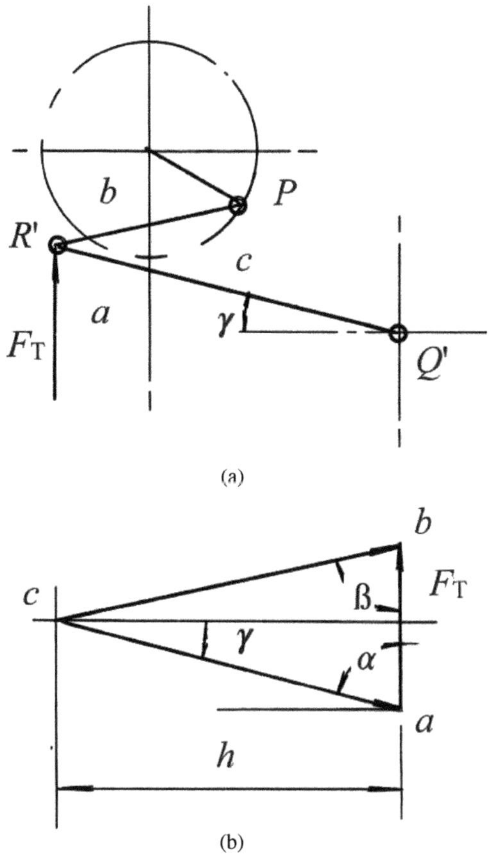

Fig. 14.4. (a) ‘Space’ diagram for displacer drive. (b) Corresponding force diagram.

the crank-pin load in the mechanism of Fig. 14.1 exceeds the corresponding SHM value at any given crank angle.

Figure 14.4 is a specimen Bow’s construction for the displacer mechanism. The corresponding algebra is a simple and obvious modification to that applied via Fig. 14.1 to the piston mechanism.

14.3. Loads in Piston Drive

Peak gas load along the line of action of F_T (not necessarily coincident with peak load internal to the mechanism) occurs in the vicinity of piston top dead-centre (TDC_P). This in turn is close to the point at which crank radius *OP* and coupler arm *PR* align. For future reference:

$$\phi_{TDCP} \approx \mathrm{a}\sin\frac{e}{r+PR}. \tag{14.2}$$

Piston acceleration peaks near TDC_P. Local inertia force is in the opposite sense to that of peak gas load, suggesting that forces in the piston drive-train calculated from gas load in isolation will be conservative.

Polar diagrams (Figs. 14.5(a) and 14.5(b)) have been generated by replacing F_T with instantaneous p/p_{min} as a function of crank-angle[1] ϕ. Feeding the thermodynamic calculation with volume variations generated by this very mechanism to a design with the provisional gas-path specification of Chapter 11 makes Fig. 14.5 'universal' — i.e., applicable to any p_{min}, to any piston area A_p (and, of course, to any rpm).

With p_{min} in units of bar (i.e., atm.) A_p can be entered in terms of diameter D_p in mm provided the outcome is divided by 10 [(Pa/bar) × m²/mm²] to give a result in Newton, N. With 7.0 for amplification factor (from Fig. 14.5) and D_p = 50 mm, peak force F_{PR} in coupler *PR* is

$$F_{PR} = 7.0\ [-] \times 1\ [\text{bar}] \times 1/4\pi \times 50^2\ [\text{mm}^2]/10 = 1375\ \text{N}.$$

The value is exceeded by the dynamic load rating *C* [N] of all needle-roller bearings of size (nominal inner-race diameter) greater than 3 mm (see tables in Appendix). Although this does *not* render the formal selection criterion redundant, it justifies proceeding in the expectation of being able to specify a number of hours' service and selecting a compact needle-roller from the standard range for duty at the crank-pin.

14.4. Mechanical Configuration

Mechanical design is frequently about reconciling competing criteria. The present undertaking is no exception. It calls for mechanism forces to be confined as far as possible to a common plane. This can minimise deflections under load — particularly important in the context of needle-roller bearings, which have a limited tolerance to load asymmetry.

[1] At the present stage the 'isothermal' analysis serves.

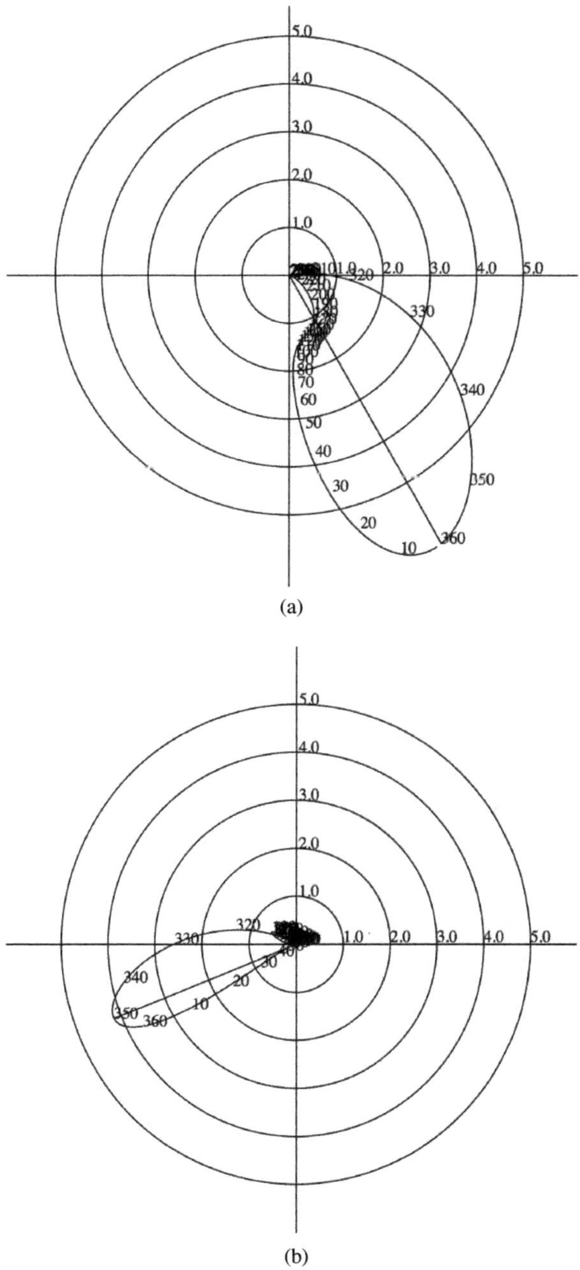

Fig. 14.5. Gas loads per $p_{\min}$ and per A_p. (a) Load in PR — and therefore on crank-pin. (b) Load in PQ.

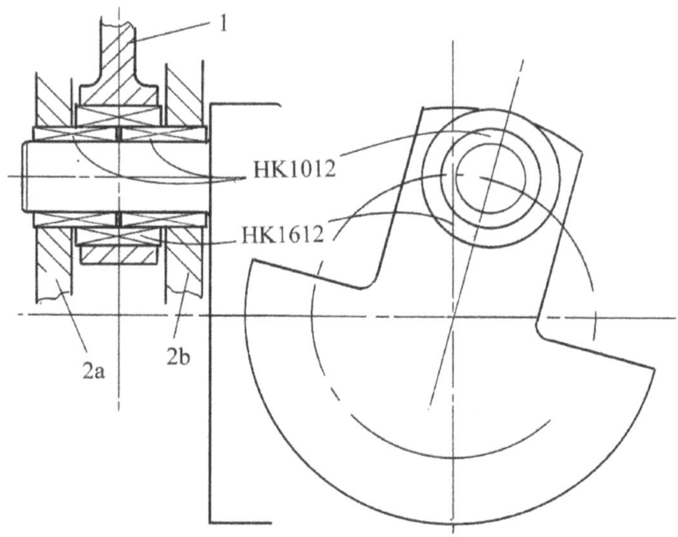

Fig. 14.6. Scheme to explore possibility of running needle-roller of displacer drive on outer diameters of needle rollers of work-piston drive. Bearings indicated are to SKF designation. (Most manufacturers use a recognizably similar code.)

Compactness is also a criterion, and at the risk of replacing one load-asymmetry with another,[2] a configuration being explored is that shown in the sub-assembly sketch of Fig. 14.6. Coupler *PR′* which drives the displacer (labelled 1 in the sketch) sits mid-way between twin couplers *PR* (labelled 2a and 2b) driven by the work piston. The latter function as housings for their own respective needle-rollers running on the crank-pin, while the former accommodates a needle-roller whose nominal *internal* diameter is the nominal *external* diameter of the inner pair. It evidently straddles the axial gap between them.

Several questions arise — all leading back to the matter of loads.

Using the drawn-cup outer ring as the inner track of the displacer bearing is unconventional — quite probably un-tried. On the other hand (a) material specification, hardness and precision are appropriate and (b) the outer bearing is over-specified (see earlier load calculation) and, to this extent, lightly loaded.

[2] SKF units embody 'modified line contact' between rollers and raceway(s) allowing minor misalignment (SKF General Catalogue, 1989).

The displacer load applied through coupler *PR′* at *P* adds to piston loads transmitted to the main bearing via coupler *PR*. The two contributions to crank-pin load are not in phase. This threatens to upset the scaling approach, because magnitudes cannot be added arithmetically. In principle the problem can be circumvented by scaling *x* and *y* components of net force vectors independently, summing respective components and forming the modulus of the sums.

This, however, may not be necessary if one load (displacer inertia or piston gas force) dominates, since a bearings selection process satisfactory for present purposes can be based on peak load. Comparing peak specific gas load in *PR* (Fig. 14.5(a)) with peak specific inertia load in displacer link *PR′* (Fig. 14.3(b)) reveals that occurrences are at 360° and 240°, respectively, viz, separated by some 120° of crankshaft rotation. Moreover, at the 360° of peak gas load, the inertia contribution is negligible.

On this basis the earlier estimate of peak crank-pin load serves for the provisional check of suitability of the pair of HK1012 bearings.

15

A Lagrange Formulation

15.1. What Level of Detail?

As computing power, speed and storage increase without apparent limit, the answer may well be in terms of the staying-power of the analyst. In 1987 Gleick was in no doubt:

Only the most naïve scientist believes that the perfect model is the one that perfectly represents reality. Such a model would have the same drawbacks as a map as large and as detailed as the city it represents, a map depicting every park, every street, every building, every pot-hole, every inhabitant — and every map! Were such a map possible, its specificity would defeat its purpose: to generalize and to abstract.' (There are no apologies for reuse (Organ, 1997) of this skilfully-articulated insight).

The picture has since been muddied by appearance of solid-modelling software, of 3D flow-modelling packages and — above all — the increasing ability of such packages to 'talk' to each other. An overriding consideration nevertheless prevails: Building a geometric and kinematic model in, say SolidWorks™, involves commitment to specific dimensions for crank kinematics, for exchanger layout, etc. — and not merely lengths, internal diameters and numbers of tubes, but the respective path traced out by each in 3D space. The process also demands specifying method of fabrication and assembly — otherwise what is the point? A prediction by Fluent™ for specific cycle work of the completed design will doubtless be impressive — but if that cycle work remains of the same old order, viz, $0.1 \leq N_B \leq 0.2$ then, by the criteria of this text, nothing has been achieved.

The commitment required to generate a viable 3D model calls for an appropriate starting specification — an outline design promising drastic change in working-fluid utilisation and corresponding vastly-increased specific cycle work N_B. Where is this to come from?

The cost-effective design of a Stirling engine is beginning to look like a process in at least two steps.

Where compressibility effects are a possibility, the reference frame appropriate to exploring the boundaries of the performance envelope is the Characteristics (MoC) coordinates system — appropriate since by travelling at particle velocity u and at $u \pm a$, respectively, temperature and pressure disturbances propagate in the numerical model as they do in reality (Shapiro, 1954). Insights uniquely available from this approach have been illustrated in the author's 2007 title. The characteristics option has, however, failed to attract a following.

An alternative with apparently unlimited potential involves tracking individual molecules according to the kinetic theory of gases. Animated display of a computer-coded model (Organ, 2007) shows, in slow motion, heaps and troughs of molecules corresponding to pressure and rarefaction waves. Attractions are the two-dimensional picture (quasi-3D if axial symmetry applies) and *absolute* conservation of mass (there being no mechanism for escape of even a single molecule!). On the other hand, application to first-principles design awaits evolution of digital storage through two or more generations.

This chapter accordingly proceeds by compromise: the Lagrange reference frame is relied upon to handle moving temperature disturbances. Embodiment of $dp/p = f(Re, Ma, ¶_v)$ deals with the steady-flow component of regenerator pressure gradients up to any combination of p_{ref} and *rpm* leading to choking.

15.2. Starting Assumptions

The working fluid is an ideal gas obeying $p/\rho = RT$.

The gas path is visualised as being 'straightened out' into the usual horizontally-opposed layout with expansion space to the left, compression to the right.

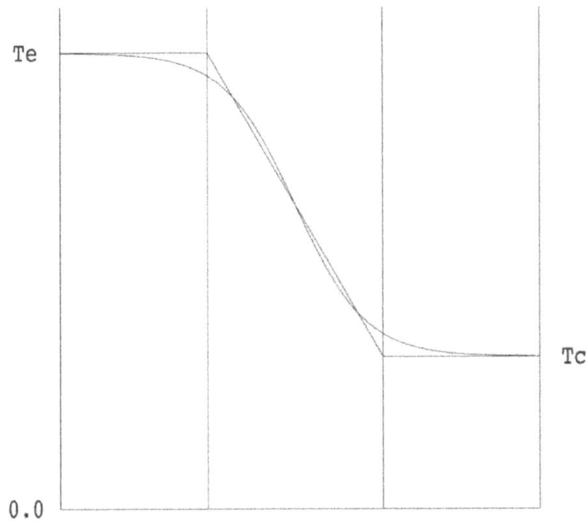

Fig. 15.1. Options for specifying length-wise distribution of regenerator matrix temperature — piece-wise linear or continuous as per Eq. (15.1).

Mass M of working fluid enclosed within the gas path is constant, i.e., leakage is zero. (The Lagrange formulation exploits this to ensure — and to help confirm — convergence.)

Flow is one-dimensional ('slab' flow): a fluid particle initially lying in a plane perpendicular to the direction of motion (x-direction) remains in that plane, and its instantaneous properties characterise those of all particles in the same plane.

Distribution of enclosure temperature T_w in the flow direction does not vary with time. It may be either piece-wise linear as indicated in Fig. 15.1, or may vary continuously with axial location according to a suitable function, e.g.:

$$T_w(x) = ½(N_T + 1) - ½(N_T - 1)\tanh(x). \qquad (15.1)$$

The x origin for Eq. (15.1) is at the regenerator mid-plane. The equation is super-imposed on Fig. 15.1.

Gas in contact with the piston faces at left- and right-hand extremities remains at the temperatures assumed for those faces, viz, T_E and T_C, respectively.

15.3. Independent Variables

Working fluid occupies not physical length, but *volume*. This reveals the appropriate independent variable to be *cumulative volume, v*. In the present formulation v is reckoned rightwards from the outer (left-most) position of the expansion piston face. Time (equally crank angle) elapses up the y-axis — i.e., vertically.

Gas mass M is divided into n sub-volumes containing equal fractions dm/M of the total. The number of particle tracks includes the bounding faces of the opposed pistons, and is $n + 1$. Under the 'slab flow' assumption, following the motion of a bounding face (of a particular v_j) over a cycle describes a particle trajectory:

$$\textbf{\textit{Mass conservation}}: \quad dm = p\Delta V/RT. \tag{15.2}$$

In Eq. (15.2) ΔV is the volume enclosed between two adjacent bounding faces j and j+1 — *not* a change in v with time (or crank-angle).

The 'obvious' approach to forming the finite-difference approximation to Eq. (15.2) involves expressing volume element ΔV as $v_{j+1} - v_j$ and other terms as mean values, e.g., $\frac{1}{2}(p_{j+1} + p_j)$, leading to:

$$\tfrac{1}{2}(p_{j+1} + p_j)(v_{j+1} - v_j) = \tfrac{1}{2}dmR(T_{j+1} + T_j). \tag{15.3}$$

However, temperature gradient $\partial T/\partial x$ within the regenerator is some 20°C/mm (20,000°C/m!), meaning that use of mean temperature $\frac{1}{2}(T_{j+1} + T_j)$ is insufficiently accurate to prevent Eq. (15.3) drifting over the millions of arithmetic operations envisaged. Appendix H offers an accurate alternative. For convenience of presentation *only*, the present account proceeds in terms of Eq. (15.3).

Each variable is expanded about its current value in the fashion of $p_j + Dp_j$ etc. After an increment of time dt, measured here in terms of an increment in crank angle $d\varphi$ at uniform angular speed ω, local p_j, v_j and T_j have changed to $p_j + Dp_j$, $v_j + dv_j$ and $T_j + DT_j$. Substituting, cancelling the factor ½ and subtracting Eq. (15.3) from the result:

$$\begin{aligned}-dv_j(p_{j+1} + p_j) + dv_{j+1}(p_{j+1} + p_j) - dmRDT_j - dmRDT_{j+1}Dp_j(v_{j+1} - v_j)\\ + Dp_{j+1}(v_{j+1} - v_j) = -(Dp_{j+1} + Dp_j)(dv_{j+1} - dv_j).\end{aligned} \tag{15.4}$$

The process has effectively linearised the original equation. The nonlinear product of small quantities at the RHS succumbs to iteration.

A thermodynamic cycle is assumed to start with gas stationary and at uniform pressure, so that $p_1 = p_2 \cdots = p_{n+1}$. Gas temperatures T_2, T_3, ... , T_n are those of the immediately adjacent enclosure. Initial axial locations are set in terms of the p_j, T_j so that successive pairs of locations v_{j+1} and v_j enclose equal masses of gas $dm = M/n$. When the pistons are given their first increment of motion, the problem amounts to finding the various v_j, T_j and p_j, after time increment dt and corresponding changes dv_j, DT_j and Dp_j.

Confidence of correct functioning calls for checks at steps intermediate to full formulation (gas law plus three conservation laws). Checks need to be contrived on the basis of number of available equations vs number of unknowns.

Allowing T_1 and T_n to be determined at all times by the T_E and T_C (of piston faces with which they respectively remain in contact) leaves $n - 1$ values of DT_j unknown. The number of dv_j is the same, since piston-face increments are prescribed. Pressure change Dp_j applies along each of the $n + 1$ particle paths, giving a total of $3n - 1$ unknowns per crankshaft increment.

There are n sub-volumes to which Eq. (15.4) applies. Provisionally neglecting pressure drop yields n equations of the form Dp_{j+1} = Dp_j, i.e., a total so far of $2n$ equations, leaving $n - 1$ to be specified. With appropriate, respective definitions of the changes DT_j in path-line temperatures T_j, formulation to this point can be tested for ability to replicate (a) adiabatic conditions throughout ($T_j/p_j^{(\gamma-1)/\gamma}$ = const, i.e., $\mathrm{Dd}T_j/T_j = (\gamma - 1)/\gamma\, \mathrm{D}p_j/p_j$) and/or (b) the 'isothermal' assumption ($T_j = T_w(v_j)$) throughout. The latter case is implemented as $\mathrm{D}T_j = (\mathrm{d}T_w/\mathrm{d}v_j)\mathrm{d}v_j$.

The $(3n - 1) \times (3n - 1)$ array of unknowns is processed by SIMQX — a somewhat dated item of Fortran library software for the solution of simultaneous equations. It functions by Gauss–Seidel elimination. Should the formulation eventually prove worthy of being invoked by an optimisation algorithm it will be worth trading SIMQX for a more efficient routine. The NAG™ library offers options.

Figure 15.2 is a relief in quasi-perspective of temperature distribution over an initial cycle under the 'adiabatic' assumption. Number of subdivisions of a crankshaft revolution is 180, with 60 sub-division of mass *M*.

The anticipated overall rise and fall in temperature can be seen be comparing $T_{j,1}$ (left piston face) to the adjacent $T_{j,2}$ and $T_{j,n}$ to $T_{j,n+1}$

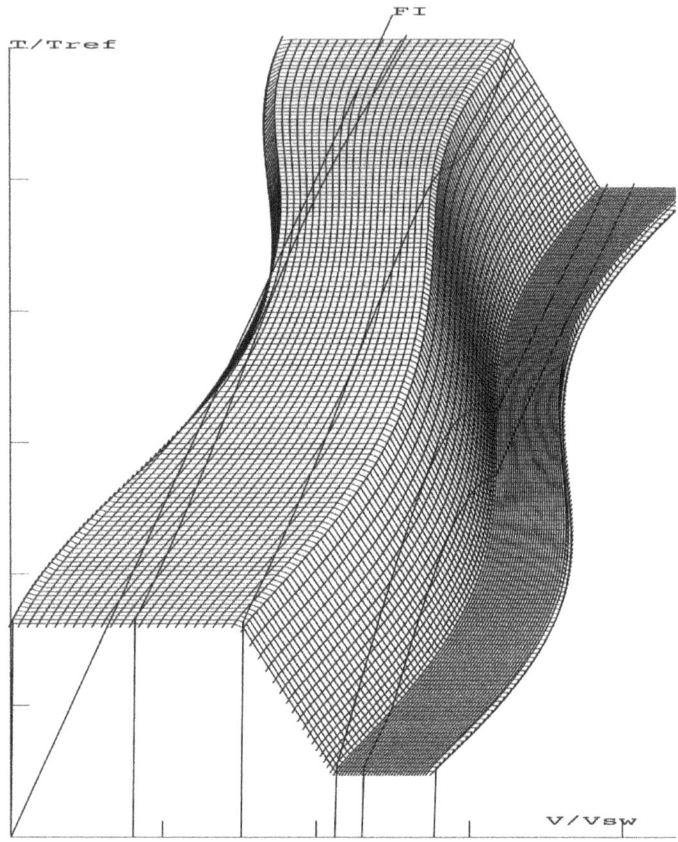

Fig. 15.2. Temperature solution for start-up cycle under the 'adiabatic' assumption. Oblique lines ruled on surface are outlines of heat exchangers and regenerator.

(right piston face). As expected, the starting gas temperature profile washes back and forth unaffected by the temperature of the adjacent solid enclosure.

Figure 15.3 is the corresponding 'isothermal' case which again complies with expectation — T adapting to local enclosure temperature T_w at all locations and for all crank angles.

A criterion of robust functioning would be welcome at this stage. The response of the model to an unrealistically-small number of sub-divisions should be revealing. Figure 15.4 shows the initial cycle with number of subdivisions of M reduced to $n = 7$, but otherwise under the 'isothermal'

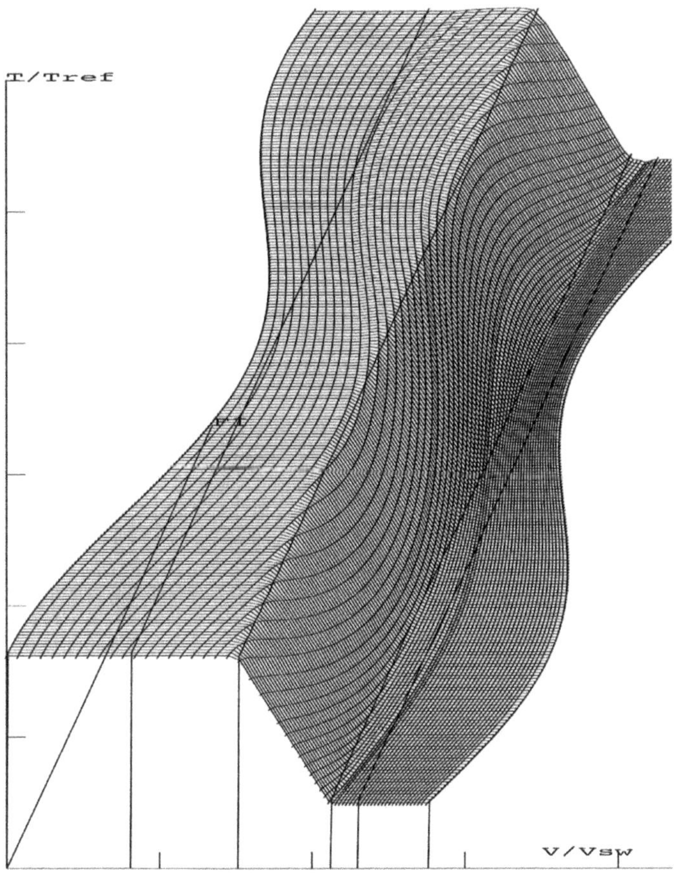

Fig. 15.3. 'Isothermal' counterpart of Fig. 15.2.

assumption. There is no hint of imminent distress in the integration process.

Figure 15.5 represents a tentative step towards modelling the internally-focusing solar concept. Simple-harmonic volume variations have been replaced by those generated by the 'pseudo-rhombic' mechanism (Chapter 10) but the gas-path proportions behind Fig. 15.4 have been retained. The formulation is still able to cope with an unrealistically small number of subdivision, although the more drastic rates of volume change now call for a minimum of 12. For the illustration 15 sub-divisions have been used.

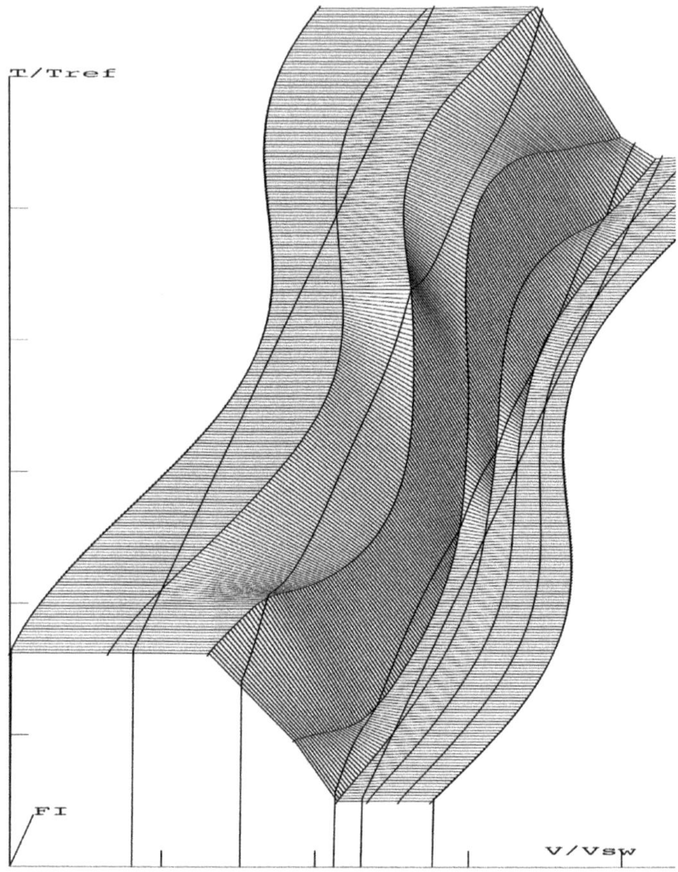

Fig. 15.4. Number of subdivisions n of M reduced to 7. Otherwise as for 'isothermal' case.

The isometric depiction causes an element of optical illusion: at first glance, the expansion space (outlined by the left-side curve) both increases and then decreases while the compression space (outlined by the lower-right curve) is held close to zero — evidently not what is required. To see that this is not the case, follow the compression-piston curve from lowermost right to the point where it begins to more rightwards (slightly less than half-way through the cycle), pick one of the lines ruled on the temperature surface and follow it leftward to meet the expansion piston. The latter will be seen to have commenced its path rightward which, in

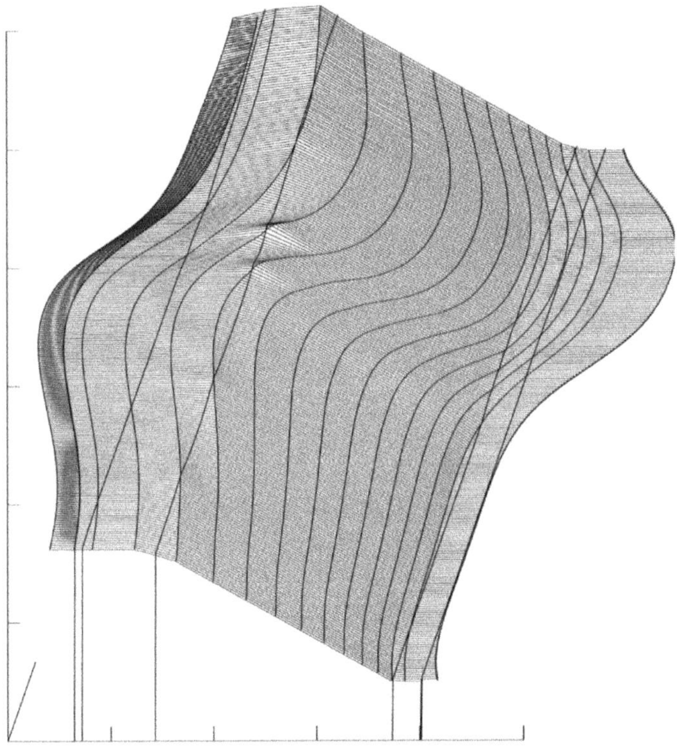

Fig. 15.5. Volume variations derived from 'pseudo-rhombic' mechanism, but gas path proportions (large dead space) unchanged from those of Fig. 15.4. Lagrange set-up can still cope with unrealistically-small number of subdivisions *n*, although the minimum is now 12.

unison with the compression piston, brings about the constant-volume transfer process.

15.4. Discussion

Little difficulty has been encountered is setting up the framework of a gas-process model in Lagrange coordinates. With appropriate step sizes for the integration intervals, adiabatic and 'isothermal' conditions are convincingly replicated. Formulation to this point is sufficiently robust to function with unrealistically-large subdivisions dm of working fluid mass M.

16

Path-Line Equation

16.1. Priority

The numerical model to this point might be described as being of 'single conservation law' scope — mass conservation or gas law. But for its potential for greater things, it is no more powerful nor revealing than the 'adiabatic' model of Chapter 3.

Extension to reflect conservation of energy (and eventually momentum) will be carried out with the overriding priority of providing for *verifiability*. With the energy equation embodied, but with heat transfer suppressed (h = zero throughout) it must return the temperature relief of Fig. 15.1 generated via the adiabatic relationship. It must handle heat transfer coefficients near to infinity without numerical problems, returning the 'isothermal' solution of Fig. 15.2.

With a view to making these checks as conclusive as possible, the dummy momentum law $p_{j+1} = p_j$ (implemented as $Dp_{j+1} = Dp_j$) is retained. (The usual expression for pressure-drop in terms of friction factor is not directly compatible with the Lagrange structure.)

16.2. Path-Line Equation

Whereas an Euler formulation sticks with its fixed 'control' volumes, the Lagrange counterpart integrates along selected particle tracks. The relevant equation is — unremarkably — the path-line equation. Its delightful simplicity is offset somewhat by the tricky context — that of the severe temperature gradient in the regenerator: dT_w/dx of 20,000 C/m requires special treatment if it is not to provoke numerical oscillation — or worse.

With this in mind, a change of entropy Ds while following a particle of unit mass can be expressed in terms of changes Dp and DT in properties p and T.

$$Ds = c_p DT/T - RDp/p. \qquad (16.1)$$

In Eq. (16.1), operator D corresponds to 'total' or 'substantial' derivative $\partial/\partial t + u\partial/\partial x$.

Two effects contribute to Ds (a) heat exchange $q'dt$ per unit mass and (b) friction. The thermal component $q'dt$ can be an increase or a decrease. Friction always contributes an increase. The gas element flows in a duct of free-flow area A_{ff}, wetted perimeter p_w and local wall temperature T_w. Length element Δx relates to independent variable Dv as $\Delta x = (v_{j+1} - v_j)/A_{ff}$. In terms of coefficient of convective heat transfer h and local temperature difference $(T_w - T)$, instantaneous rate of heat acquisition by the mass element is $hp_w(T_w - T)\Delta x$. The element has mass $dm = \rho\Delta V = \rho(v_{j+1} - v_j)$. Ratio A_{ff}/p_w is hydraulic radius r_h, so q' may be written

$$q' = h(T_w - T)/\rho r_h. \qquad (16.2)$$

The friction component is $|u|F$dt. Shapiro's (1957) account evaluates in terms of friction factor C_f:

$$F = \frac{1}{2}u^2 C_f / r_h. \qquad (16.3)$$

When applied in the context of the wire gauze regenerator, Eq. (16.3) must be considered a 'plug-in' pending substitution of a term better reflecting dependence on both *Re* and *Ma*.

Entropy change in terms of fluid properties (Eq. (16.1)) is equated to heat transfer and friction effects causing those changes. Invoking the ideal gas equation, and recalling that $c_p/R = \gamma/(\gamma - 1)$ there results:

$$TDp/p - [\gamma/(\gamma - 1)]DT = (|uF/R| + h\Delta T/\rho R r_h)dt. \qquad (16.4)$$

In Eq. (16.4), $\Delta T = T_w - T$, where both temperatures are local, instantaneous values.

16.3. Preliminary Commissioning

This account had evolved against the following background: The 'real' gas path is three dimensional. The numerical model, by contrast, is one dimensional. Exact correspondence would be a coincidence — or a symptom of erroneous formulation — rather than grounds for celebration. 'Verification' by comparing simulated and measured power and efficiency is not the way forward. Instead, checks of correct functioning will involve exercising the code in one-dimensional situations having known solutions. The initial focus is accordingly on temperature in isolation from the effect of pressure transients. As already anticipated, this is achieved by provisional use of a 'dummy' momentum law $Dp_{j+1} = Dp_j$ applied to all n adjacent pairs of particle paths. Pressure non-uniformity will be addressed at a later date.

In the examples which follow, the number of sub-divisions n has been chosen in the interests of display: the number required for acceptable solution results in plots which are too dense for pictorial display.

Figure 16.1 shows the temperature solution achieved by forcing NTU = zero throughout. Despite a coarse integration interval (crank-angle increment of 2.5°), the solution is indistinguishable from that of the earlier Fig. 15.1 — an indication of correct functioning at the 'nuts-and-bolts' level.

Figure 16.2 is the temperature relief with NTU set to the artificially high value of 10^6 throughout, and may be compared with the 'real' isothermal solution of Fig. 15.2.

Figure 16.3 shows how the 2-conservation-law formulation copes with an impracticably small number of subdivisions n of M, and can be compared with Fig. 15.3.

Figure 16.4 is the result when NTU value is calculated from local, instantaneous Re. The 'ridge' developing after flow reversal at the expansion end between exchanger outlet and regenerator inlet reflects the assumption of fixed matrix temperature distribution: the incoming gas is below T_E after expansion, and responds suddenly to the high NTU of the regenerator.

For Fig. 16.5, the original volume variations (simple-harmonic) are replaced by those generated by the 'pseudo-rhombic' mechanism. The start-up cycle illustrated. The earlier gas-path proportions are retained. so the potential for near-total purging is not demonstrated. However,

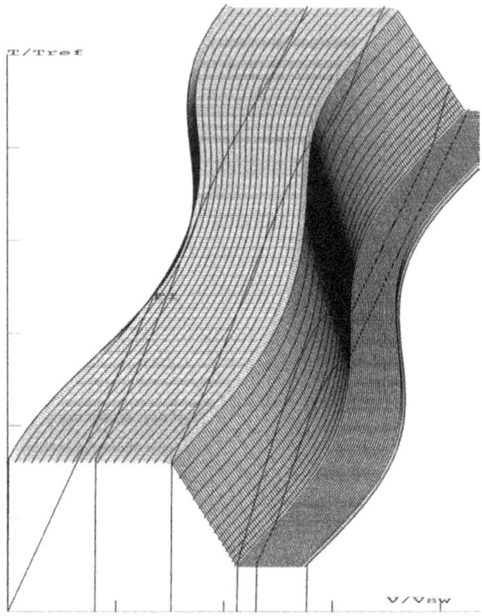

Fig. 16.1. 'Adiabatic' case synthesised by setting *NTU* = zero throughout. Indistinguishable from the explicit adiabatic solution of Fig. 15.1.

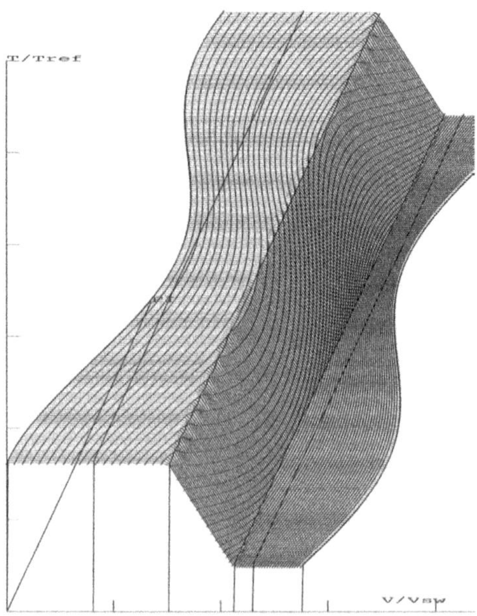

Fig. 16.2. 'Isothermal' solution retrieved by setting $NTU = 10^6$ throughout.

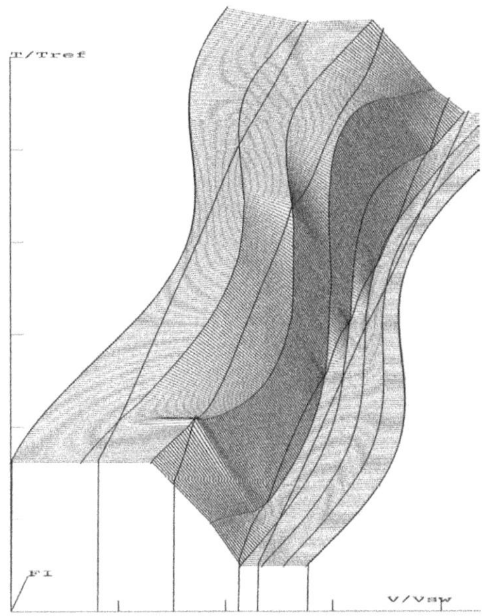

Fig. 16.3. Result of setting number of subdivisions of M to unrealistically low value $n = 7$. Response comparable to that of earlier test of robustness (Fig. 15.3) — but now reflecting influence of finite heat transfer.

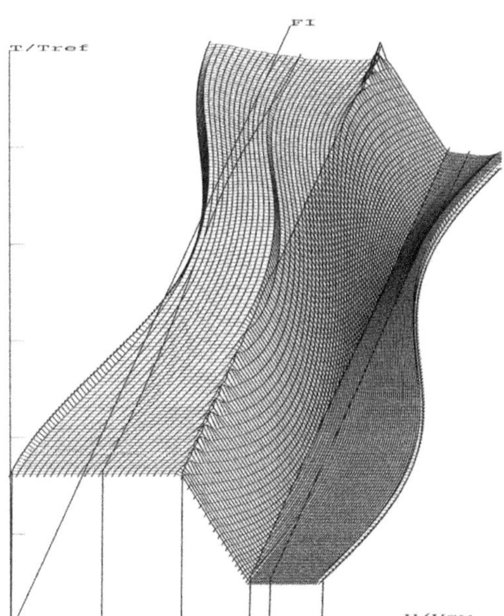

Fig. 16.4. Temperature solution for start-up cycle with St (and thus NTU) calculated from local instantaneous Re ($\partial p/\partial x = 0$ throughout).

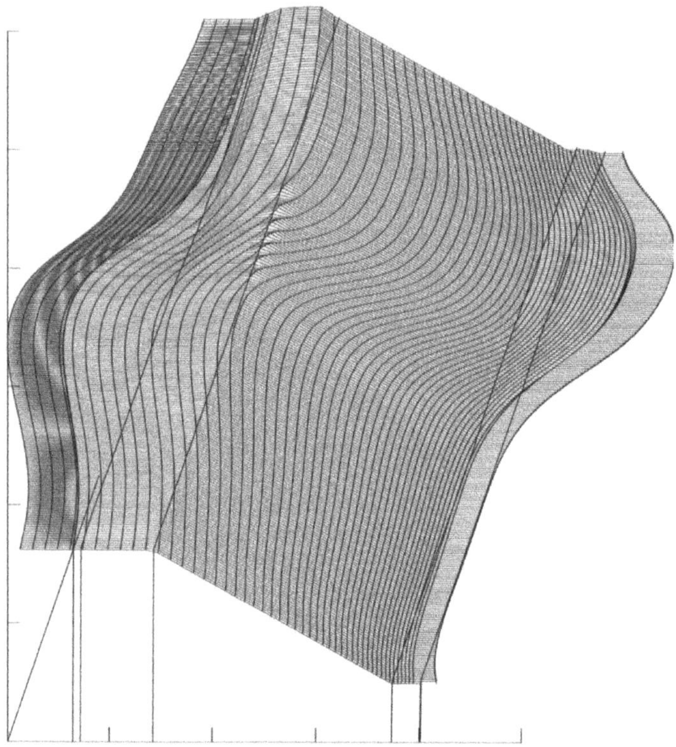

Fig. 16.5. Temperature solution for start-up cycle as for Fig. 16.4, but with SHM volume variations replaced by those generated by 'pseudo-rhombic' mechanism. (Gas path proportions unchanged.)

temperature swing is increased over the earlier SHM case, and shows up in the expansion exchanger at the uppermost point of the diagram. The *NTU* in the regenerator remains sufficiently high to hold gas temperature close to the linear matrix temperature throughout the cycle.

16.4. Discussion

The Stirling engine does not function according to the so-called *ideal cycle*. The several departures from the ideal can be traced to failure of the real machine to process more than a fraction of working fluid mass *M* through the full temperature range.

Progress is going to be dependent upon availability of flow and heat transfer correlations reflecting the realities of viscous, unsteady, compressible flow. The availability of real-time data processing promises a vastly easier task than that which Kays and London overcame in the 1950s using DC electrical apparatus. However, nothing short of a long-term programme, adequately-funded, will lead to the necessary coverage of experimental conditions.

In the meantime, the Lagrange model goes some way to bridging the gap. Two simple benchmark tests suggest that it is robust and well-suited to the capture and display of the interactions underlying its formulation.

17

Think Before You Compute

17.1. In Case You Are Wondering

The chapter heading is the title of a 2020 title by Cambridge University Press. It is re-used here with the blessing of author E. J. Hinch.

Although Hinch does not address directly the combination of gas processes[1] which make up the practical Stirling cycle, many of the issues he confronts arise in formulating a viable numerical simulation.

Efforts to date have, with one known exception, been based in Euler (fixed-grid) coordinates. The exception arises because all attempts by this writer to work in that system have failed to yield a simulation which satisfies any benchmark test of correct functioning. The failure has given rise to numerical models in alternative coordinate systems — Lagrange and Characteristics — and even to an animated model of molecular interactions based in the kinetic theory of gases (Organ, 2007).

With the assurance of a full-length title preaching essentially the same message, it becomes worthwhile to restate the objection to use of Euler: (a — in general) — under-estimating and over-simplifying the subtlety and complexity of the interactions and (b — in particular) falling into the trap of representing a differential operator such as $\partial/\partial t$ in terms of finite differences $\Delta/\Delta t$. The latter is an abuse, one symptom of which is numerical diffusion and dispersion absent from the defining equation(s).

---

[1]Cyclically-reversing, unsteady, compressible flow with friction, heat transfer and abrupt area-change simultaneous with the notoriously-challenging conjugate heat exchange problem (regenerator).

The commonly-fielded claim that numerical diffusion somehow compensates the diffusion element of the natural flow does not hold water, since the former is arbitrary and the extent of the latter unknown.

With hindsight it is remarkable that the advice *think first, compute later* has apparently never been offered before. Now that it has, Hinch's new title is essential reading for anyone proposing to enter this minefield.

Hinch's focus is rightly on appropriate matching of solution algorithm to the specific flow case and to its defining equation(s) and boundary condition(s). Study of the thermal-lag engine calls for the same, but introduces another equally pressing need. The latter arises from the total absence of design guidelines for the type. At the scale of the table-top demonstrator, the thermal-lag performs on a par with the Stirling counterpart. However, neither can be said to have been 'designed' (unless evolution counts as design, in which case the latter has had two centuries to evolve while the former has had no time at all). As far as is known, a thermal-lag engine has never been subjected to pressurisation, so it must be a consideration that specimens have functioned on a one-atmosphere charge whereas the modern 'serious' Stirling engine never has.

Evaluating any untapped potential calls for the evolutionary process to be artificially speeded up. The only known way requires the support of computer simulation — but simulation suitable for serving as the core of mechanised optimisation. Considering the need to explore a range of rpm and charge pressures in addition to the myriad combinations of flow-passage geometry, systematic optimisation will call for tens of thousands of perturbations of the cycle model (the 'object function'), each of which must be allowed to run to cyclic equilibrium. An object function taking more than a few minutes per cycle hardly qualifies.[2]

This chapter introduces a thermal model of which Hinch would approve. Its Lagrange coordinate system allows heat transfer and friction effects to act on individual particles, as reality — and Shapiro (1954) — insist they must. But it is also fast: with one-degree crank-angle increments and 200 subdivisions of the gas path, cyclic equilibrium is achieved

[2]The 17th International Stirling Engine Conference (Newcastle, 2016) included papers and discussion on the use of proprietary packages such as Fluent™ — and taking hours rather than minutes to deal with a gas-path geometry barely recognizable as a Stirling engine, thereby going a long way towards defeating the entire purpose!

in fewer than 10 cycles, and these take less than 1 sec of CPU[3] time. To distinguish from other formulations later in the book this one is nick-named *Turbo2×2*.

17.2. TURBO2 × 2

Figure 17.1 represents an arbitrary point in the gas-path and a pair of adjacent particle paths. Gas occupies volume rather than length. The horizontal coordinate is accordingly cumulative volume v reckoned from the left-hand end which, in the case of the thermal-lag type, is closed and stationary.

Successive paths lines divide total working fluid mass M into n equal sub-masses $\mathrm{d}m = M/n$ and into generally different individual volumes $V = v_{j+1} - v_j$. Each $\mathrm{d}m$ is subject to the ideal gas equation $pV = \mathrm{d}mRT$. After a short interval with the system in motion p, T and V have generally changed by $\mathrm{d}p$, $\mathrm{d}T$ and $\mathrm{d}V$ so that $(p + \mathrm{d}p)(V + \mathrm{d}V) \approx \mathrm{d}mR(T + \mathrm{d}T)$. Multiplying out the latter expression and subtracting the earlier one:

$$p\mathrm{d}V + V\mathrm{d}p - \mathrm{d}mR\mathrm{d}T \approx -\mathrm{d}p\mathrm{d}V. \tag{17.1}$$

As well as changing in magnitude, volume V has moved bodily, carrying with it its temperature distribution at local velocity u. The

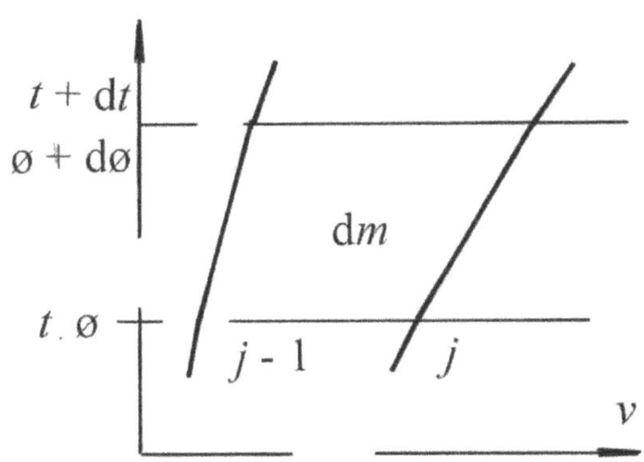

Fig. 17.1. A section of the gas-path showing adjacent particle paths. Horizontal coordinate is cumulative volume. Time or crank-angle elapses vertically.

[3] 1.8 GHz dual-core Intel i5.

appropriate notation is then *total* difference $\mathrm{D} = \partial/\partial x + u\partial/\partial t$. With $\mathrm{d}m$ chosen suitably small, effective temperature T is $1/2(T_{j+1} + T_j)$. Substituting:

$$p(\mathrm{D}v_{j+1} - \mathrm{D}v_j) + (v_{j+1} - v_j)\mathrm{D}p - \mathrm{d}mR(\mathrm{D}T_{j+1} + \mathrm{D}T_j) \approx - \mathrm{D}p(\mathrm{D}v_{j+1} - \mathrm{D}v_j)\,.$$

Any given time step gives rise to n such equations. Anticipating the need for simultaneous solution, unknowns $\mathrm{D}v$, $\mathrm{D}T$ and $\mathrm{D}p$ are organised into an array:

$$-p\mathrm{D}v_j + p\mathrm{D}v_{j+1} + \cdots + \mathrm{d}mR\mathrm{D}T_j + \mathrm{d}mR\mathrm{D}T_{j+1} + \cdots + (v_{j+1} - v_j)\mathrm{D}p. \qquad (17.2)$$

In terms of fluid properties p and T, entropy change $\mathrm{D}s$ while following a particle is $c_p\mathrm{D}T/T - R\mathrm{D}p/p$. Expressed as irreversible processes of heat transfer and friction, rate of entropy change is $(q' + uF)/T$, where q' is heat rate per unit mass and F is a wall friction term. Using Shapiro's $F = 1/2u^2C_f/r_h \times \mathrm{sign}(u)$ and with $T_w - T$ for local temperature difference, $q' = h(T_w - T)/\rho r_h$.

Equating the two expressions for $\mathrm{D}s$ yields the path-line equation. It will be seen shortly that only the right-hand path (Fig. 17.1) is of interest:

$$\mathrm{D}p/p = \gamma/(\gamma - 1)\mathrm{D}T_{j+1}/T_{j+1} - (q' + uF)\mathrm{d}t/(RT_{j+1}). \qquad (17.3a)$$

Ordering unknowns and their coefficients as previously:

$$-0 \times \mathrm{D}v_j + 0 \times \mathrm{D}v_{j+1} + \cdots - 0 \times \mathrm{D}T_j - [\gamma/(\gamma - 1)]\mathrm{D}T_{j+1}/ T_{j+1} + \cdots + p^{-1}\mathrm{D}p = -(q' + uF)\mathrm{d}t/(RT_{j+1}). \qquad (17.3b)$$

If temperatures T_1 and T_{n+1} are assumed to take the (known) values of the opposed piston faces with which they are in permanent contact, an integration step in time (or crank angle) involves $n - 1$ temperature increments $\mathrm{D}T$, $n - 1$ increments $\mathrm{D}v$ in cumulative volume v and the increment Dp in p, making a total of $2n - 1$ unknowns. Solution requires of $2n - 1$ equations, and just this number is available: n equation (17.1) for mass conservation, and $n - 1$ equation (17.3) along the interior path-lines (the paths of the enclosing end-faces being defined by piston motions).

Previous solutions (e.g., Organ 2016b) have been solved simultaneously by numerical methods for linear simultaneous equations. SIMQX (anon, undated) is an unsophisticated but serviceable and freely-available example. However, with $n = 200$ and a crankshaft revolution divided into 360 steps of 1°, simulating a single cycle takes about two minutes on a machine equipped with 1.8 GHz dual-core Intel i5 processor. This suggests at least 10 minutes per solution — depending on the criterion for cyclic equilibrium.

This is not a promising basis for exploring the large number of combinations of design variables which characterise the thermal-lag engine: a mechanised optimisation run would involve some tens of thousands of perturbations and some hours per run.

The specific case cited (200 × 360 sub-divisions) can be speeded up by a factor of 100 by a sleight-of-hand. Unusually for an advantageous short-cut, there is no compromise of accuracy.

17.3. Quasi-Explicit Integration Step

The method takes advantage of the availability of an approximation to dp at any given time step. Since dp is small, zero will serve!

Figure 17.2 represents the first gas element in the sequence. The left-hand face is the closed end — which is stationary, so Dv_1 is zero. Moreover, DT_1 is also zero, since T_1 is assumed known and invariant. The estimated value for Dp is transferred to the right-hand sides of Eqs. (17.1) and (17.2), leaving only two unknowns — Dv_2 and DT_2. Two equations and two unknowns means — solution!

The equations may be abbreviated:

$$a_{1,1}\mathrm{D}v_{j+1} + a_{1,2}\mathrm{D}T_{j+1} = b_1,$$
$$a_{2,1}\mathrm{D}v_{j+1} + a_{2,2}\mathrm{D}T_{j+1} = b_2,$$

with solutions:

$$\mathrm{D}v_{j+1} = (b_1a_{2,2} - b_2a_{1,2})/(a_{1,1}a_{2,2} - a_{2,1}a_{1,2}),$$
$$\mathrm{D}T_{j+1} = (a_{1,1}b_2 - a_{2,1}b_1)/(a_{1,1}a_{2,2} - a_{2,1}a_{1,2}).$$

The calculation now shifts one step to the right. With the same approximation for Dp there are again two unknowns Dv_{j+2} and DT_{j+2} and the same two equations.

The process repeats, eventually yielding a Dv_{n+1} corresponding to the initial guess for Dp. The resulting v_{n+1} must be identical to piston position after its specified motion increment. If less, then total gas mass M has been compressed into too short a total volume by the assumption of too

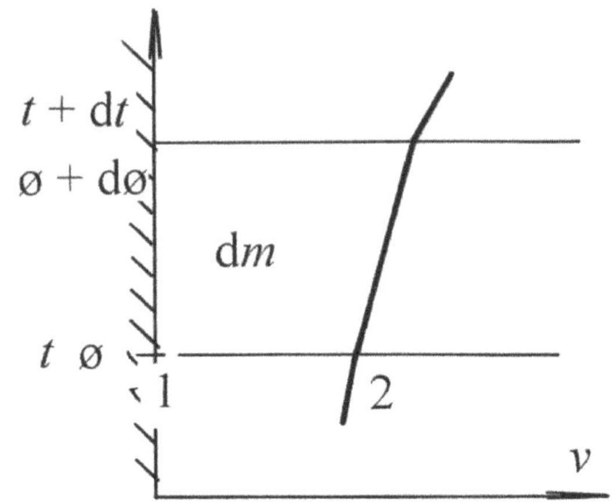

Fig. 17.2. Path lines 1 and 2 at left-hand (closed and stationary) end of system.

high a D*p*. If greater, then the assumed D*p* was too low. Acquiring the correct D*p* is a simple case of pro-rating in terms of the calculable error in computed v_{n+1} and iterating to convergence. Denoting by $Dv_p^{\phi+\Delta\phi}$ the total volume enclosed by the piston after crank-angle increment $\Delta\phi$, the ratio r by which the estimate of D*p* should be modified is $(v_{n+1} + Dv_{n+1})/Dv_p^{\phi+\Delta\phi}$:

$$Dp^{new} = -p + (p + Dp^{prev}) \times r. \tag{17.4}$$

The convergence criterion can be that embodied in the library routine for simultaneous solution (SIMQX in this case), so the short-cut method does not compromise accuracy.

17.4. Test Cases

The graphical illustrations which follow are temperature reliefs. In all cases the vertical coordinate is T, the horizontal coordinate cumulative volume v. Time or crank angle elapses from lower left to upper right. To avoid cluttered presentation a reduced number of integration steps is used: 200 time (or crank-angle) divisions per cycle with gas mass M subdivided in 100 equal sub-masses dm. These are fewer than the minimum required for satisfactory solution, the latter being 360×200. The penalty for use of the coarse grid shows up in the 'isothermal' and 'adiabatic' cases, where the p–V diagram which ought to be a single curve in both cases actually encloses a small but finite area.

17.4.1. 'Isothermal'

The term is a misnomer. It arises in connection with the Schmidt cycle analysis which assumes that the temperature T of the working gas follows the temperature T_w of the enclosing surface with which it is in contact, regardless of relative motion. According to that assumption $DT = \partial T_w/\partial t + u\partial T_w/\partial x$. Noting that the first term $\partial T_w/\partial t$ is zero, this translates in terms of the present coordinate system to

$$DT = \left(v + \frac{1}{2}Dv\right) \times \partial T_w/\partial v\left(v + \frac{1}{2}Dv\right). \tag{17.5}$$

In Eq. (17.5), $\partial T_w(v + 1/2Dv)/\partial v$ is wall temperature gradient at mid-integration interval. It states that, over a cycle, the temperature distribution $T(v)$ of the gas should remain unchanged at that of the enclosure $T_w(v + 1/2Dv)$. The change is implemented by substituting terms of the solution algorithm. Coefficients $a_{1,1}$, $a_{1,2}$ and b_1 remain unaltered, but $a_{2,1}$ becomes $-1/2\partial T_w(v + 1/2Dv)/\partial v$, coefficient $a_{2,2}$ becomes unity and right-hand side b_2 is replaced by $v \times \partial T_w(v + 1/2Dv)/\partial v$.

Figure 17.3 shows the solution for the data of Table 5.1 of the earlier chapter and confirms expectation.

17.4.2. Adiabatic

In this special case, the path-line equation is the standard isentropic relationship between Dp and DT. In the notation of the present study:

$$Dp/p = \gamma/(\gamma - 1)DT_{j+1}/T_{j+1}. \tag{17.6}$$

The case is implemented by again leaving coefficients $a_{1,1}$, $a_{1,2}$ and b_1 unaltered, but $a_{2,1}$, becomes zero, $a_{2,2} = \gamma/(\gamma - 1)DT_{j+1}/T_{j+1}$ and right-hand side b_2 is replaced by the approximation Dp/p.

The solution (Fig. 17.4) is as anticipated: the original temperature profile is shunted back and forth un-modified but for a small rise and fall corresponding to the adiabatic pressure swing induced by cyclic volume change.

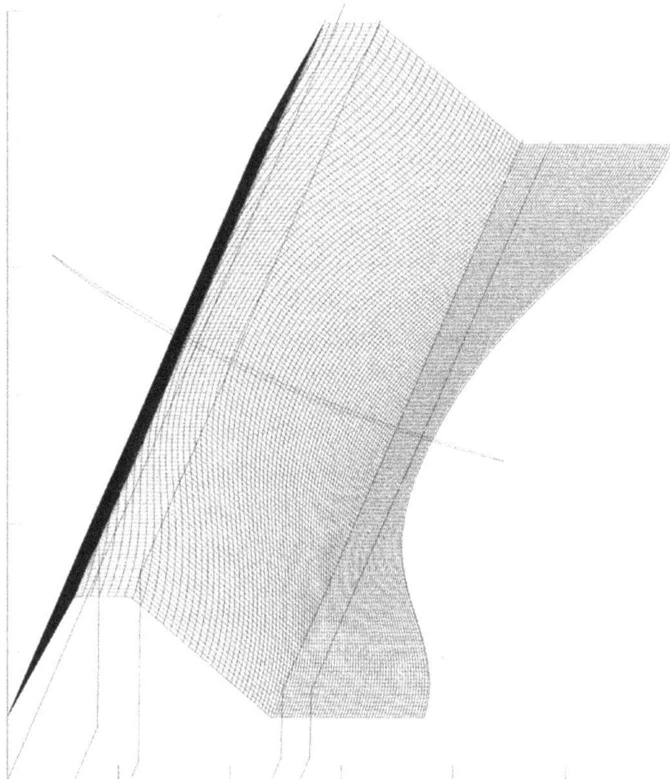

Fig. 17.3. Time-temperature history over the initial cycle for the explicit 'isothermal' assumption (Eq. (17.5)).

17.5. Convective Heat Transfer Evaluated Locally

On any path-line j and at any given crank-angle ϕ, increment Dv_j in cumulative volume to that point can be converted to speed u (or to mass rate m') and thence to local instantaneous Reynolds number Re. In turn, Re can be converted to Stanton number St (or to Nusselt number Nu) *via* the preferred experimental correlation. The corresponding value of q' used in Eq. (17.3) leads to local, instantaneous temperature difference $T_w - T$ and thus to the variation of gas temperature profile over a cycle.

Figure 17.5 displays the start-up cycle corresponding to the specification and operating conditions of Table 5.1. The distribution at the end of the cycle is markedly different from that under static equilibrium at the

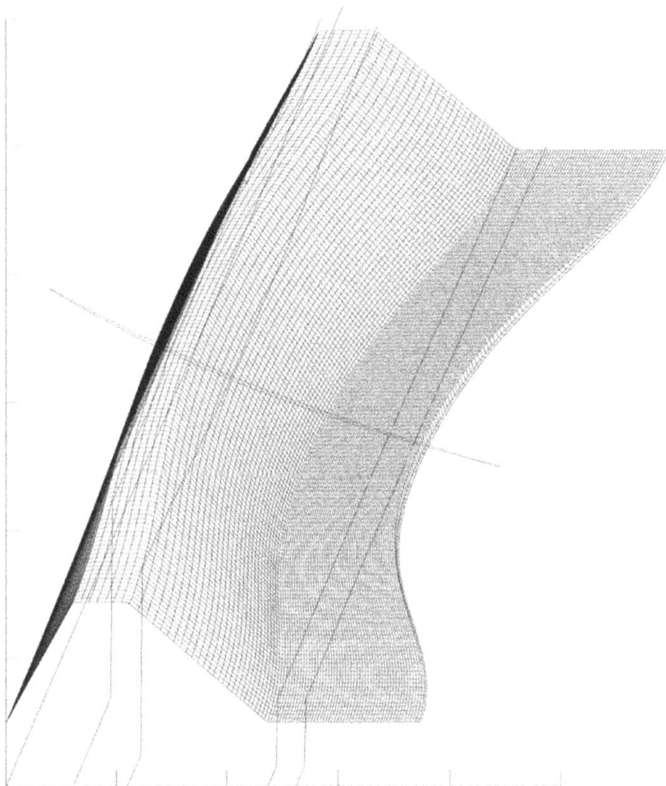

Fig. 17.4. Time-temperature history over the initial cycle under the explicit 'adiabatic' assumption (Eq. (17.6)).

start — consistent with the qualitative explanation of functioning offered in the previous chapter. It can be compared with Fig. 17.6 showing the profile at cyclic steady state after 10 cycles.

Setting engine speed to the impractically-low value of one rpm can be expected to give ample time for heat transfer, and a solution similar to the 'isothermal'. Figure 17.7 confirms expectation.

A counterpart test might generate the earlier explicit 'adiabatic' case by reverting to 1500 rpm and supplying data which cause *St* to be zero at all times. This would be unrevealing, since the path equation (Eq. (17.3)) with *St* set equal to zero is precisely the isentropic relationship (Eq. (17.6)), and the adiabatic solution is the only possible outcome.

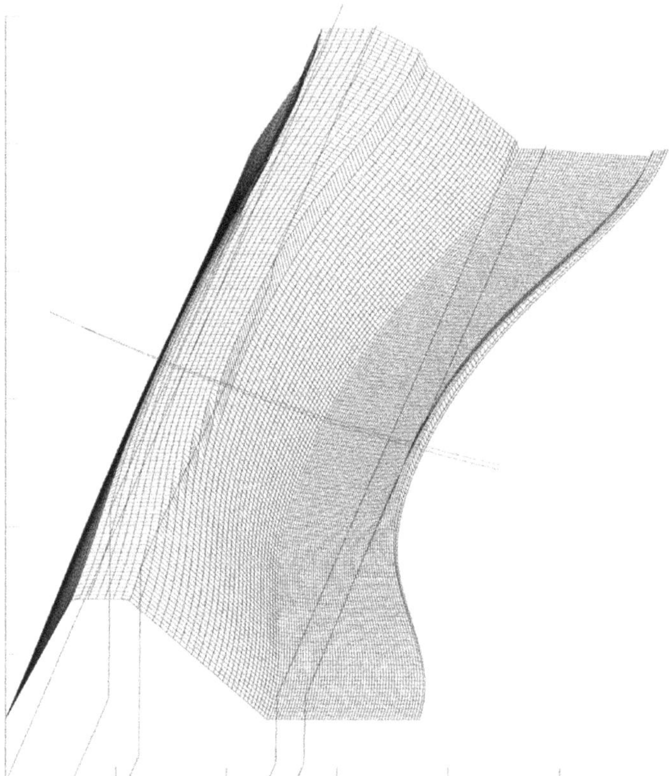

Fig. 17.5. Time-temperature history over the initial cycle with convective heat transfer evaluated locally.

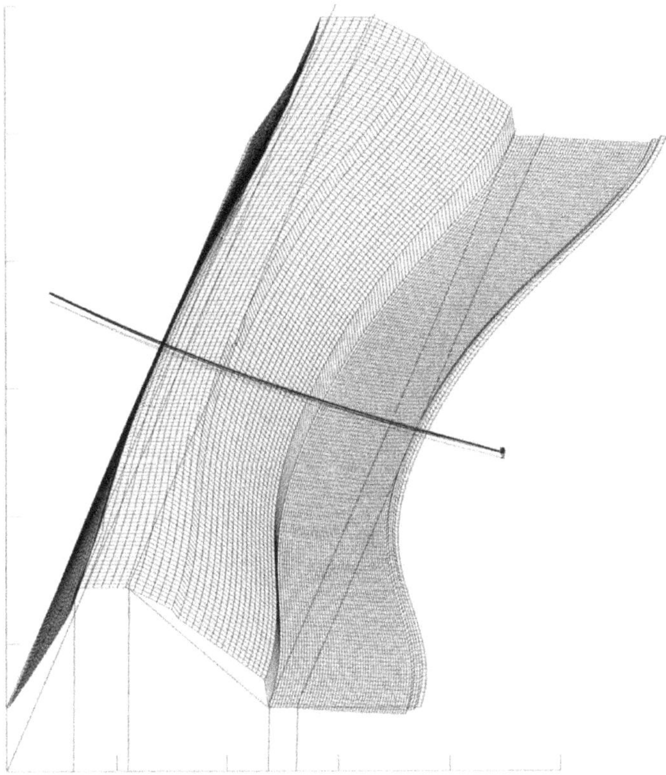

Fig. 17.6. Similarity of temperature profiles at the beginning and end of 10th cycle consistent with approach to cyclic equilibrium.

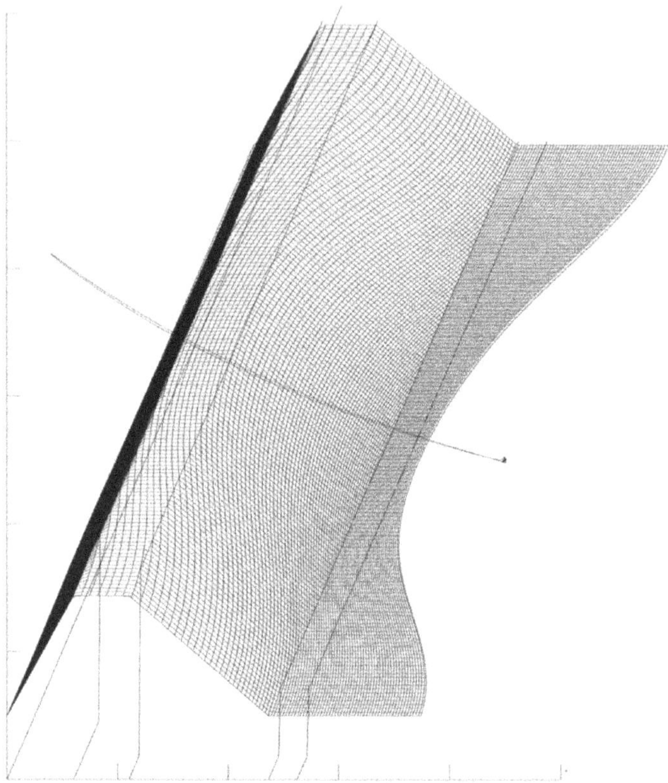

Fig. 17.7. Setting cycle speed to rpm = 1 gives ample time for heat exchange — and for evident attainment of the conditions of the 'isothermal' assumption.

17.6. Prelude to Optimisation

There are two good reasons for thinking ahead to optimisation when formulating the numerical solution to a flow problem. The first is obvious. The second is this: the numerical algorithm perturbs the object function mindlessly in its search for a maximum (or minimum). Any quirks or weaknesses are exploited with indifference. Fortunately, this usually shows up in an 'optimum' which is obviously anything but. If the core formulation has been subjected to a searching test and found wanting, a negative result is far from wasted.

18

Lagrange Meets Euler

18.1. Precautions

Experiments by Vaizey suggest that performance of the thermal-lag engine benefits from pressure lag caused by the flow resistance of the heat exchangers. This is not a confusion with the thermo-acoustic phenomenon: the latter relies on pressure information propagating at a rate of $u \pm a$, where u is local particle velocity and a is local acoustic speed. In contrast, a pressure difference set-up across a flow restriction takes finite time to equalise.

Pursuing the matter calls for a formulation taking account of flow resistance. The latter is commonly expressed as pressure gradient $\mathrm{d}p/\mathrm{d}x = -1/2\rho u^2 C_\mathrm{f}\mathrm{sgn}(u)/r_\mathrm{h}$. However, this is only the spatial component in the $t-x$ field. Of no consequence in a one-dimensional Euler formulation, the matter is important here. The appropriate rendering is now in terms of the partial derivative:

$$\partial p/\partial x = -\frac{1}{2}\rho u^2 C_\mathrm{f}\mathrm{sgn}(u)/r_\mathrm{h}. \tag{18.1}$$

In Eq. (18.1), C_f is a friction factor. The approach which prevails is to acquire C_f from 'correlations' determined in wind-tunnel tests as a function of Reynolds number *Re* and flow-passage geometry. Having been determined under conditions of steady, uniform-temperature, constant-density flow at constant pressure, these are the only conditions under which the correlations are valid.

The proposed context, in contrast, is unsteady flow with rapid swings in pressure, density and temperature, both locally and temporally.

One might question the wisdom of proceeding. It is not a matter of wisdom, but of pragmatism: there is not alternative. On the other hand, an opportunity arises — that of stealing a march: the steady-flow correlations can serve as 'place-holders', allowing formulation to progress pending availability of definitive replacements.

Formulation proceeds by completing the path equation (Eq. (16.4) of the earlier chapter) by addition of the term for entropy generated in friction. In the notation of one-dimensional 'slab' flow, this is in terms of $\partial p/\partial x$. The path equation becomes:

$$\mathrm{D}p/p = \gamma/(\gamma - 1)\mathrm{D}T/T - (RT)^{-1}(q' + u/\rho \times \partial p/\partial x)\mathrm{d}t. \qquad (18.2)$$

Expanding as in the earlier treatment of Chapter 16 reveals an equation linear in the three unknowns, $\mathrm{D}T$, $\mathrm{D}p$ and $\mathrm{d}x = u\mathrm{d}t$ of the earlier formulation, this last being the lateral progress of the path-line during time $\mathrm{d}t$. Accounting for pressure gradient now calls for as many values of $\mathrm{D}p$ per time increment as there are path-lines, viz, p_{j-1}, p_j, p_{j+1}, etc.

At first sight Eq. (18.2) has introduced an additional unknown, $\partial p/\partial x$. This is not so, since $\partial p/\partial x$ is expressible in terms of u (Eq. (18.1)). It has, however, raised the need for an additional equation — the unsteady momentum term:

$$\mathrm{D}p/\mathrm{d}t = \partial p/\partial t + u\partial p/\partial x. \qquad (18.3)$$

The need to invert a $(3n - 2) \times (3n - 2)$ matrix per time increment is averted by the same ploy as previously. At the left-hand (stationary) end of the thermal-lag engine, u is zero and, by definition, $\mathrm{D}p/\mathrm{d}t = \partial p/\partial t$. The approximate values which kick-start solution ahead of every left-to-right sweep are those of $\partial p/\partial t$ at this point.

18.2. Friction Correlation

The simplest curve which fits experimental points of C_f vs steady Re for flow though stacks of rectangular wire screens is

$$C_f = c/Re + d, \qquad (18.4a)$$

where, for given volume porosity $\P_v$, c and d are constants. Substituting into Eq. (18.1), recalling that $Re = 4\rho u r_h/\mu$ and multiplying through yields:

$$\partial p/\partial x = -\frac{1}{2}\{cu\mu/4r_h + du|u|p/RT\}/r_h. \tag{18.5}$$

The counterpart to Eq. (18.4a) for the parallel duct is Eqn. (18.4b) in which the first term of the right-hand side reflects purely viscous effects, dominant at low *Re*, while the second takes over in the turbulent region. Conversion to an expression for $\partial p/\partial x$ is as for Eq. (18.5).

$$C_f = c'/Re + d'Re^{0.2}Pr^{-0.6}. \tag{18.4b}$$

18.3. Abrupt Change in Free-Flow Area A_{ff}

Figure 18.1 shows two sections of duct, respectively of uniform free-flow areas A_{ff1} and A_{ff2}, meeting at a discontinuity. In terms of ratio r_a of the smaller area to the larger, the approach is that of Kays and London (1964): with flow from larger to smaller as indicated they describe the pressure drop by their Eq. (5.1):

$$\Delta p = \frac{1}{2}\rho u^2\{1 - r_a^2 + K_c\}. \tag{18.6a}$$

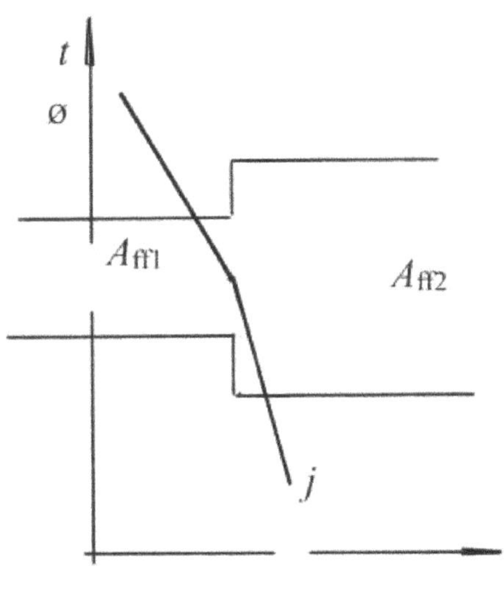

Fig. 18.1. Path-line crossing junction between two elements of parallel duct. (In the cumulative-volume coordinate of the present formulation it is a straight-line.)

In Eq. (18.6a) u is evaluated in the smaller cross-section. With flow from smaller cross-section to larger:

$$\Delta p = \frac{1}{2}\rho u^2\{1 - r_a{}^2 - K_e\}. \tag{18.6b}$$

Kays and London offer plots of K_c and K_c determined experimentally as a function of r_a with Reynolds number *Re* as parameter. The complex dependency means that a large number of graphs would be required to characterise equations (18.6a) and (18.6b) visually. It appears necessary to make the appropriate provision for area change in the numerical flow model and to note the effect of different r_a. Recalling the priority of processing speed, the provision should integrate without non-essential IF-test switches. An option is to code in such as way that the pressure step is treated in the same way as $\partial p/\partial x$.

Increment Δp is achieved over the distance travelled by a path-line which crossing the area change during integration interval dt, namely udt or, in the x-coordinate of the formulation (cumulative volume, v) of the increment dv_j. Progress along the path-line will be subject to local rate of pressure rise $\partial p/\partial t$. The momentum equation in 'real' coordinates becomes:

$$\mathrm{D}p_j = (\partial p/\partial t)\mathrm{d}t + u \times \text{Eq. (18.6)}/\Delta x. \tag{18.7}$$

18.4. Solution Algorithm

The matrix of unknowns is now 3 × 3. Coefficients of the unknowns are grouped in three columns:

$\mathrm{D}v_j$	$\mathrm{D}T_j$	$\mathrm{D}p_j$	RHS
a_{11}	a_{12}	a_{13}	b_1
a_{21}	a_{22}	a_{23}	b_2
a_{31}	a_{32}	a_{33}	b_3

The coefficients embody terms in $\mathrm{D}v_{j-1}$, $\mathrm{D}T_{j-1}$ and $\mathrm{D}p_{j-1}$, whose values at any given right-ward step in the sequence are known. The significant change in row 2 (path-line equation) is inclusion of the term from Eq. (18.2) for friction re-heating, viz, $(RT)^{-1}(u/\rho \times \partial p/\partial x)\mathrm{d}t$, duly converted to the $v - \phi$ coordinate system of the algorithm. In the third row, b_3 holds the approximation $\partial p/\partial t$ (updated at each iteration). Coefficient of Dv_j at a_{31} is Eq. (18.5) for $\partial p/\partial x$. Change DT_j in temperature along path line j does not feature, so a_{32} is zero.

18.5. Specimen Output

Figure 18.2 is the temperature solution corresponding to the specification and operating conditions of the earlier Table 5.1. Friction reheating is

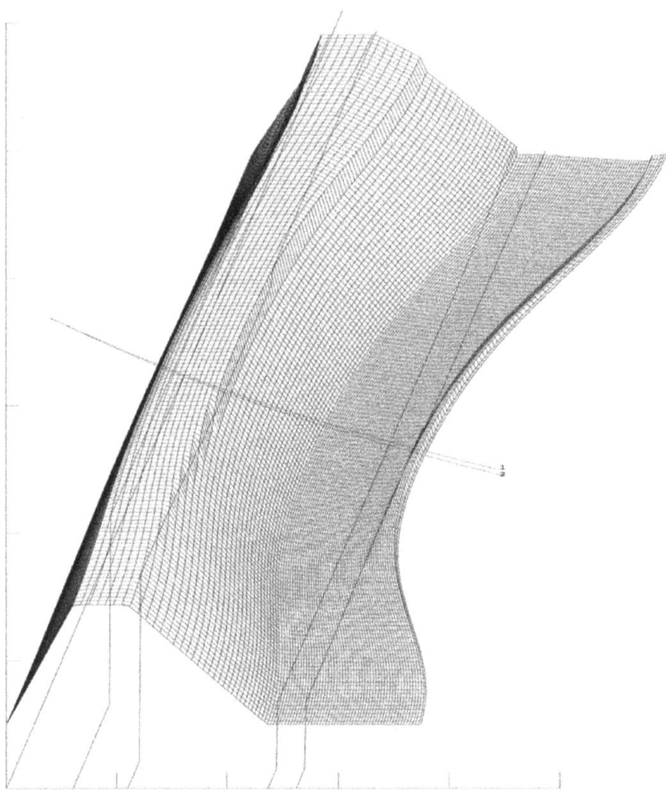

Fig. 18.2. Temperature solution for the start-up cycle under the conditions of Table 5.1. Comparison with the earlier uniform-pressure solution reveals no effect of friction heating subsequently embodied in the path-line equation.

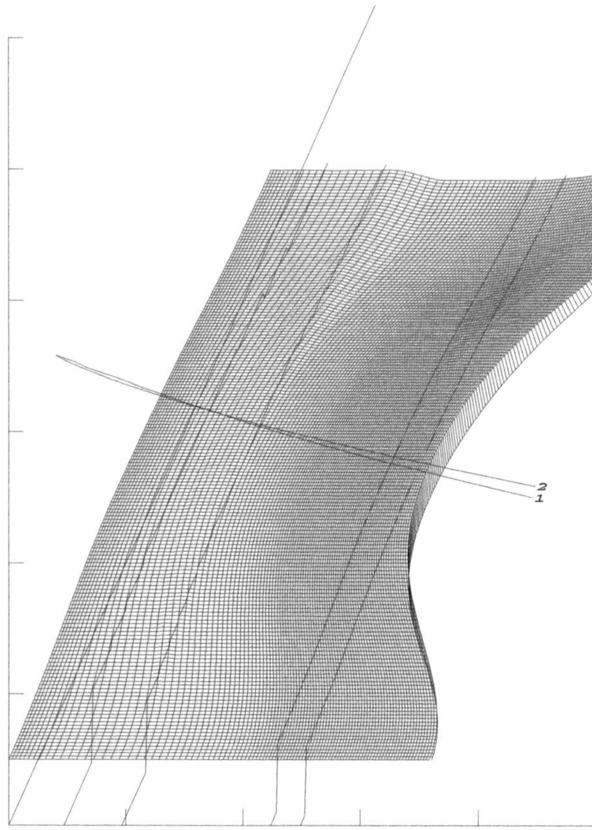

Fig. 18.3. Pressure profile over start-up cycle at rated rpm (1500). Superimposed trace 1–2 (to different scale) is cycle history of pressure at r.h. piston face vs enclosed volume, confirming cyclic equilibrium some way off!

evidently too slight to show up (cf. the uniform-pressure solution of Fig. 17.5).

Insight of particular value in design is provided by the pressure-time history. Figure 18.3 is for the start cycle, again under the conditions of Table 5.1. Flow at the left-hand (closed) end of the regenerator is zero throughout the cycle. Instantaneous flows are correspondingly less than those in the Stirling regenerator at all points in the cycle. All things being equal (particularly volume porosity $\P_v$ and hydraulic radius r_h), local instantaneous pressure gradient can be expected to be proportionately less. Figure 18.4 maps $\partial p/\partial x$ over the start-up cycle. Input data again

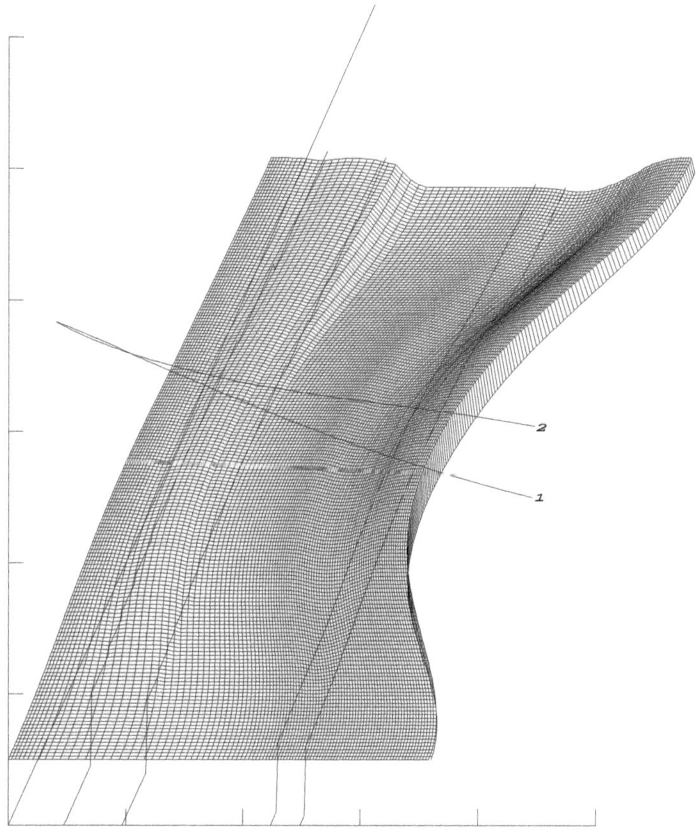

Fig. 18.4. As for Fig. 18.3, but machine heavily over-driven. Re-scaled to fit, but reflecting the changed balance of viscous and inertia contributions.

correspond to the specification of Table 5.1, but with rpm raised to an unrealistic 5000. Even with this measure, observable $\partial p/\partial x$ is confined to the cold exchanger at the driving end where free-flow area is reduced and u is highest.

18.6. Optimisation

The regenerator of the Stirling engine functions by storing heat over one phase of the cycle and releasing a similar amount over another. It seems safe to assume that the matrix of the thermal-lag engine performs a similar role — but under markedly different conditions. Although there are instances among Stirling engines where no one fluid particle traverses the entire length of the regenerator, hot and cold ends nevertheless experience both inflow and outflow.

By contrast, the closed end of the thermal-lag engine (left-hand in upper Fig. 5.1) sees zero flow. The phenomenon — or lack of it — raises the question as to what drives heat exchange in this otherwise dead region. At first sight the driver can only be the adiabatic component of temperature difference accompanying pressure swing. But this in turn re-opens the question tackled in Chapter 5: just how does the device function?

In the context of the Stirling engine there is an argument (any one of the author's earlier texts) for grading the matrix to compensate for — and/or to take advantage of — differences in end-to-end flow conditions. Any such argument must be even stronger in the case of the thermal-lag type, where the difference between the work of compression and that of expansion in devices demonstrated to date is clearly marginal.

On the assumption that it must be possible to do better than the usual regenerator provision of randomly-packed wire wool, Turbo2×2 has been modified to incorporate end-to-end variation in regenerator flow passage geometry — i.e., in the parameters determining friction factor C_f and NTU namely, hydraulic radius r_h and volume porosity $¶_v$. Recalling that $¶_v = ¶_v(r_h, m_w)$ it is evident that there is considerable scope: any one — or two — of $¶_v$, r_h, m_w can be varied while the remaining two (or one) are/is held constant. Moreover, the lengthwise variation can be linear or any function of x. The number of options suggests difficulty in identifying a manageable way of displaying the dependency of any optimum.

Contrary to expectation, computed cycle work has shown negligible dependence on any of the linear variations so far attempted. Like many others, this aspect of thermal-lag behaviour remains under investigation!

19

Regenerator Flow Impedance — A Diffusion Formulation

19.1. Origins

The Lagrange formulation is pursued in response to lessons learned during earlier exercises in cycle modelling, mostly using other reference frames. The resulting approach — to which the Lagrange option particularly lends itself — is incremental: using 'dummy' energy and momentum equations, the mass-conservation element can be coded and checked before proceeding to see that it does, indeed, conserve mass. Checking for correct implementation of the energy equation (path-line form in the present case) involves more 'circumstantial' options, as recently illustrated.

The next step is replacement of the dummy momentum equation ($p_j = p_{j+1} = p_{j+2} = \cdots$ for all ϕ) with $Dp/d\phi$ computed in terms of local conditions along the path-line. Implementation was undertaken with the dummy energy equation (prescribed temperature distribution) again standing in for the path-line equation. The target was not a model of physical reality — since that is out-of-reach — but accurate solution of the one-dimensional defining equations. Despite use of various discretisation schemes, this was not achieved, one symptom being a persistent saw-tooth pressure profile.

Transformation of the momentum law to Lagrange coordinates is not an obvious process, and an apparent explanation for the difficulty came from an unlikely source: a Wikipedia entry asserting that the Lagrange reference frame deals in *scalars* — at first sight disqualifying the momentum equation. The Wikipedia position is open to question, however, since the Method of Characteristics — the no-compromise numerical discretisation scheme for all three conservation laws — is dependent upon integration along particle paths.

The only way forward appeared to be a fresh start from a trouble-free formulation. The sole formulations on record without reports of similar difficulty have been in Euler coordinates.

The return to square-one was a ruthlessly-simplified, two-conservation-law model (mass and momentum) with fixed, linear temperature distribution between two variable volumes — but the same, saw-tooth pressure 'solutions'!

Something fundamental is at stake.

19.2. The Momentum Equation

A form expected to cover all aspects explored here is that used by Shapiro (1954), which reflects prior subtraction of the continuity equation (mass conservation):

$$\partial u/\partial t + u\partial u/\partial x + \rho^{-1}\partial p/\partial x + F = 0. \tag{19.1}$$

Symbol F stands for Shapiro's 'wall-friction' term:

$$F = \frac{1}{2}u^2 C_f/r_h \times u/|u|. \tag{19.2}$$

Nothing of present relevance is lost by postponing consideration of the acceleration terms, leaving

$$\partial p/\partial x = -\frac{1}{2}\rho u^2 C_f/r_h \times u/|u|. \tag{19.3}$$

Warning bells should be sounding: Eq. (19.3) differs from others used in the formulation in that it cannot be inverted, e.g., to express u in terms of ρ and $\partial p/\partial x$. Equally, it does not linearise unconditionally for use in an implicit formulation.

Preparation for numerical integration calls for expansion in terms of time. Whereas the ideal gas equation lends itself as $p\mathrm{d}V + V\mathrm{d}p \approx R(T\mathrm{d}m + m\mathrm{d}T)$, Eq. (19.3) in contrast is resolutely steady-state (but see later section). This suggests applying it at mid-time-step. A discretised form based on mid-time-step values of ρ and u might be,

$$(p_{j+1} - \Delta p_{j+1} - p_{j-1} - \Delta p_{j-1})/2\Delta x \approx -\frac{1}{2}\rho u^2 C_f / r_h \times u/|u|.$$

This and all variants have displayed the same oscillatory characteristic.

Plausible pressure profiles have nevertheless resulted from two formulations. The first applies Eq. (19.3) retrospectively, i.e., after solution has been advanced by a time step in which Δu and $\Delta\rho$ have been determined on the basis of mass conservation alone.

The second involves perturbation, in which 'true' values of the variables are treated as the p, u, and ρ of an ideal, lossless solution adjusted by respective 'errors', ε_p, ε_u and ε_ρ, viz, as $p + \varepsilon_p$, $u + \varepsilon_u$ and $\rho + \varepsilon_\rho$. The result of an integration step is a set of new values of the 'errors'. These are added to the known p, u, and ρ to give the required solution.

Appendix C gives further details, illustrated by pressure solutions for possible use for other investigators committed to Euler formulations.

19.3. A Philosophical Dimension

Earlier chapters questioned application of the friction-factor concept to the matrix regenerator. The doubts are now reinforced from a different direction.

The Lagrange formulation is essentially implicit, calling for the momentum law to be linear in the unknowns of the formulation — or capable of being linearized. As an algebraic exercise, there is no problem. A first venture saw Eq. (19.3) put into difference form equivalent to:

$$\Delta(\partial p / \partial x) = -\frac{1}{2} u / r_h |u| \{(u^2 C_f \Delta\rho + \rho C_f \Delta u + \rho u^2 \Delta C_f\}. \tag{19.4}$$

Provisionally accepting $C_f = c/Re + d$,

$$\Delta C_f = -c/Re^2 \times \Delta Re. \tag{19.5}$$

It soon becomes apparent that an expansion is required for ΔRe. Not forgetting that $Re = |4\rho u r_h/\mu|$ is itself is always a modulus:

$$\Delta Re = 4r_h\{u\Delta\rho + \rho\Delta u - \rho u\Delta\mu\}. \tag{19.6}$$

The simulation computes in terms of m', i.e., of the group $\rho u A_{\mathrm{ff}}$. The equations are coded accordingly.

Not only does the added complexity fail to repay the effort — it is doomed to failure. The reason, which has been slow to come to light, is relevant to one-dimensional modelling regardless of coordinate framework — Euler, Lagrange or Characteristics.

The normalisation by which numerical values — or curves — for C_{f} are reduced from experimental measurement is the arbitrary division by $1/2\rho u^2$ of the sum of two,[1] distinct effects, one viscous and proportional to $|u|$ (the c/Re of the usual curve-fit), the other inertial and measured in terms of u^2 (the constant d of the fit). The underlying picture is retrieved by multiplying out the original $1/2\rho u^2 C_{\mathrm{f}}$ with the curve-fit $c/Re + d$ substituted and recalling the definition of Mach number Ma:

$$\frac{1}{2}\rho u^2 C_{\mathrm{f}} = \frac{1}{2}\gamma p(c/Sg + Ma^2). \tag{19.7}$$

The original normalisation was inherently questionable. The problem it provokes is that there is any number of combinations of Sg and Ma which return the same numerical value for $\partial p/\partial x$. Solutions can be artificially stabilised — or smoothed — but would attract greater confidence if doctoring were unnecessary.

Unsteady flow through the regenerator encounters not just resistance but *capacitance*: at only two instants over a cycle is outflow equal to inflow. In other contexts, this would be recognised as a diffusion phenomenon and analysed accordingly. When the same approach is followed here, earlier problems disappear — strongly suggesting that this would have been the appropriate course all along.

19.4. A Diffusion Approach

Embodying the usual C_{f}-Re relationship in Eq. (16.3) gives

$$\partial p/\partial x = -\frac{1}{2}\rho u^2 (c/Re + d) r_{\mathrm{h}}^{-1}.$$

To keep the algebraic development uncluttered, provisionally consider only the viscous term, c/Re. Multiply both sides by free-flow area

[1] Actually three — but two make the point.

A_{ff}, recall the definition of Re as $4\rho|u|r_h/\mu$, noting that $\rho u A_{ff}$, $= m'$ and substitute:

$$A_{ff}\partial p/\partial x = -c\mu/(8\rho r_h^2)m'. \tag{19.8}$$

Inverting:

$$m' = -kA_{ff}\partial p/\partial x. \tag{19.9}$$

In Eq. (19.9), $k = 8\rho r_h^2/\mu$ is 'conductivity' with units kg/mPas.

The diffusion equation (in Euler coordinates) is set up between two equal control volumes, each ΔV (Fig. 19.1), by equating the difference between flows in and flow out to rate of mass accumulation $\partial m/\partial t$. Under present assumptions (fluid temperature fixed locally and equal to that of the immediately adjacent matrix), $\partial m/\partial t = (\mathrm{d}V/RT)\partial p/\partial t$:

$$kA_{ff}\{\partial p/\partial x_{right} - \partial p/\partial x_{left}\} = (\Delta V/RT)\partial p/\partial t. \tag{19.10}$$

In a flow-passage of uniform cross-sectional area and elemental length Δx, volume element $\Delta V = A_{ff}\Delta x$. In terms of finite differences Δx and Δt, Eq. (19.10) becomes

$$kRT\frac{\{p^i_{j+1} - 2p^i_j + p^i_{j-1}\}}{\Delta x^2} = (p^{i+1}_j - p^i_j)/\Delta t. \tag{19.11}$$

The term kRT has units of m^2/s, i.e., of diffusivity. Superscript i labels a point on the elapsing time scale, i being current time at which values are known for p^i at all j, and $i + 1$ the time Δt ahead for which the difference equation has to be solved for p^{i+1}_j.

Finally:

$$p^i_{j+1} - 2p^i_j + p^i_{j-1} = N_F^{-1}(p^{i+1}_j - p^i_j). \tag{19.12}$$

In Eq. (19.12) N_F is Fourier modulus, evaluated as $kRT\Delta t/\Delta x^2$ and is dimensionless.

Provided N_F is $\leq 1/2$ an approximate solution to the pressure field is obtained by calculating the unknown p^{i+1}_j explicitly across successive groups $p^i_{j+1} - 2p^i_j + p^i_{j-1}$ of 'known' values:

$$N_F^{-1}\, p^{i+1}_j \approx p^i_{j+1} + p^i_j(-2 + N_F^{-1}) + p^i_{j-1}, \tag{19.13}$$

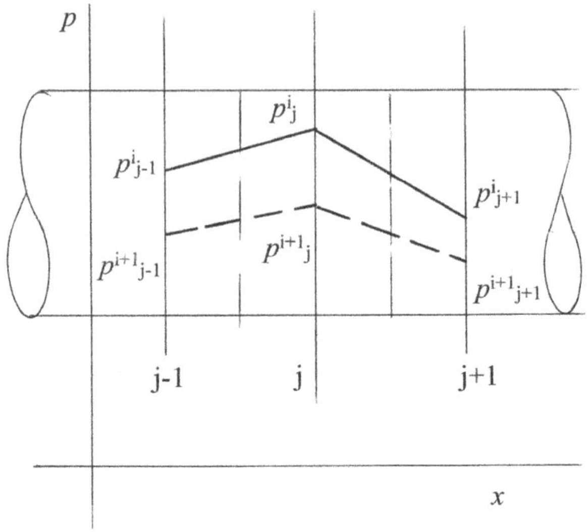

Fig. 19.1. Notation for setting up diffusion equation.

approximate because *all* p^i_j are varying concurrently, and not just the token unknown p^{i+1}_j. The matter will be picked up again towards the end of this chapter.

19.5. Interim Graphical Output

There are no known accounts of directly-relevant laboratory measurements (momentum effects excluded and only the *Re* term effective). However, all that would be required in principle is a tube containing a stack of gauze, a piston at either end and a mechanism to synchronise and drive both. To this extent it is a horizontal, opposed-piston Stirling engine minus heat-exchanger dead-space and minus the problems of high temperatures and high-pressure sealing.

Low peak piston speed minimises the momentum term and eases the matter of achieving low *Re* (low pressure level and/or small r_h).

The algebraic development to this point is used to simulate an experiment with such an apparatus. Operating conditions (experimental and simulated) can be described in terms of phase angle α and volume ratio κ, of *rpm*, of working fluid, of charge pressure p_{ref}, etc. Checks behind the illustrations which follow confirm accurate conservation of mass. The pressure (vertical) scale is arbitrary, because the focus is on the

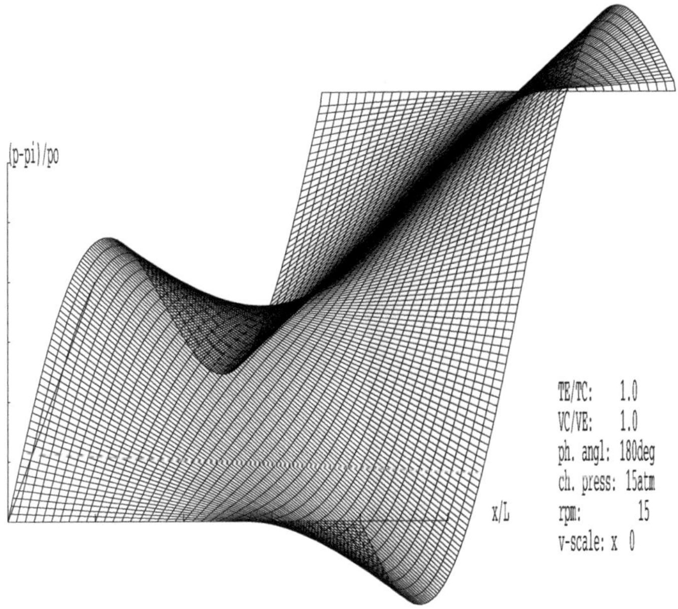

Fig. 19.2. Pressure surface generated from diffusion formulation with rpm and hydraulic radius chosen so as to isolate viscous component of flow resistance.

continuity of the surface, and whether the profile is a plausible outcome of the volume variations which form the boundary conditions.

Working fluid in all cases is air. Volume variations are simple-harmonic with zero dead-space at inner dead-centre. The initial complete cycle from rest is illustrated. For Fig. 19.2 $\alpha = 180°$, i.e., the air is shuttled back and forth with no change in overall volume. To confine flow to the purely viscous region, rpm = 15 and hydraulic radius is given the unrealistically low value of 0.0004 mm. The isometric plot contains an element of optical illusion: to see that peak and trough occur some 180° apart requires following a surface trace from left to right. Cyclic equilibrium is not achieved, so neither is perfect symmetry. However, drawing a straight line between lower left and upper left, doing the same between lower right and upper right and counting particle tracks reveals that maximum and minimum pressure excursions left and right are of similar magnitude.

For Fig. 19.3 phase angle is reset to zero. The pressure surface meets the expectation of symmetry. The integration process has again conserved mass accurately, and there is no indication of the 'saw-tooth' problem which beset earlier formulations based on the steady-flow expression for dp/dx (Eq. (19.3)).

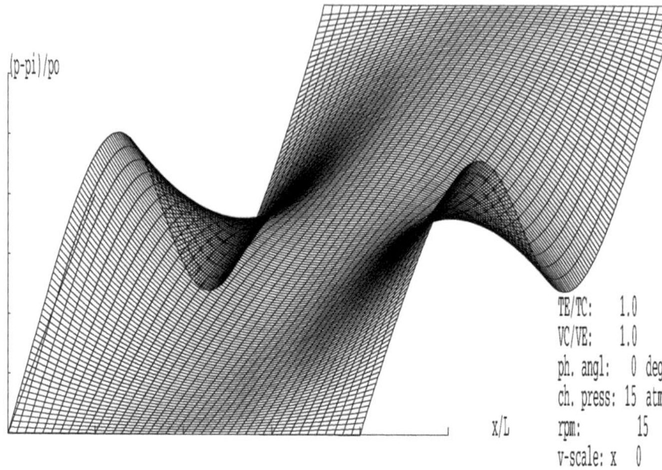

Fig. 19.3. As for Fig. 19.2, but with volume phase angle α reset to zero.

With algebraic framework demonstrated it is possible to proceed to a case for which there is an obvious experimental counterpart.

19.6. Contribution of Momentum Term ($Ma \ll 0.1$)

The contributions to pressure gradient of viscous and momentum terms in Eq. (19.7) add linearly. Isolating the latter and treating as for the former:

$$A_{ff}\partial p/\partial x = -\frac{1}{2}\, dRT/(pA_{ff}r_h)m'^2. \tag{19.14}$$

The addition process has resulted in a quadratic equation in mass-rate m':

$$am'^2 + bm' + c = 0,$$

$$a = \frac{1}{2}\, dRT/(pA_{ff}^2 r_h), \qquad b = c\mu/(8\rho r_h^2 A_{ff}),\ c = \partial p/\partial x.$$

Numerator and denominator of the standard solution for the quadratic are divided by b. Inspection reveals the relevant root to be positive:

$$m' = \frac{-1 + \sqrt{(1 - 4ac/b^2)}}{2a/b}. \tag{19.15}$$

Provided the absolute value of $4ac/b^2$ is less than unity, the root term may be replaced by the series expansion:

$$(1 - x)^n \approx 1 - x + n(n - 1)x^2/2! - n(n - 1)(n - 2)x^3/3! + \cdots.$$

Substituting into Eq. (19.15) and recalling the definition of coefficient b of the quadratic equation gives:

$$m' = -kA_{\mathrm{ff}}\partial p/\partial x + \text{higher-order terms } (h.o.t).$$

With $(4a/b^2)A_{\mathrm{ff}}\partial p/\partial x$ abbreviated to x,

$$h.o.t. = (2a/b)^{-1}\{-x^2/8 - x^3/16 - \cdots.$$

The diffusion equation is formed as before by eliminating m'. The Fourier modulus is no longer constant, and now requires to be evaluated locally at each step in the integration process. The $h.o.t.$ succumb to iteration. Fig. 19.4 was acquired for the flow conditions for which the earlier Fig. 19.2 was generated.

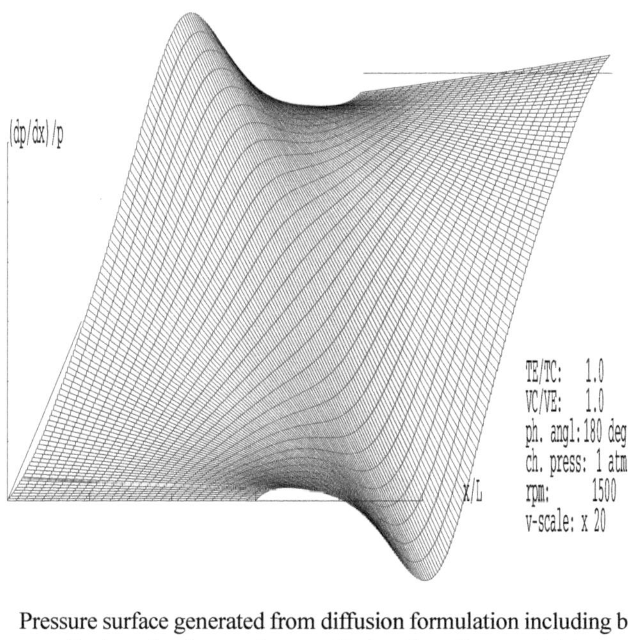

Fig. 19.4. Pressure surface generated from diffusion formulation including both viscous and momentum contributions for flow resistance. Hydraulic radius $r_h = 0.04$ mm, rpm = 1500.

A form of the momentum law is now available which one might be happy to embody in a simulation in Euler coordinates. The context, on the other hand, remains the Lagrange formulation which, it will be recalled, is implicit, whereas the integration step just employed is explicit. 'Upgrading' to implicit is essential — and for two reasons:

19.7. Implicit vs Explicit

The diffusion equation formulated around pressure p in compressible flow ($p/\rho = RT$) is one step more involved than, say, the statement of unsteady thermal diffusion with constant, uniform thermal conductivity, k. The best illustration of the potential stumbling block is in terms of the better-known transient-temperature problem:

$$\partial T/\partial t = \alpha\partial^2 T/\partial x^2. \tag{19.16}$$

In Eq. (19.16), α is thermal diffusivity, $k/\rho c$ [m²/s].

Only a limited number of combinations of boundary and initial conditions allow analytical solution. In general a numerical process is required, calling for Eq. (19.16) to be recast in finite-difference form. The commonest might be called 'explicit'. With reference to Fig. 19.5:

$$(T_j^{\text{new}} - T_j)/\Delta t \approx \alpha(T_{j+1} - 2T_j + T_{j-1})/\Delta x^2.$$

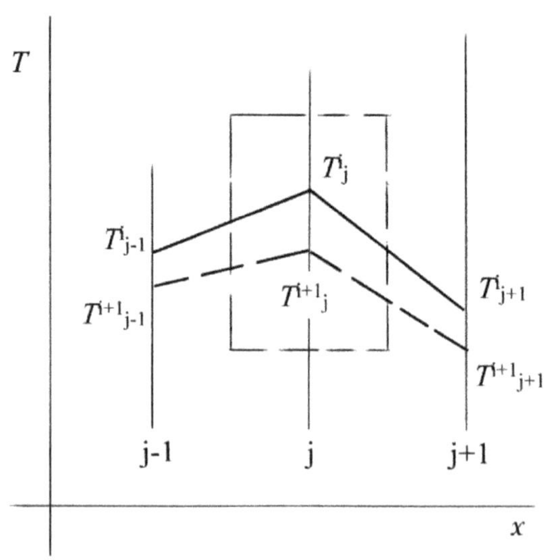

Fig. 19.5. Notation for setting up thermal diffusion equation.

Collecting terms in T_j^{new} and forming Fourier Modulus $N_F = \alpha\Delta t/\Delta x^2$,

$$T_j^{new} \approx N_F\{T_{j+1} + T_{j-1} + T_j\,(1/N_F - 2)\}. \tag{19.17}$$

For the special case of $N_F = 1/2$, Eq. (19.17) reduces to

$$T_j^{new} \approx \frac{1}{2}\{T_{j+1} + T_{j-1}\}. \tag{19.17}$$

Equation (19.17) amounts to the celebrated Binder-Schmidt graphical algorithm, whereby T_j^{new} is achieved graphically by joining T_{j+1} to T_{j-1} with a straight line.

The method has involved the arbitrary manoeuvre of treating T_j on the left-hand side of Eq. (19.17) as unknown, but known and fixed on the right. The reality is that, over time increment Δt, all three temperatures, T_{j-1}, T_j and T_{j+1}, are liable to variation of comparable magnitude.

Of greater concern in the context of regenerator flow, where diffusivity generally takes a high numerical value, is that Eq. (19.17) is unstable for $N_F > 1/2$. Both problems disappear on returning to the derivation of the diffusion equation (Eq. (19.16)), on basing heat flow rates on respective mean values over the integration interval — and on treating all three temperatures increments of the discretized expression as unknowns. With ΔT to represent a change over Δt,

$$\Delta T_j/\Delta t \approx \alpha\left(T_{j+1} + \frac{1}{2}\Delta T_{j+1} - 2T_j - \Delta T_j + \Delta T_{j-1} + \frac{1}{2}\Delta T_{j-1}\right)/\Delta x^2.$$

Grouping unknowns on the left-hand side:

$$-\frac{1}{2}\Delta T_{j-1} + \Delta T_j(1 + 1/N_F) - \frac{1}{2}\Delta T_{j+1} \approx T_{j+1} - 2T_j + T_{j-1}. \tag{19.18}$$

A case with prescribed variation in boundary temperatures ΔT_1 and ΔT_n left and right and divided by n equally spaced nodes has $n - 2$ unknown temperatures — and the same number of equations of type (19.18). Solution is thus possible by any of the standard methods for solution of linear, simultaneous equations.

The two numerical algorithms, explicit and implicit, are compared here by applying each to the case of a flat plate of thickness b, initially at uniform temperature T_0 and having its vertical faces suddenly raised to, and maintained at, a higher temperature. The internal temperature response is acquired by considering one symmetrical half of the plate (of thickness $b/2$) and making use of the fact that dT/dx = zero at the mid-plane ($x = 0$ in the figures).

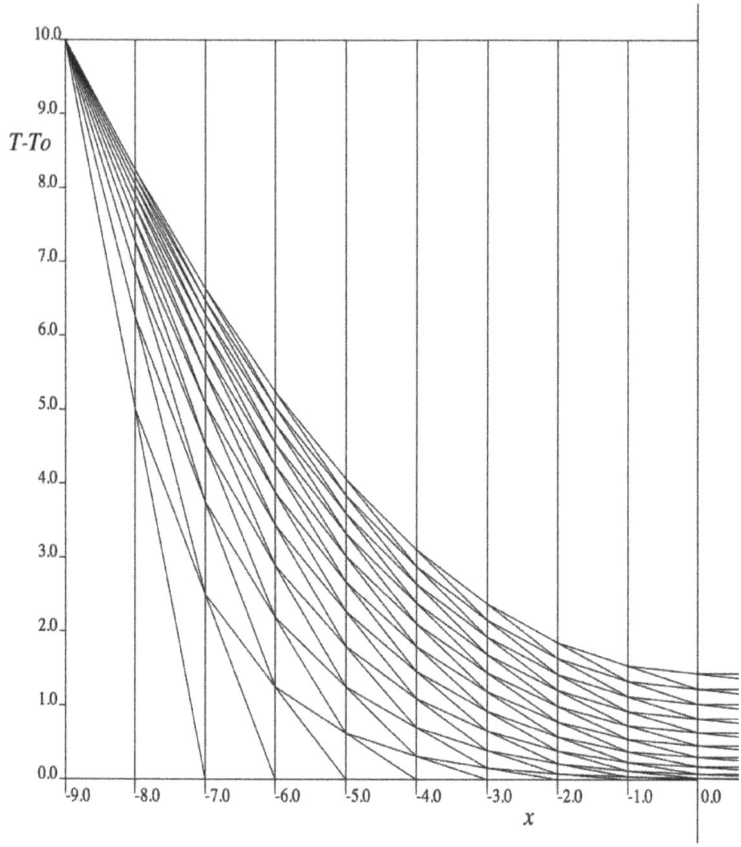

Fig. 19.6. Temperature solution generated by (explicit) Binder–Schmidt construction.

Figure 19.6 illustrates the (explicit) Binder–Schmidt method (Eq. 5.17). Figure 19.7 is the more general solution of the same problem (Eq. (19.17)), but with N_F set to $\frac{1}{2}$. Figure 19.8 is the same problem dealt with in the implicit scheme. As anticipated, the solutions are far from identical. But is either correct? One test of appropriate formulation is satisfied when solutions acquired with different combinations of integration step Δx and Δt (obviously retaining the value of N_F i.e., of $\Delta t/\Delta x^2$) are substantially the same. The implicit scheme achieves this where the explicit scheme does not.

The Lagrange formulation must proceed on the basis of an implicit method.

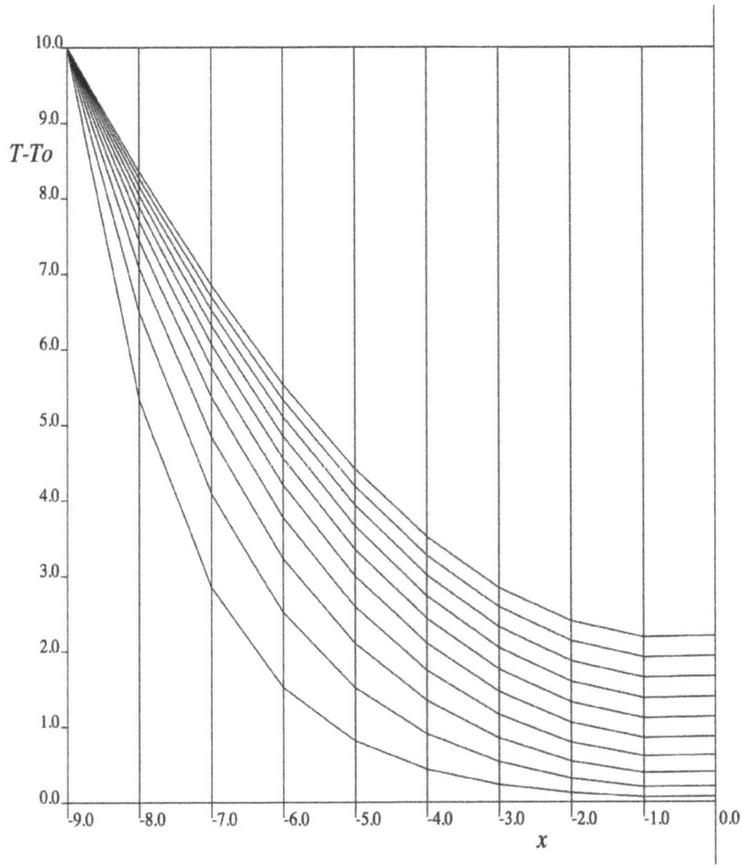

Fig. 19.7. Problem as for Figs. 19.6 solved by application of explicit scheme (Eq. (19.17) with $N_F = 1/2$).

19.8. Lagrange and Discretization

The numerical scheme must reflect the fact that the coordinate frame is in motion at the very u which gives rise to flow resistance. A number of formulations have been coded and explored. All have involved interpolation — and unacceptable compromise.

A breakthrough appeared imminent on re-examining the momentum equation in relation to the path-line equation. The latter derives from equating the two expressions for entropy increment Ds, the first expressed

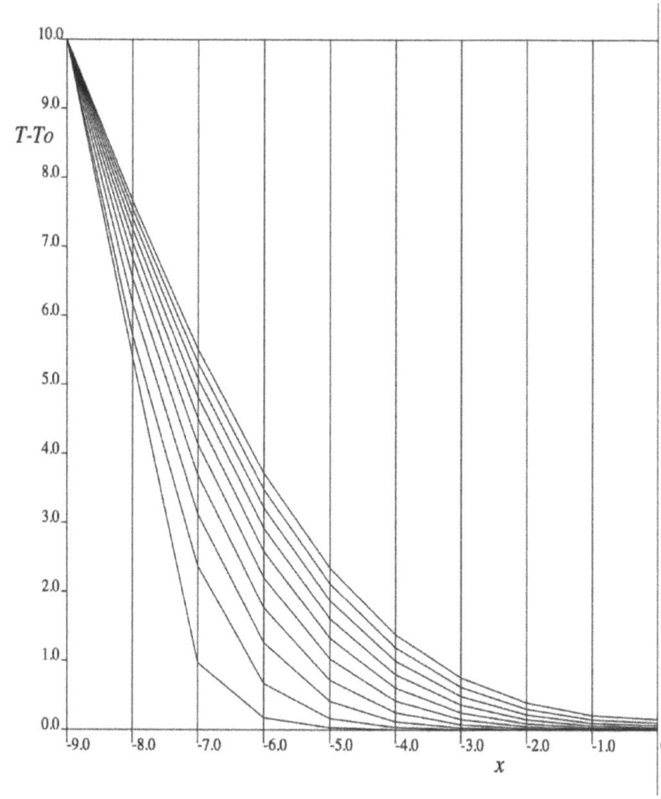

Fig. 19.8. Problem as for Figs. 19.6 and 19.7, but solved by application of implicit scheme based on Eq. (19.18).

in terms of state-change, viz $\mathrm{D}s = c_p\mathrm{D}T/T - R\mathrm{D}p/p$ the second in terms of the individual contributions to irreversibility, viz to $T\mathrm{D}s = (q' + uF)\mathrm{d}t$, where F stands for Shapiro's 'wall friction' term $\frac{1}{2}u^2C_f/r_h\mathrm{sgn}(u)$. Total derivative $\mathrm{D}p$ expands to $\partial p/\partial t + u\partial p/\partial x$ and the wall friction term can be expressed in terms of pressure gradient. Term $u\partial p/\partial x$ thus appears on both sides of the equality.

The initial temptation is to cancel. This, however, would be misguided. The momentum law, after subtraction of the continuity equation (mass conservation), is:

$$\partial u/\partial t + u\partial u/\partial x + \rho^{-1}\partial p/\partial x + F = 0. \tag{19.19}$$

The acceleration term $\partial u/\partial t$ is responsible for compressibility (wave) effects, and local pressure gradients which can be of *opposite sign* to those suggested by the friction term $\partial p/\partial x = -1/2\rho u^2 C_f/r_h \, \mathrm{sgn}(u)$. It is clear that both terms must be retained.

An impasse threatens. The Euler framework is unsuited to the following temperature disturbances; the Lagrange appears to be an unpromising choice for pressure solutions. Stirling engine design calls for a picture of both — preferably as they interact. One interim approach involves two, complementary simulations working in parallel, unsteady heat transfer modelled in Lagrange coordinates, flow resistance in Euler. By the principles of this title, two complementary simulations, each of limited scope, but each of which is a verifiable solution of its respective defining equations algebra is to be preferred over a nominally more comprehensive approach involving arbitrary numerical compromise.

Another possibility arises: the Method of Characteristics (MOC) is widely recognised as the 'no-compromise' algorithm for solution of the equations defining friction, and heat transfer in one-dimensional flow. While useful in probing the limits of the rpm envelope it is not well-suited to the normal speed range involving thousands of wave traverses of the gas path per cycle.

On the other hand, there has apparently been no previous attempt to rework the MOC omitting the time-dependent term of the momentum equation responsible for pressure-wave phenomena. Considerable effort has been expended applying Shapiro's algebra of determinants (his 972 pp) to the set of conservation laws with acceleration term $\rho\partial u/\partial t$ suppressed. Unsurprisingly, the characteristic direction $\Delta x/\Delta t|_{\mathrm{path}} = u$ emerges as before. However, the process does not yield plausible propagation directions for other properties.

Again, enquiries continue!

20

From Specification to Reality

20.1. Mechanical Design

It was proposed earlier that the cycle embodiment most likely to benefit from improved working-fluid distribution is the internally-focusing solar variant. It does not require a multi-tube heater, and the extent to which working fluid is excluded from the expansion space during compression depends only on the imagination of the designer.

This chapter will begin the process of converting specification to reality. The former provides for air (un-pressurised) as working fluid, so the drive mechanism can be exposed. Several features compete for priority. The overriding aim is to achieve the volume variations of the earlier Fig. 12.4 in a mechanism sufficiently robust to allow an hour or so on test. High on the list, however, is that, as far as possible, sub-components should lend themselves to fabrication from stock materials.

The final result is going to be somewhat Meccano-like. This, however, would not be the first time[1] that the result of working 'outwards' from a specification has challenged expectation. 'Outwards' in the present case means starting from unorthodox volume variations.

The description and pictorial illustration of an individual component will be accompanied by an estimate of its 'embodied energy' [kWh] and

[1] A spectacular instance was the first NASA lunar landing module. At first sight, too fragile to make it from assembly shop to launch pad — and about as far from popular expectation (maybe the Dan Dare comic strip) as it is possible get! On the other hand, perfect for its unique task, which it performed to schedule and with barely a hitch.

Table 20.1. Data used in estimates of embodied energy.

	Density (kg/m^3)	Energy (kWh/kg)	CO_2 (kg/kg)
'Aluminium' (general and incl. 33% recycled)	2700	43.05	8.24
'Steel' (avg. recycled content)	7800	5.58	1.37
Stainless steel	7850	15.75	6.15
Kovar™ (Ni 29%, Co 17%, Si 0.2%, Mn 0.8%, Fe bal.)	Assumed as for stainless steel		
Glass	2500	4.17	0.85

equivalent carbon cost [kg/CO_2]. The arithmetic is based on an estimate of mass of the raw-material blank from which machining starts. An exact value would require figures for clamping and machining allowance. These have to be the result of guesswork. Table 20.1 gives the figures used for the energy and CO_2 estimates. These are clearly 'generic' figures which will no doubt be refined as their rôle in a fully-costed design becomes recognised.

There is an interesting contribution from the French *Institut du développement durable et des relations internationales,* according to whom '... *we consume more embodied energy in our expenditure on transportation than direct energy*.' This suggests that the energy taken to produce, transport and sell cars, trains and buses exceeds that consumed in travelling around in them! In the UK some 60% of the retail price of highway fuel is tax. How fiscal intervention should enter the energy balance is not clear. The calculation promises to be more straightforward in the case of a solar energy converter where the 'fuel' does not (so far) attract tax. The matter will be returned to when there are figures to discuss.

The descriptions of selected components which follow are complemented by views in isometric projection generated by CADStudios (UK) Ltd from orthographic engineering drawings supplied by the author. The CAD software allows the finished assembly to be rotated (in the sense that the 'real' engine rotates) to check for clashes. The design offered here as 'complete' has benefited from the process.

Readers interested in constructing the engine can contact CADStudios direct[2] for advice on acquiring machining files, animation and isometric

[2]CADStudios, +44 (0) 1726 74304. e-mail: duncan@cadstudios.co.uk.

illustrations. Sets of traditional engineering drawings (first angle, orthographic projection) carrying material specification and toleranced dimensions, together with an inventory of bought-out components (rolling-element bearings, etc.) are available from the author.[3]

20.2. Drive Mechanism

Figure 20.1 shows the mechanism sub-assembly in isometric view generated by CAD software. Figure 20.2, on two facing pages, is the corresponding orthographic drawing. In conjunction, isometric and orthographic illustrate how the kinematics of the patent drawing (earlier Fig. 12.1) are realised in practice.

20.2.1. Crankshaft (details #4, 5, 6)

For ease of fabrication this is a sub-assembly with main journal and crank-pin clamped in location and alignment at a split crank web (Fig. 20.3).

The material of main journal and pin is Stubbs-grade silver steel of stock diameters 14.0 and 10.0 mm, respectively. Needle-roller bearings

Fig. 20.1. Isometric view of mechanism sub-assembly. Built by CADStudios from individual detail drawings referred to by number system of sub-assembly.

[3] See preliminary pages for contact details.

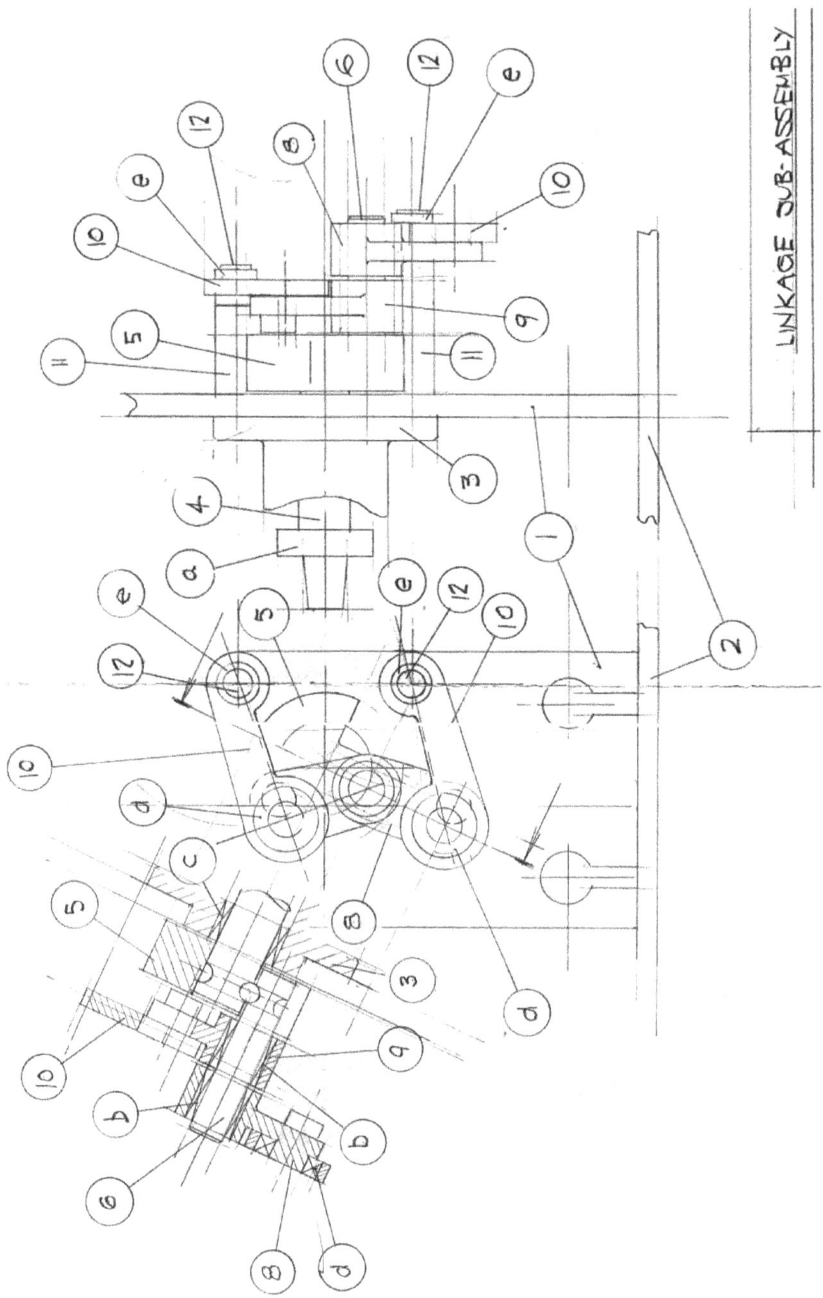

Fig. 20.2. Mechanism sub-assembly in orthographic projection of original kinematic study.

Fig. 20.3. Crank web showing 1 mm slit to allow clamping (by M5 cap-head screw not illustrated).

run on these diameters. In view of likely time in service, no further surface or heat treatment is specified.

Thread-cutting is reduced to use of an M8 tap by providing for a cap-head screw to secure the flywheel to a tapered section of the main journal.

Tolerances on the internal diameters of the mild-steel (e.g., BS En-1A) web can be achieved by reaming (in advance of the 1.0 mm slit-saw operation!), provided both diameters are reamed at a common setting to ensure parallelism.

The embodied-energy estimate starts from the envelope volume of the steel blank for the web. Conversion factors are from Table 20.1. Dimensions are from respective detail drawings:

$$V = 1/4\pi \times 0.05^2 \times 0.02 \text{ [m}^3\text{]} = 3.927 \times 10^{-5} \text{ m}^3,$$

$$M = \rho V = 7800 \text{ kg/m}^3 \times 3.927 \times 10^{-5} \text{ m}^3 = 0.3063 \text{ kg.}$$

Embodied energy = 0.3063 × 5.58 kWh/kg = **1.71 kWh**,
CO_2 equivalent = 0.3063 × 1.37 kg/kg = **0.42 kg.**

Estimates for the main journal (detail #4) and crank pin (detail #6) are obtained in the same way, a minor difference being that these items are simply parted-off to length, meaning that machining and clamping provision is minimal, and that the mass of the finished component differs little from that of the blank:

	Embodied energy [kWh]	CO_2 equivalent [kg]
Main journal	**0.54**	**0.132**
Crank pin	**0.19**	**0.046**

The embodied-energy approach offers a common denominator for comparison over a range of energy converters — particularly informative where comparison is between converters operating from a common energy source.

p-v panels are available as follows: mono-crystalline: 1319 kWh/m^2; poly-crystalline: 1130 kWh/m^2; thin-film 362 kWh/m^2. The area of the solar collector of the current study is one m^2, so at this premature stage in the comparison 1319.0/2.44 = 540 crankshafts equal a single *p-v* panel of one m^2!

20.2.2. Main bearing housing (detail #3 and isometric Fig. 20.4)

This carries the main journal on deep-groove ball-bearing SKF 6001 (bought-in item *a*) at the outboard end and needle-roller bearing SKF HK 1412 (bought-in item *b*) at the inboard end, the former providing axial

Fig. 20.4. Main bearing housing. Embodied energy: **31.3 kWh** CO_2 equivalent **5.99 kg**.

location. Suitable material is BS HE-30 in the WP condition. The component is produced by straightforward turning and drilling, the sole precautions being that (a) both internal diameters should be machined at the same setting and (b) that the axis should be perpendicular to the mounting face. This may require skimming the latter while the housing is mounted on a mandrel machined to an appropriate fit in the finished bore.

20.2.3. Binary, or 'dog-bone' links (detail #10 on sub-assembly, isometric Fig. 20.5)

These are identical, and can be machined 'as one' from stock 5 mm rolled plate. The ideal material would be aluminium alloy BS 2014A (cf. old designation BS L-64). Supplies in small quantities are becoming scarce, so it might be necessary to use BS HE-30 in the WP condition.

The smaller end accommodates needle-roller bearing SKF HK0810 (bought-in item *e*) the larger holds deep-groove ball race SKF 61800 (b/i item *d*).

20.2.4. Ternary links (details #8 and 9, isometrics at Fig. 20.6)

These are essentially mirror-images. Choice of material is as for the binary links. The bores accommodate SKF HK1015 needle-roller bearings (b/i item'*b*') in both cases. The 10 mm diameter stubs each carry a deep-groove ball-race *d* (SKF 61800), one housed at the larger-diameter end of the associated binary link. The other sits in the big-end of the

Fig. 20.5. Binary link. One of identical pair. Embodied energy: **1.02 kWh** CO_2 equivalent **0.195 kg**.

Fig. 20.6. Ternary link. Details #8 and 9 and of sub-assembly. Embodied energy: **2.79 kWh** CO_2 equivalent **0.53 kg (NB *each*)**.

respective vertical connecting rod (piston or displacer). The 10 mm diameter stubs each carry a deep-groove ball-race *d* (SKF 61800), one housed at the larger-diameter end of the associated binary link. The other sits in the big-end of the respective vertical connecting rod (piston or displacer).

20.2.5. Chassis plate (detail #1 on sub-assembly) and base-plate (detail #2)

The mechanism sub-assembly is mounted *via* the bearing housing to a 'chassis' comprising vertical chassis-plate and base plate. Both are machined from 6 mm rolled plate stock, BS2014A being preferred, BS HE-30 (WP) being acceptable. The two are located and secured together at the 'keyhole' provisions in the base of the chassis plate by cylindrical nuts (detail #2) tensioned by M5 countersunk-head socket screws.

The four holes at the top of the plate are elongated to allow the cylinder unit to be located after a degree of vertical adjustment (Fig. 20.7).

Mounting holes are provided for the 12 mm dia. holders (detail #11) for the 8 mm diameter cantilever pins (details #12) on which run the HK0810 needle-roller bearings of the binary links (b/i components *e*). Pin holders can be parted-off to length from BS HE-30 WP 12 mm round bar stock. The pins are parted-off to length from 8 mm stock-size

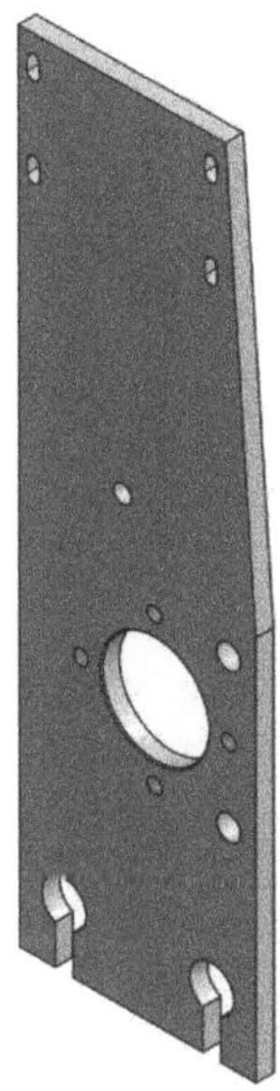

Fig. 20.7. Chassis plate. Embodied energy **20.92 kWh** CO_2 equivalent **4.0 kg**.

Stubbs silver steel and bonded (after trial assembly) into the holders with epoxy resin or the appropriate grade of Loctite™.

It may be found desirable to stabilize the cantilever pivot pins against flexure by supporting at the outboard ends. Machining the chassis plate clamped to a blank of 3 mm. thickness would automatically generate the correct outline and facilitate the retrospective modification at virtually no extra effort. Extended pivot pins (detail #12) are easily prepared in anticipation.

Pivot-shaft holders (not illustrated) combine to contribute **0.92 kWh** embodied energy and **0.176 kg** equivalent CO_2. Between them the two silver-steel pivot shafts add **0.153 kWh** and **0.038** equivalent CO_2.

20.3. Piston Connecting Rod/Bypass Link (Fig. 20.8, Detail #15)

Piston and displacer connecting rods are also labelled 'jumpers' on the basis that they bypass the rotating members of the crank mechanism.

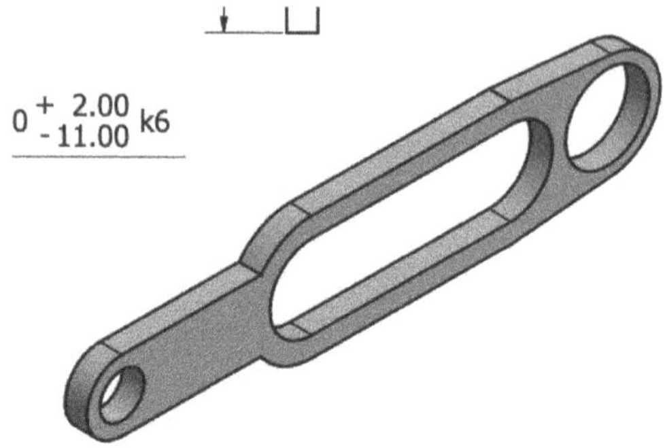

Fig. 20.8. Piston connecting-rod/bypass link. Embodied energy **2.61 kWh,** CO_2 equivalent **0.5 kg** (piston); **2.9 kWh** CO_2 equivalent **0.56 kg** (displacer).

Both operate with lateral offset from the vertical axis, but the angular swing either side of the vertical in both cases is 2 degrees or so, promising minimal side loads on guides, seals and piston. Stock material is 5 mm plate, preferably BS 2014A. Mounting to the respective ternary link is via SKF 61800 deep-groove ball races (b/i item *d*).

Displacer connecting-rod (Fig. 20.8) bypasses the mechanism and pivots at swivel pin (detail #14 not illustrated here) on proprietary dry-running wrapped bush Glacier DU0810. The pin secures to displacer drive-rod 16 by a pinch-bolt, an arrangement facilitating set-up.

Figure 20.9 is an example of the alternative output from the CAD model — a traditional orthographic drawing.

The 6 mm diameter displacer drive rod (detail #16, also not illustrated in isometric) is parted-off from 6 mm stainless-steel tubing to AISI 316 grade available exstock in the UK from Coopers Needles Ltd. Wall thickness is 1.5 mm. Tolerance on the outside diameter is ± 0.005 inch (0.127 mm), so depending on received condition it may require polishing to size. If so, target surface finish would be 0.4 micrometre Ra. Otherwise, no machining is required apart from tapping one end M4 × 0.6, inserting a 25 mm length of M4 studding (stainless-steel Allthread™) and pinning. A stainless-steel pin 0.5 mm diameter is suitable.

The drive rod is guided at dry-running bush Glacier 0610 DU which is the appropriate interference fit in the floating guide (detail #13) mounted on the chassis plate.

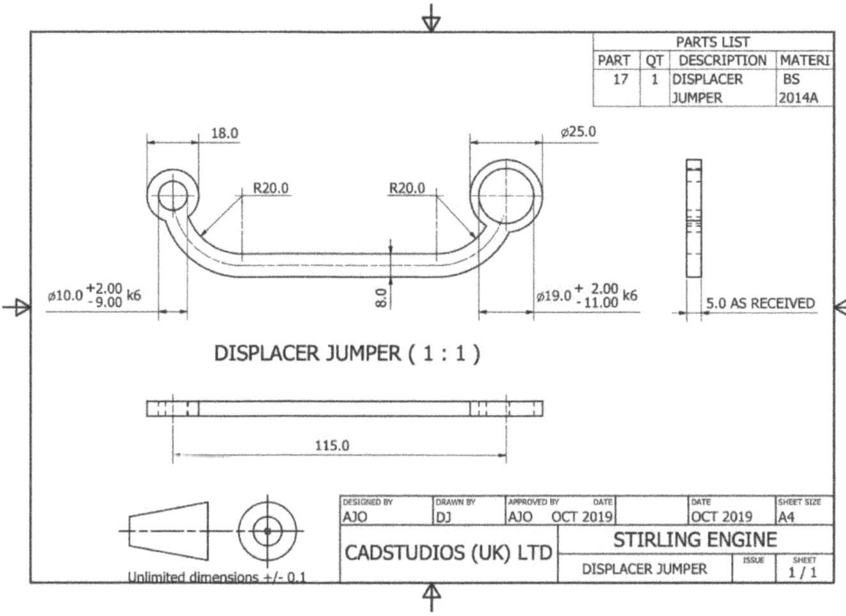

Fig. 20.9. Specimen orthographic output from CAD modelling process. Item illustrated is displacer connecting-rod/jumper.

20.4. Cylinder Liner (Detail #25 — Not Illustrated)

The cylinder liner and piston seal combination is technology borrowed from Airpot™ Corp. (Airpot, undated) and embodied in their dry-running pneumatic actuators, damper/dashpot assemblies. An item from their range of 'displacement products' (Fig. 20.10) has already been used successfully in a number of 'thermal-lag' engines. Units comprise a length of precision-bore glass tubing and a carbon/graphite piston ground to size.

The present application calls for service at elevated temperature, for which precision, clear fused quartz tubing is routinely cut to length by suppliers such as Heraeus Conamic.

The cylinder head/window (detail #26) is a Plano-Concave lens of the type used to diffuse laser beams and available in clear fused silica. Lenses come in stock sizes (Newport, undated) the choice here being the stock diameter of 55.8 mm and (negative) focal length of 100 mm. The supplier is able to weld lens to cylinder — not, apparently, invisibly, but at a distortion-free, structurally-sound bead of parent material.

Fig. 20.10. Specimen 'displacement products' from the Airpot™ range of pneumatic actuators and dashpots. With one end sealed the piston bounces, indicating ultra-low friction combined with excellent sealing.

The liner accounts for **0.089 kWh** embodied energy and CO_2 equivalent of **0.0018**. Corresponding figures for the lens are **0.29 kWh** embodied energy and equivalent CO_2 of **0.0601.**

20.5. Cylinder Barrel (Detail #20, Isometric at Fig. 20.11)

Material is BS HE-30 WP bored slightly over-size relative to the quartz-tube liner to take a film of epoxy resin after satisfactory trial assembly. The fins are token, as it is anticipated that the majority of heat rejection will be via the combined piston/heat sink.

20.6. Piston/Heat Sink (Detail #18, Isometric View at Fig. 20.12)

The piston is the conduction path for heat rejection. Vertical oscillation of fins extending laterally beyond the bore diameter is equivalent to an alternating flow of air. Machining is preferably by NC wire erosion, although 'additive machining' by laser melt appears a possibility. If by wire erosion, material suggestion is BS 2014A. HE-30-WP appears acceptable.

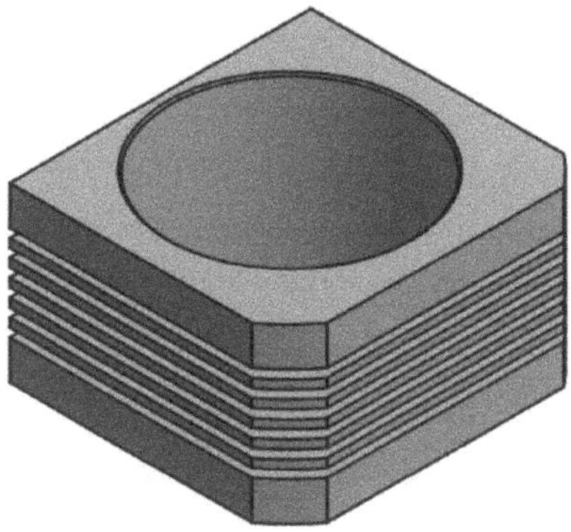

Fig. 20.11. Cylinder barrel. Embodied energy **32.5 kWh** CO_2 equivalent **6.23 kg**.

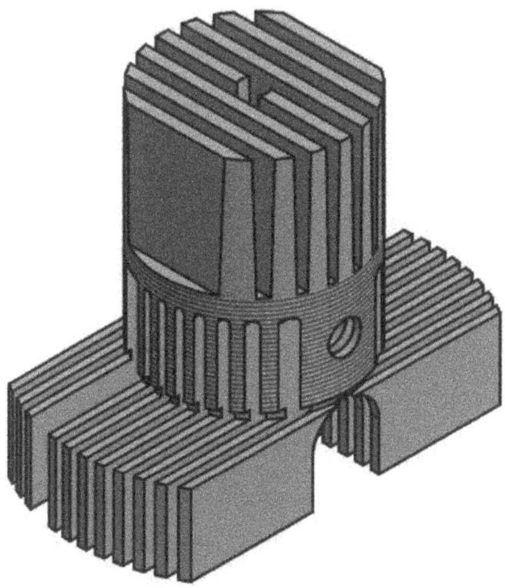

Fig. 20.12. Piston/heat sink. Embodied energy **58.1 kWh;** CO_2 equivalent **11.12 kg**.

The tapered fins on the upper face intercalate with matching fins on the underside of the displacer body. Lateral clearance between adjacent fins at any given point in the stroke depends on the value specified for the fully-engaged (minimum dead-volume) position and the taper angle. Experiment is the only way of determining any optimum.

External and internal diameters are achieved first (at a common setting to ensure concentricity), and the required alignment achieved by positioning the blank with these diameters as reference.

Sealing and (minimal) side-thrust are dealt with by the technology borrowed from Airpot™: a plain carbon/graphite sleeve (detail #22/23) machined to a close fit in the fused-silica liner. Grinding would be the preferred way of achieving the required surface finish. This follows bonding to the finished piston, and is by reference to the 6.0 mm internal diameter of the displacer guide/seal (bought-in item *g*) — not illustrated).

20.7. Displacer/Regenerator Housing (Detail #19, Isometric View at Fig. 20.13)

Proprietary alloy Kovar™ combines coefficient of thermal conductivity $k = 17$ W/mK and low coefficient of thermal expansion similar to that of glass (α = mm/mmC) with dimensional stability. Turning is completed before wire erosion of the fins, for which it should be possible to adapt the NC code used for the matching feature of the piston.

The regenerator (detail #24, not illustrated) is a stack of discs punched from proprietary woven-wire mesh. G Bopp and Co are able to supply ready-punched. For initial evaluation mesh material will be stainless steel. Eventually, the requirement will be for gauze woven from material of higher temperature specification. Wire diameter and mesh number are chosen elsewhere.

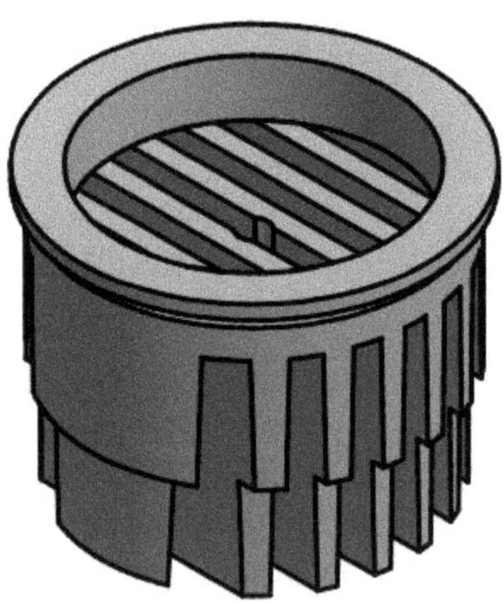

Fig. 20.13. Displacer body/regenerator housing. Embodied energy **12.11 kWh;** CO_2 equivalent **4.74 kg**.

The stack requires sintering, as the means of restraint is the circular nut (detail #16c) which secures it to the displacer drive-rod (detail #16a). On the provisional basis of 90% volume porosity $\P_v$ (see under Regenerator Design for choice of high $\P_v$) embodied energy **0.124 kWh;** CO_2 equivalent **0.0485 kg**.

20.8. Flywheel, Rim Detail #28a, Disc Detail #28b, Hub Detail #28c: Isometric of Assembly at Figs. 20.14 and 20.15

A way of producing a 'one-off' rim with minimal material wastage (minimum embodied energy) would be to roll a 630 mm length of 25 mm × 30 mm cold-rolled BS En-1A into an annulus, to trim and weld the joint and to clean up by turning to finished diameter and cross-sectional dimensions.

The disc (detail #28b) is cut over-size from stock 3 mm aluminium-alloy plate (BS HE-30 WP) turned to specified outside diameter, bored to take the hub, and drilled 4.1 mm diameter to take the pan-head retaining screws.

The hub can be from 45 mm diameter HE-30 WP bar stock. Tapered bore and 18 mm location diameter and corresponding location face require to be turned at the same setting (or other suitable measure taken to ensure true running of the assembly).

20.9. Tentative Comparison

The raw materials estimate added to that for bought-in items (Table 20.2) adds up to 13.3 kg (Table 20.2). Embodied energy is 149.57 kWh. The figures are per m^2 of solar receiver area, and include the flywheel provision. Applications involving direct drive of a load having sufficient mechanical inertia do not require a separate flywheel[4], in which case the figures reduce to 9.65 kg and 129.07 kWh respectively.

[4]The characteristic open indicator diagram of the Stirling allows satisfactory operation function with less flywheel inertia than required by an internal-combustion engine of the same number of cylinders.

Table 20.2. Embodied energy and CO_2 equivalent of bought-in components — *on basis of material content only*. Respective finishing processes (heat treatment, wire-weaving, etc.) could well account for several hundred percent inflation of that basic figure. (Arbitrarily) assuming an average of 300% gives the estimate for bought-in items as **3.68 kWh** and **0.852 kg CO_2**.

	Mass/unit [kg]	× No. req'd	kWh	CO_2 [kg]
Deep-groove ball bearings				
SKF 61800	0.0055	×4		
SKF 6001	0.022	×1		
Needle-roller bearings				
SKF HK 1412	0.013	×1		
SKF HK 1015	0.006	×2		
SKF HK 0810	0.003	×2		
			0.418	0.103
Sub-sub-total	0.015			
Glacier DU bushes				
0610 DU		×2	?	?
0810 DU		×2		
Lens — clear fused silica		×1	0.29	0.06
Cylinder liner — clear fused silica		×1	0.089	0.002
Regenerator stack — as pre-punched discs		×1	0.124	0.048
Sub-totals			0.921	0.213

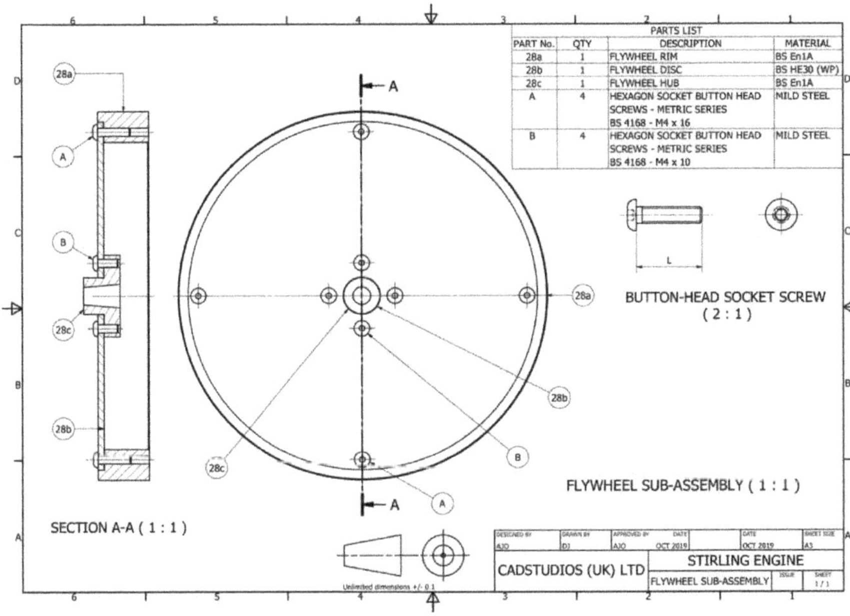

Fig. 20.14. Flywheel assembly. Embodied energy **21.16 kWh** (total); CO_2 equivalent **8.92 kg** (total).

The superficial comparison is with established p-v technology, for which estimates of embodied energy range from 1130 to 1320 kWh/m^2. On these figures, the Stirling alternative, assuming 15% thermal efficiency (150 W shaft power from one m^2 of collector area) begins to operate carbon-free after about 1000 hours — or 250 days total assuming 4 hours/day of suitable sunshine.

The 1–m^2 *p-v* panel of 40% efficiency begins carbon-free operation after $1130 \times 10^3/400 = 2825$ hours' operation — or 706 suitable days. (At the higher estimate of embodied energy the figures are 3300 hours or 825 days.)

The comparison requires qualification: the Stirling option requires a mirror or a lens. Plastic Fresnel lenses (imagine a transparent 33-1/3 vinyl record) were available in Canada by mail order for a couple dollars and arrived rolled-up in a cardboard tube.

If output is to be electrical, then provision of a generator has to be taken into account — as does its efficiency. Figures for embodied energy

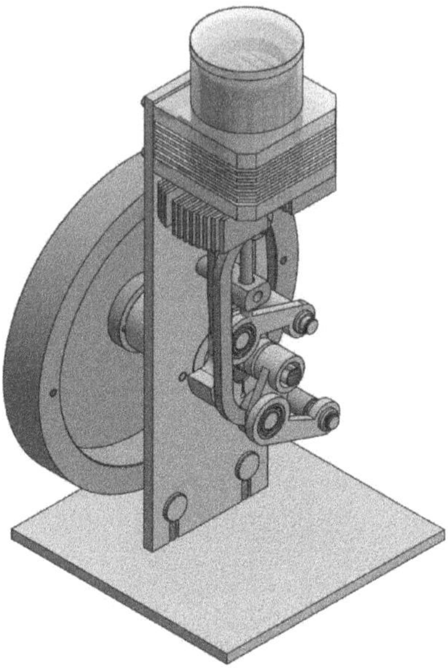

Fig. 20.15. Complete concept as constructed by CADStudios. The regenerator stack is omitted.

appear not to be generally available, but a Delco 'MotorSport' alternator of 12 volt, 60 amp rating has a mass of 3.0 kg and is advertised in the UK for £195 + VAT.

Likewise, if output from the solar panel is to be converted to mechanical power, then provision for an electric motor has to be embodied into the sums. Electric motors are now available in a bewildering range of technologies, power-to-weight values and cost, so the embodied energy burden has to be left open.

Figure 20.15 is the CAD rendering of the complete concept.

20.10. Provisional Conclusion

If CO_2 is bad for the environment, then continuing to extract and burn fossil fuels at the current rate seems highly inadvisable. Accepting the most pessimistic predictions, the increases in exploitation and consumption recently approved appear nothing short of suicidal.

20.11. Looking Ahead

This has been rather in the nature of a design study than a design — but detail has been sufficient to identify an undemanding exercise in machining and modest outlay in terms of materials. The author's experience in a different area of design, construction and endurance testing (Organ, 1967, 1973; Tobias *et al.*, 1969) justifies proceeding to the construction stage. The interim gives time to speculate on the way forward — pre-supposing promising initial trials.

To a production engineer this would probably mean looking for scope for simplification. One opportunity combines a reduction in component inventory with potential for improved[5] heat rejection.

20.11.1. Component reduction

A technology which has revolutionised cost, size, weight and range of applications of the reciprocating air compressor has been the tilting (or rocking) piston. Gudgeon-pin and associated bearings are eliminated, and sealing reduces to a single, easily-replaced off-the-shelf item. The latter is frequently an O-ring, but a lip-seal combines sealing with service as a simple one-way valve.

Restricting the angle of tilt to values which do not compromise sealing normally calls for connecting-rod centre-distance l several times greater than crank semi-throw r. The Leybold Stirling engine, marketed as a teaching tool, may have been the first of the *genre* to embody the tilting piston. Figure 20.16 gives an idea of the high l/r ratio necessary when a conventional crank/connecting-rod is involved. If this seems like a shortcoming, the 'max-purge' crank mechanism of present concern does not suffer from it: minimal 'tweaking' is required to the optimum proportions in order to confine angularity of the power-piston connecting-rod to less than ± one degree.

[5] 'Potential' because the heat rejection path comprises at least three elements in series — forced convection from working fluid, thermal conduction through a solid enclosing body and convection between the latter and the environment. Achieving a measurable improvement involves tackling the most resistive element in the series — which first of all has to be identified.

Fig. 20.16. The Leybold teaching engine may have been the first Stirling engine to make use of the rocking piston. Illustration courtesy of Leybold Heraeus GmbH.

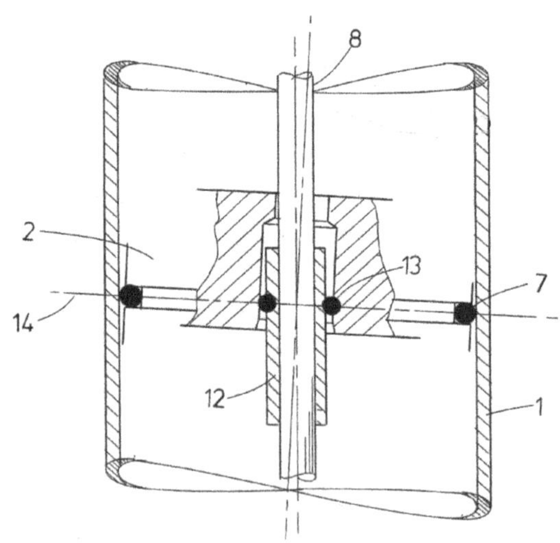

Fig. 20.17. Resiliently-mounted displacer-rod seal shown with angularity exaggerated (3°) relative to the calculated maximum (< 1°). From the drawings accompanying the patent application (Appendix F).

On the other hand, *any* departure from the vertical — however slight — has to be accommodated by the seal between displacer drive-rod and piston. A possibility which might serve for initial experimentation would be a further O-ring lying in the same plane as the outer piston seal (Fig. 20.17).

A solution with potential for extended service life is a resiliently-mounted parallel bush. A further O-ring, suitably pre-loaded by an appropriate interference fit, and again sitting in the plane of the main piston seal, provides static sealing while accommodating the minimal angular mis-alignment (Fig. 20.17 again). Most of the current generation of machinable 'self-lubricating' non-metallic materials are candidates for the bush, and determining the best choice would be an inexpensive and undemanding experiment.

20.11.2. Heat transfer enhancement

The 'max-purge' mechanism with conventional gudgeon pin holds the internal fins of displacer and piston in close engagement over an extended fraction of a cycle. During this interval, motion of the working fluid between the fins is minimal — a condition generally unfavorable to convection. Pressure is nevertheless varying, heat still needs to be exchanged, and all variants of the cycle require that convection should be the achievable maximum.

It is over this same interval of crank rotation that the rocking component of piston motion is concentrated. Slight modification to the taper of the fins (of piston or of displacer or of both) would accommodate the left-to-right displacement. Internal combustion engine practice uses the term 'squish' to describe the anticipated motion of the fluid, which would be either around the lateral extent of the fins, or through horizontal holes — or via both.

Net change in performance would be a balance between improved heat rejection, additional shearing work and the (marginal) increase in dead-space to accommodate tilt of the piston fins. The inevitable increase in dead-space detracts from the minimum achievable by use of the 'max-purge' mechanism, so that the change will not necessarily be an improvement.

An alternative implementation turns the interior fins through 90° (Fig. 20.17), so that the rocking motion now induces cyclic shearing of the working fluid between gaps of unaltered clearance. Dead-space is now un-changed but for the effect of a slight taper to the lateral extremities. Viscous losses promise to be less than for the earlier 'paddle' motion.

Experiment appears to be the only way of quantifying the benefit (or penalty) of either scheme.

21

Optimisation: A Preliminary Foray

21.1. The Challenge

There are numerous accounts of attempts to turn Tailer's patented thermal-lag principle into a power-producing engine. In one respect they are unanimous: each lacks evidence of design at any level. Regenerator material is routinely 'wire wool', but with no apparent concern for density (or even uniformity) of packing, for wire diameter or for porosity in relation to rpm and swept volume. When such accounts have made it to print, it is because the engines have functioned — at least unloaded.

Like the Stirling, the thermal-lag engine relies on the interplay of gas motion and heat exchange. But unlike that of the Stirling, the interplay is sufficiently inscrutable to defy attempts to define an 'ideal' cycle. If the thermal-lag type can function in the complete absence of guidelines — let alone of calculation — what comes to mind is scope for performance improvement. It thus becomes a promising candidate for an exercise in mechanised optimisation — but with due regard for the pitfalls of this seductive option.

21.2. Aspects of Optimisation

For present purposes, a search for the optimum means systematically perturbing (slightly increasing and/or slightly decreasing) the numerical data fed to an 'object function' (numerical model of work per cycle) and noting any improvement or deterioration. When all relevant parameters have been perturbed, each is incremented (or decremented) by the respective perturbation amount which yielded the improvement. The cycle repeats, with individual perturbation amounts reducing as necessary, until no further improvement results.

Specialist texts offer a choice of more sophisticated approaches. Ironically, the greater the sophistication, the greater the tendency to illustrate with a pedestrian example. A frequent choice is the simple cantilever clothes-peg!

It is possible, and occasionally desirable, to apply the sequence (perturbation, inspection and adjustment) manually. More commonly it is achieved by an algorithm integral with the code which invokes the cycle model. And herein lies a potential problem. In the hands of the algorithm the process is mindless: Weaknesses as well as strengths in the object function are exploited and amplified with indifference. The resulting 'design' can be closer to the worst than to the optimum. But all is not lost: in the early stages (and these are early stages in understanding the thermal-lag engine) optimisation is arguably an effective debugging tool!

What makes it worth pressing on at this point is the fact that cycle simulation and optimisation algorithm are decoupled. The latter can be exercised and illustrated pending development of a simulation sufficiently tested to serve as a realistic object function.

21.3. Case Study

The process requires a start-point in the form of a numerical specification. An arbitrary design will serve. (Specialists prefer the adjective 'sub-optimal' to arbitrary.) The Vaizey engine (Fig. 5.1) is particularly suitable: it demonstrated remarkable performance while owing nothing to 'formal' thermodynamic design. The regenerator was 'wire-wool' which, however, does not have any defined heat-transfer or flow characteristics. It is substituted retrospectively here by close-packed, square-weave mesh, for which wire diameter d_w and mesh number m_w define hydraulic radius and volume porosity. And thus arises an opportunity: Flow velocity varies between a maximum at the hot end to *zero* at the closed end. If ever there were a case for a grading the regenerator packing, this is surely it! Linear grading will get the ball rolling, and can be specified in terms of hydraulic radii and volume porosity at the extremes of a packing of overall length L_r. With x for distance from the closed (left-hand) end:

$$r_h = r_{h0} + (r_{h1} - r_{h0})x/L_r, \tag{21.1a}$$

$$¶_v = ¶_{v0} + (¶_{v1} - ¶_{v0})x/L_r. \tag{21.1b}$$

(Varying $¶_v$ in a duct of uniform cross-sectional area causes length-wise variation in effective free-flow area A_{ff}. Depending on the sensitivity of performance to regenerator specification it may eventually be worth grading free-flow area A_{ffr} independently — indeed, nonlinearly.)

Table 21.1 gives start values r_{h0}, r_{h1}, $¶_{v0}$ and $¶_{v1}$.

Also provided for in the specification is variable displacement between the axes of cylinder and crankshaft (désaxé). This generates asymmetry between compression and expansion phases. In an engine relying on asymmetries of one sort and another, this offers two further 'degrees-of-freedom'. All parameters are non-dimensional, and the pair which govern volume variations are e/r and l/r, where e is axis offset, l is connecting-rod centre distance and r is crank-pin offset.

Any improvement in simulated performance achieved by tweaking a given parameter is likely to be cumulative. This engine, with its inherently low compression ratio, requires every possible tweaking opportunity.

It has been taken for granted to this point that the 'object function' will be Z – specific cycle work W divided by swept volume V_{sw} and a characteristic cycle pressure p_{ref}. W is evaluated as $\int p\mathrm{d}V$, where p is measured at the piston-face, having been computed by cycle simulation *Turbo2×2* described elsewhere.

For any given set of parameter values, the resulting Z will be dependent upon rpm. Table 21.1 takes 1500 rpm. Alternatively, working with Z multiplied by rpm (i.e., power/$p_{ref}V_{sw}$) as object function reveals instead the parameter values which maximise power under the conditions of Table 21.1. However, exploring, say, 10 values of rpm increases an already heavy computational load[1] by an order-of-magnitude.

A sub-topic requires attention before biting the bullet.

[1]50 perturbations of 10 parameters suggests 500 cycles to equilibrium. Each cycle of the numerical model used here is iterated 5 times, and 10 cycles are required for equilibrium — i.e., 25,000 cycles. This increases to 250,000 if power is explored rather than specific work.

Table 21.1. Outline specification of Vaizey's impressive single-cylinder thermal-lag engine.

Swept volume $V_{sw} = V_{ref}$	7.6 cc
Charge pressure $p_{ref} = p_{min}$	0.66[a] atm./10^5 Pa
rpm	variable
T_C (= T_{ref}), T_E	30°C, 600°C
Volumetric compression ratio, V_{max}/V_{min}	1.645
Pressure ratio at rated temperatures p_{max}/p_{min} (simulated)	2.037
(Reference length $L_{ref} = V_{ref}^{1/3}$: **<u>19.66</u>** mm, Reference area $\boldsymbol{A}_{\mathbf{ref}} = V_{ref}^{2/3}$: **<u>38.6</u>** mm^2)	
Regenerator:	
Packing length	65 mm
Diameter	10.0 mm
Material	'wire wool' volume porosity ¶$_v$: 0.85 (assumed)
Pulse-tube:	
Outer diameter 17.5 mm	inner diameter 10.0 mm
Heater:	
Length	2 × 20 mm (reverse-flow);
No. flow channels	30 rectangular 2.0 mm × 2.0 mm

Cooler:					
Length	20 mm				
No. flow channels	123 holes 1.0 mm dia.				
	exp.exch.	regen.	comp. exchanger	comp. cyl.	pulse tube
Dead-space ratio					
V_d/V_{ref}	0.188	0.4418	0.0548	0.100	0.766
Length L_x/L_{ref}	1.474	2.947	0.736	0.100	2.947
Area/A_{ref}	0.128	0.149	0.074	1.000	0.259
Hyd. rad r_h/L_{ref}	0.0184	0.021	0.007	0.250	0.300

[a]The figure for p_{min} is arrived on the basis that, without provision to the contrary, mean cycle pressure probably settles to atmospheric. Pressure-ratio of about 2 then suggests $p_{max} = 2 \times p_{min} = 1.33$, giving $p_{mean} = 1/2(0.66 + 1.33) =$ unity.

21.4. False Optima

It would be equally possible to kick-off the optimisation sequence from a totally different specification — for example, one derived by linear scaling from Tailer's original patent sketch. (This would again call for imposing a regenerator specification in terms of flow-passage geometry.) However, one could be confident that the resulting 'optimum' would be quite distinct from that deriving from the Vaizey specification. This is one of the reasons why a reader contemplating constructing a thermal-lag engine should not be bound by the results of the present exercise.

By the same token there would be little point in exhaustive perturbation of the specification of Table 21.1. Instead, the section which follows will outline the numerical algorithm and illustrate progress as it deals with a specimen number of parameters.

21.5. A Basic Optimisation Search

Figure 21.1 is the search algorithm coded in Fortran. For reasons already given, the sophistication of golden-ratio, Hessians, etc., is not justified until means are available for circumventing all false optima.

The 'object function' is coded separately as Fortran function Z(PARAM, N, NCYC). *Z* is computed iteratively over NCYC consecutive cycles, where NCYC has been found by inspection to lead to equilibrium. PARAM is a one-dimensional array at least *n* long containing the values of the parameters to be perturbed.

The initial computational step is to allocate to simple variable Z0 the value of Z corresponding to the start values of all parameters. Provision of integer values JSTART and JFIN for the subsequent loop allows the search to be focused between parameters PARAM(JSTART) and PARAM(JFIN) rather than arbitrarily perturbing the lot.

The perturbation amount, DPAR, is set before each perturbation at a fraction FRACT(J) of the current value of the respective parameter. Fraction 0.1 has been found suitable for the initial stages of the search. Values of Z resulting from perturbation by + DPAR and by −DPAR are stored temporarily in ZR and ZL, respectively. The values are compared. If ZR is greater than ZL, the fact is recorded by setting integer array element ISGN(J) to +1. If the opposite, the value −1 is stored. If the absolute value of the difference ZR – ZL is less than a pre-set limit, ISGN is set to zero and multiplier FRACT(J) halved.

```
!
! PERTURBATION SEQUENCE
!
  Z0 = Z(PARAM, N, NCYC)
  WRITE(16,*)Z0
!!GO TO 555
  DO JJ = 1,10
    DO J = JSTART,JFIN
      DPAR(J) = FRAC(J)*PARAM(J)
      DPARJ = DPAR(J)
      PARAM(J) = PARAM(J) + DPARJ
      ZR = Z(PARAM, N, NCYC)
      PARAM(J) = PARAM(J) - 2.0D+00*DPARJ
      ZL = Z(PARAM, N, NCYC)
      IF(ZR .GT. Z0 .AND. ZL .LT. Z0)ISGN(J) = 1
      IF(ZR .LT. Z0 .AND. ZL .GT. Z0)ISGN(J) = -1
      IF((ZR .GT. Z0 .AND. ZL .GT. Z0) .OR. (ZR .LT. Z0 .AND. ZL .LT. Z0))THEN
        FRAC(J) = 0.5D+00*FRAC(J)
        DPAR(J) = FRAC(J)*DPARJ
IF(PARAM(2) .LT. 0.65)DPAR(J) = ZERO
        ISGN(J) = 0
      ENDIF
!!WRITE(16,*)J, ZR - Z0, Z0 - ZL, Z0
!
! RETURN VALUE OF PARAM(J) TO UN-PERTURBED STATE
!
      PARAM(J) = PARAM(J) + DPARJ
    END DO
!
! ADD OR SUBTRACT INCREMENT AS APPROPRIATE
!
    DO J = JSTART,JFIN
      PARAM(J) = PARAM(J) + ISGN(J)*DPAR(J)
    END DO
    Z0 = Z(PARAM, N, NCYC)
    EOR = PARAM(1)
    XLOR = PARAM(2)
    XRHR0 = PARAM(3)
    XRHR1 = PARAM(4)
    XRHXE = PARAM(5)
    XRHXC = PARAM(6)
    XLR = PARAM(7)
    XLPT = XLR
    ALFFPT = PARAM(8)
```

Fig. 21.1. Search algorithm coded in Fortran. In numerical order, parameters are e/r, l/r then the following normalised by L_{ref}: r_{hr0}, r_{hr1}, r_{hxe}, r_{hxc} and L_r. Pulse-tube length L_{pt} is assumed to remain equal to L_r as L_r is perturbed.

When a value of ISGN has been registered for all parameters, current perturbation amounts are multiplied by respective ISGN and added to start values to give results leading to the calculated performance improvement.

There is obvious scope for optimising the algorithm itself!

21.6. Numerical Output

Table 21.2 is an extract from the printout generated as the search proceeds. The parameters chosen for perturbation at listed in the caption — but not in raw form: numerical values are respective current value pro-rated by starting value. This together with the fact that initial perturbation increments are uniformly the same faction of parameter value, explains numerical similarities.

The results suggest that the initial value of Z (0.0849) is increased by a factor of 2 by altering the design parameters to the fractional values at the bottom of each vertical list. The fact that simulated indicated power is increased by the identical percentage is explained by the fixed value (Table 21.2) of rpm.

The apparent improvement is achieved to the accompaniment of a considerable reduction in hydraulic radii r_{hr0} and r_{hr1} at both ends of the regenerator relative to respective start values (columns 3 and 4). The hydraulic radii of hot and cold exchangers decrease by 60% and increase by 25%, respectively (columns 5 and 6).

It is tempting to look at columns 1 and 2 and indulge in a bit of self-congratulation for the earlier prediction of a strong influence of volume variations. The matter can be probed by perturbing kinematic parameters

Table 21.2. Progress towards local optimum. Numerical values listed under *e*, *l*, etc., are actually pro-rated to respective start values, e.g., e/e_0.

	e	l	r_{hr0}	r_{hr1}	r_{hxe}	r_{hxc}	L_r	A_{ffpt}	Z[-]	Power [W]
1	1.1000	0.9000	0.9000	0.9000	0.9000	1.1000	1.0000	1.0000	0.00846	0.08492
2	1.2100	0.9000	0.8100	0.8100	0.8100	1.2100	1.0000	0.9500	0.01023	0.10258
3	1.3310	0.8550	0.7290	0.7290	0.7290	1.3310	0.9750	0.9025	0.01416	0.14205
4	1.4641	0.8123	0.6561	0.6561	0.6561	1.3310	0.9750	0.9025	0.01501	0.15060
5	1.6105	0.7716	0.5905	0.5905	0.5905	1.3975	0.9750	0.9025	0.01642	0.16471
6	1.6105	0.7716	0.5905	0.5905	0.5314	1.3975	0.9750	0.9025	0.01644	0.16497
7	1.6105	0.7716	0.5610	0.5610	0.4783	1.3976	0.9750	0.9025	0.01368	0.13725
8	1.6105	0.7716	0.5610	0.5610	0.4305	1.3975	0.9750	0.9025	0.01369	0.13733
9	1.6105	0.7716	0.5610	0.5610	0.3874	1.3976	0.9750	0.9025	0.01677	0.16826
10	1.6105	0.7716	0.5610	0.5540	0.3487	1.3975	0.9750	0.9025	0.01681	0.16866
11	1.6105	0.7716	0.5575	0.5470	0.3138	1.3997	0.9750	0.9025	0.01686	0.16916
12	1.6105	0.7722	0.5575	0.5470	0.2824	1.3997	0.9750	0.9025	0.01687	0.16921

Table 21.3. Progress towards local optimum with search confined to kinematic parameters (e/r and l/r of désaxé crank drive). Restricted search creates the false impression that performance is only marginally influenced by volume-crank-angle characteristic.

	e	l	Z[-]	Power [W]
1	1.1000	0.9000	0.00754	0.07563
2	1.1000	0.9000	0.00754	0.07563
3	1.1000	0.9000	0.00754	0.07563
4	1.0725	0.9000	0.00754	0.07564
5	1.0457	0.9000	0.00757	0.07596
6	1.0457	0.9000	0.00757	0.07596
7	1.0457	0.9000	0.00757	0.07596
8	1.0457	0.8986	0.00757	0.07596
9	1.0424	0.8986	0.00757	0.07599
10	1.0424	0.8979	0.00758	0.07599
11	1.0424	0.8979	0.00758	0.07599
12	1.0424	0.8975	0.00758	0.07601
13	1.0424	0.8975	0.00758	0.07601
14	1.0426	0.8974	0.00758	0.07601
15	1.0424	0.8975	0.00758	0.07601

e/r and l/r in isolation. Table 21.3 records the result: e/r and l/r are exploited *not* against the start-values of other parameters, but against current values as the 'optimised' specification evolves.

21.7. Interim Conclusions

Applied to a multi-variable problem such as the thermal-lag cycle, the machinations of optimisation are as inscrutable as the cycle itself.

Subject to appropriate conditions, optimisation is nevertheless a promising way forward. The most important such condition is provision in the study for continuous cross-validation of object function (cycle model) and counterpart experiment. Without a Z behaving as the real thing, the search for first-principles design guidelines is futile.

If mechanical simplicity is a virtue, the *genre* has a potential future, particularly in the context of renewable energy. It is fortuitous that atmospheric air has so far out-performed other candidate working fluids, including the He and H_2 essential for competitive specific power in the Stirling engine. The fact contributes to the prospect of an affordable laboratory project.

At the time of writing, there has been no published attempt to propose an 'ideal' thermal-lag cycle. Even Tailer's original patent account of functioning does not serve. Mechanical optimisation affords a systematic way of exploring cause and effect. A sufficient number of such dependencies might add up to insights which could be graphed or expressed parametrically.

Looking ahead, a superior intellect to that of this writer might then be able to design from the inside out by working directly with the interactions by which heat converts to work inside this enigmatic device.

22

Enigma Variations

22.1. Rationale

A purpose of study is to gain insight. Understanding should lead to explanation. In the field of technology, availability of explanation can be the basis for progress.

Reversing the argument affords a test: if understanding has advanced, then where explanation was previously lacking, the position should now have clarified.

22.2. Performance Phenomena in Search of an Explanation

An earlier title (Organ, 2014, p. 35) introduced Larque's 'Vari-Engine' — a small but 'serious' coaxial design with similarities to the Philips MP1002CA — slotted heat exchangers and annular, wrapped regenerator, but with drive mechanism designed specifically to allow investigation of the effect of varying volume ratio κ and phase angle α (achieved in the case of the coaxial machine by varying kinematic displacement ratio λ and kinematic phase angle β).

Figure 22.1 reproduces the original photograph showing two, parallel crankshafts as per the rhombic drive, but with the timing gears synchronised by a common idler wheel, and accordingly rotating in the same direction. Each timing gear carries a crank-pin driving an arm through a short connecting rod. Each arm pivots at its outer extremes on a pin set in the crank-case. The inner ends impart vertical motion directly to

Fig. 22.1. 'Vari-Engine'. Image provided by Ian Larque with permission to reproduce.

piston-rod and displacer rod, respectively. Both rods run in guides at their lower ends, offsetting much of the side loads on piston and displacer.

Variation of strokes S_p and S_d (and thus of λ in the parallel-bore machine) was achieved by use of interchangeable gears drilled for different crank-pin offset. Kinematic phase angle was altered by withdrawing the idler gear and advancing a crankshaft by one or more teeth. The process alters the vertical separation between piston and displacer, which is then compensated by adjusting the attachment point of one or both driving arms.

Comprehensive dynamometer tests confirmed $\lambda = 0.75$ and $\beta = 50°$ as the optimum combination. These were also the values at which no-load operation with fuel source isolated continued down to lowest T_E. (The intuitive expectation is for the optimum combination of λ and β to be dependent on temperature ratio N_T. More on this later.) This writer witnessed sustained no-load running at 100°C followed by removal of heat source and continued running for a further 5 minutes. Vaizey reported a similar event (run-down after heat removal) which he timed at 11 minutes.

Under the assumption of simple-harmonic motion, the kinematic optima translate to $\alpha = 132°$ with $\kappa = 0.77$. The literature carries accounts of several independent studies optimising α and κ for the Schmidt cycle under a range of constraints (e.g., limited mass content, limited peak pressure, etc.), including those of Finkelstein (1960a), Kirkley (1962),

Walker (1962), Takase (1972) and the present author (1990). None of these arrives at optima of κ and α anywhere near the respective 0.77° and 132°.

Setting $\lambda = 1.5$ from published 'optimisation' studies resulted in the Vari-Engine requiring very high T_E before it would start. Restoring $\lambda = S_p/S_d$ to the experimentally-determined optimum of 0.75 transformed performance, leading back to ready starting, increased no-load speed and sustained operation at dramatically-reduced T_E. In other words, experimentally-determined optimum λ is effectively the inverse of that suggested by available theory.

Kalibulut *et al.* (2009) were able to report no-load running of their engine down to T_E = 75°C with heat source removed. Although less sophisticated in terms of regenerator and heat exchanger provision, their engine was also in the 'serious' category. At 0.759 their optimum S_p/S_d was almost identical to that of the Vari-Engine. This compounds doubt in the conclusions of earlier exercises in analytical optimisation.

To confound matters, there is the well-established practice of 'low-ΔT' engines. The design target is no-load operation at minimum $T_E - T_C$. Kinematic volume ratios λ settled on after extensive competitive experiment are of the order of 1/35. In a parallel-bore machine this converts to $S_p/S_d \approx 0.0285$.

To be acceptable, an explanation must be valid over three regimes of operation: (1) the 'serious' engine (pressurised with multi-duct heat exchangers and micro-porous regenerator) operating from high-temperature heat source; (2) the same engine operating (or failing to operate) at $T_E \leq 100$°C and (3) the dedicated 'low-ΔT' type — un-pressurised and constructed of low-temperature materials.

The thermodynamic analysis which is both manageable and equally appropriate to all three has yet to be devised. This is equivalent to saying that, although the same thermodynamic and flow interactions are involved, each regime involves a different balance.

A way of approaching the much broader modelling problem which this raises is in terms of 'skeleton' analysis. The bare bones of any given aspect can be fleshed out if preliminary probing offers a promising lead.

22.3. Drive Kinematics

Comparing Fig. 22.1 with the earlier Fig.12.2 shows how much the Vari-Engine has in common with the 'max-flush' mechanism. Depending on

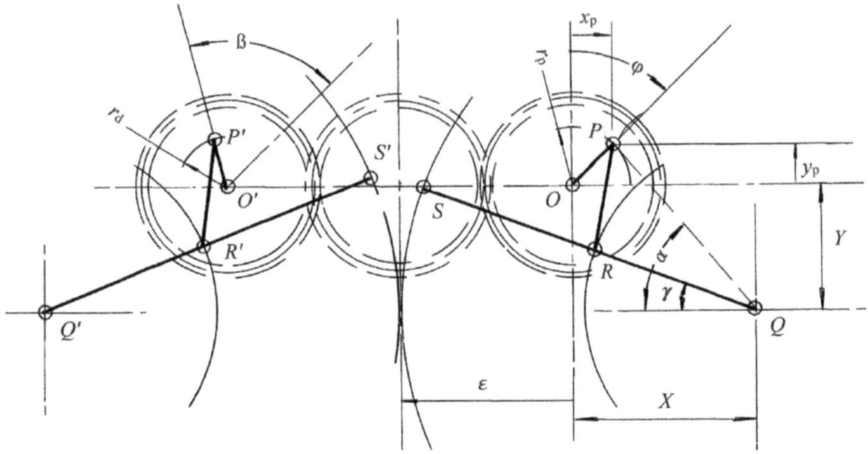

Fig. 22.2. Notation for algebra of kinematics (Fig. 22.1). Lengths *X*, *Y*, *P–R*, *Q–R*, *Q–S* scaled from the original of Fig. 22.1. *O–P*, *O′–P′* and angle β variable — as in Vari-Engine.

how the drive mechanism of the former was configured for any given run, the resulting volume variations could have departed as far from the simple-harmonic 'ideal' as do those of the 'max-flush' (recall Fig. 12.2). Values of α and κ then become irrelevant, and a cycle model of any level of sophistication, from the 'isothermal' upwards, needs to call upon the correct volume-crank-angle history.

Figure 22.2 focuses on the right-hand half of the Vari-Engine drive. From Fig. 22.1 this is the linkage controlling the work piston. Crank angle ϕ is reckoned positive as shown, although operation as a prime mover requires the anti-clockwise rotation indicated by the larger arrow.

The algebra arrives at the vertical and horizontal motions of point *S*, viz, of y_S and x_S as functions of crank angle ϕ. It starts with the instantaneous coordinates of crank-pin *P*:

$$x_p = r_p \sin(\phi), \quad y_p = r_p \cos(\phi). \tag{22.1}$$

Lengths *PR*, *QR* and *PQ* form a triangle of which all internal angles and length of side *PQ* vary with rotation of the crankshaft. The process of specifying instantaneous values in function of ø starts by determining length *PQ*. Pythagoras' theorem comes to the rescue:

$$PQ = \sqrt{(X - x_p)^2 + (-Y + y_p)^2}. \tag{22.2}$$

Angle *PQR* is available from the standard trigonometric relationship:

$$\cos(PQR) = \frac{PQ^2 + QR^2 - PR^2}{2PQ \times QR}. \tag{22.3}$$

Angle α (not to be confused with volume phase angle α) is readily obtained from the definition of the tangent function:

$$\tan(\alpha) = (-Y + y_p)/(X - x_p). \tag{22.4}$$

Angle γ follows as angle *PQR* minus α.

Re-using the earlier trigonometric formula:

$$\cos(PRQ) = \frac{PR^2 + QR^2 - PQ^2}{2PR \times QR}. \tag{22.5}$$

Relative to origin *O*, horizontal and vertical components of the instantaneous location of point *R* are as follows:

$$x_R = X - QR\cos(\gamma), \qquad y_R = Y + QR\sin(\gamma). \tag{22.6a}$$

Instantaneous coordinates of point *S* follow immediately:

$$x_S = x_R - RT\cos(\gamma), \qquad y_S = y_R + RT\sin(\gamma). \tag{22.6b}$$

y_S is the vertical location of the piston. With slight and obvious modification, the algebra applies to the counterpart linkage to the left of the centreline (i.e., to the displacer drive). The resulting $y_{S'}$ in this case is the vertical location of the displacer.

Figure 22.3 is the skeleton kinematic diagram.

Respective y_S can be plotted in function of crank angle for λ_{opt} and β_{opt} to give an indication the volume variations V_e and V_c. At first sight, Fig. 22.4 reveals nothing to write home to mother about — except that picture could be considered the very *opposite* of that achieved by the 'max-flush' mechanism (cf. Fig. 12.2). This, however, is before the corresponding indicator diagrams are plotted (Fig. 22.5). The unexpected negative loop disappears with a minor increase in λ from 0.75 to 0.80. None of these features can be reconciled with observed performance.

It could be, of course, that the explanation is staring us in the face, but waiting to emerge from enquiries on other fronts.

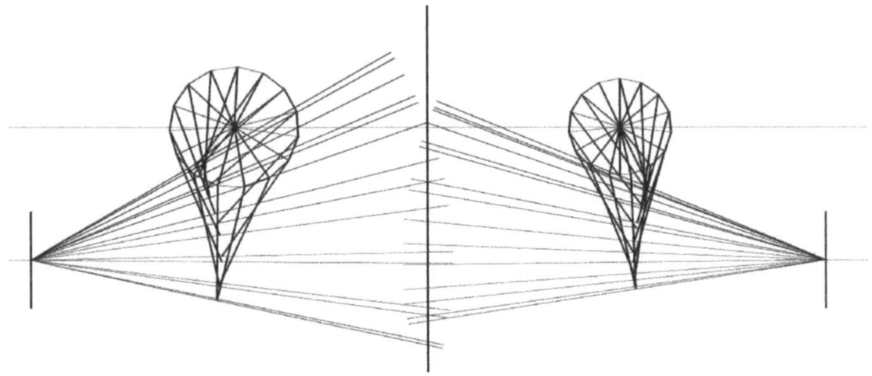

Fig. 22.3. Kinematic diagram plotted for λ_{opt} and β_{opt}. Left trace: displacer; right: piston.

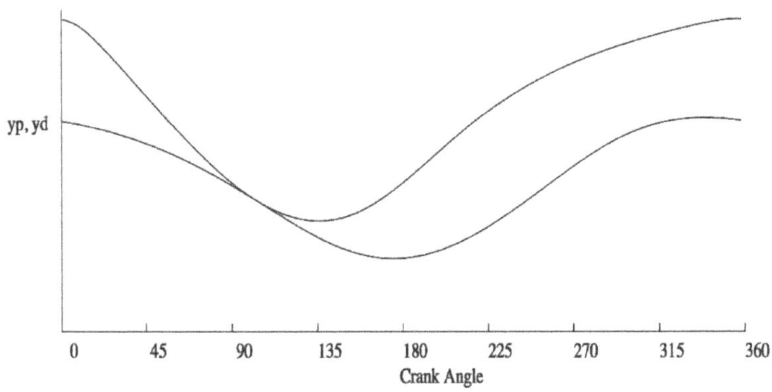

Fig. 22.4. Motion diagram plotted for λ_{opt} and β_{opt} notified by Larque. Upper trace: displacer; lower trace: piston.

22.4. Idealised Transient Behaviour

If running continues after removal of the heat supply, it is at the expense of the temperature of the components of the expansion space. In the skeleton algebraic model which follows, the mass of solid material involved is denoted m_E [kg] its instantaneous temperature (assumed uniform at any instant) by T_E and specific heat by c [J/kgK].

Each cycle of the run-down draws an amount of heat dQ_e equal to W_{cyc}/η where η is ideal thermal efficiency between current T_E and T_C, viz $(T_E - T_C)/T_E$ and W_{cyc} is ideal cycle work between the same limits.

$$\eta = (T_E - T_C)/T_C = (N_T - 1)/N_T. \tag{22.7}$$

The Schmidt expression for cycle work W_{Sch} is sensitive to κ, α and to $N_T = T_E/T_C$. To this extent it serves as a check on basic functioning of a numerical model. In terms of Finkelstein's reworking

$$W_{Sch} = pV_E Z(N_T, \kappa, \alpha, \upsilon),$$
$$dQ_E = pV_E Z(N_T, \kappa, \alpha, \upsilon)N_T/(N_T - 1).$$

On removal of heat supply, rate of temperature drop dT_E/dt of m_E is in terms of Q_E, and of thermal conduction $kA_{eff}(T_E - T_C)/L_{eff}$. A_{eff} and L_{eff} are effective thermal conduction path length and effective cross-sectional area, respectively. (There is also radiation and convection from the expansion end which will be worth estimating if the provisional formulation leads somewhere.)

Heat rate dQ_E/dt is the previous equation multiplied by rpm/60, leading to

$$m_E c dT_E/dt \approx -pV_E rpmZ(N_T, \kappa, \alpha, \upsilon)N_T/60(N_T - 1) - kA_{eff}(T_E - T_C)/L_{eff}. \tag{22.8a}$$

Equation (22.8a) has two dependent variables T_E and rpm, both are changing with time t. This is tackled empirically by noting that the Vari-Engine was equipped with slotted heat exchangers, and that the regenerator was a foil bandage. At the low rpm under investigation the assumption of laminar flow suggests local pressure drop proportional to local particle velocity — and thus proportional to rpm.

The context suggests assuming low inertia loads and mechanical constant mechanical friction. As rpm approaches zero, pumping power does the same, leaving only mechanical friction to balance indicated work Z_f. With N_{Tf} for temperature ratio at the end of run-down:

$$Z_f = Z[N_{Tf}, \kappa, \alpha, \upsilon(N_{Tf})].$$

Brake output is zero, so provisionally supposing indicated work to be dissipated in internal pumping gives $pV_E Z[N_T, \kappa, \alpha, \upsilon(N_T)]/60 = C_1 \times \text{rpm} + pV_E Z_f$.

Rearranging gives an expression from which a value of C_1 can be obtained corresponding to any pair of values of N_T and *rpm*. The values N_{T0} and rpm_0 at the start of the run-down serve, yielding

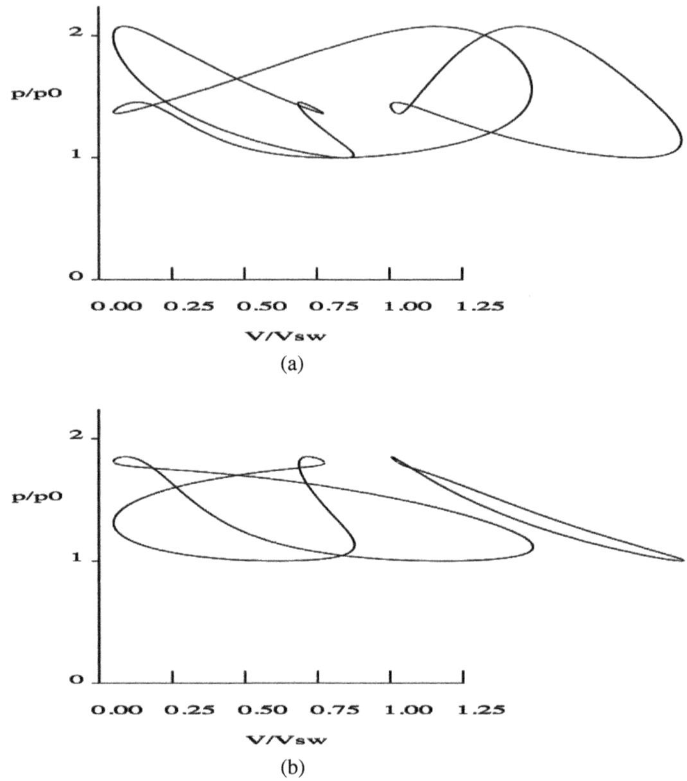

Fig. 22.5. Ideal indicator diagrams. (a) $T_E = 600°C$. (b) $T_E = 75°C$. $T_C = 25°C$ in both cases.

$$C_1 = pV_E \frac{Z[N_{T0}, \kappa, \alpha, \upsilon(N_{T0})] - Z[N_{Tf}, \kappa, \alpha, \upsilon(N_{Tf})]}{\text{rpm}_0},$$

$$\text{rpm} \approx pV_E\{Z[N_{T0}, \kappa, \alpha, \upsilon(N_{T0})] - Z[N_{Tf}, \kappa, \alpha, \upsilon(N_{Tf})]\}/C_1. \tag{22.8b}$$

The expression is divided through by $m_E c T_C$. For the κ, α, of a given test, temperature and *rpm* run down according to:

$$\begin{aligned} dN_T/dt &\approx -C_2 Z(N_T, \upsilon) \times \text{rpm} - C_3(N_T - 1), \\ C_2 &= pV_E/(60 m_E c T_C), \\ C_3 &= kA_{\text{eff}}/(m_E c T_C L_{\text{eff}}). \end{aligned} \tag{22.8c}$$

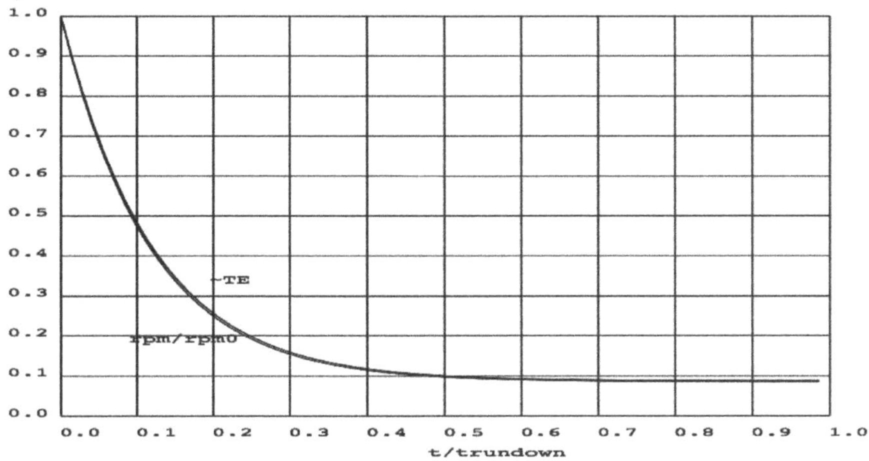

Fig. 22.6. Solution of Eqs. (22.8a) for the parameters of the Vari-Engine, mimicking a run-down test from $T_E = 100°C$ to 75°C.

Choosing a value for time increment dt and evaluating Eqs. (22.8b) and (22.8c) iteratively generates a history of N_T and rpm over the time interval between removal of heat and eventual cessation of rotation. Figure 22.6 illustrates a run-down test computed from Eq. (22.8a) using values of κ, α converted from the λ_{opt} and β_{opt} for the Vari-Engine. Curves of rpm and of T_E virtually superimpose. Not surprising, since without the heat-leak term, cyclic rate of heat removed from the expansion end and rate of temperature drop are both proportional to rpm. The curve marked ~T_E is actually $(T_E - T_{Efinal})/(T_{E0} - T_{Efinal})$ to correspond with the fractional y-axis. (The plot software refused to format the preferred expression.)

22.5. Optimisation Revisited

Until relatively recently, getting a Stirling engine to turn a crank called for a temperature ratio N_T of about 3. This writer noted the emergence of the Low-ΔT genre from the sidelines and watched in awe as Kolin (1991) and Senft (1993) coaxed no-load operation out of ever-decreasing ΔT. In 2020, executive toys activated by the heat of the hand proliferate. (They purport to demonstrate the Stirling cycle — probably outnumbering engines which actually do.) Whether gas process modelling has played a part in achievement of operation with ΔT of 1/2°C is not known.

Whatever design priorities were identified, they have evidently been satisfied on two fronts (a) structural materials: use of wood, plastic and styrofoam as appropriate and (b) adoption of 'beta' or 'gamma' configuration with universally large λ (ratio of volume swept by the displacer to that swept by the piston).

But then along came Bradley (1974), Karabulut (2009) and Larque (Organ, 2014, p. 35) and coaxed machines of conventional materials and conventional proportions to operate unloaded from unprecedentedly-low T_E.

The original enigma has become two!

(1) How do the 'high-temperature' engines of Larque and Karabulut operate so readily from near-ambient temperatures — and with phase angle and volume-ratio at odds with the optima of published studies?
(2) The gas-space proportions of conventional (i.e., high-temperature) machines able to operate from small temperature differences are unrecognisably different from those typical of the low-ΔT genre. Why?

Sherlock Holmes would probably have been in favor of exploring the obvious before becoming involved in sophisticated cycle modelling. If theory disagrees with reality, then suspicion falls on the former — which has two independent aspects (a) the cycle model and (b) the *criterion of the optimum*. Regardless of the choice of (a) the machine satisfying the criterion of maximum cycle work W per unit mass M of working fluid is different from that which maximises W per swept volume and per peak cycle pressure.

And perhaps the candidate criteria relevant to the present enquiry have never been explored! How about maximum indicated cycle work *per unit temperature difference*?

Isolating the roles of (a) above while exploring choices of (b) across both categories of engine (native low-ΔT and traditional high-temperature) suggests use of a cycle model with the minimum number of parameters over and above the essential N_T, κ, α and υ — in other words, the much-maligned Schmidt model! The choice will have the side benefit of filling gaps in earlier optimisation studies.

Before embarking it will be helpful to have a clear picture of what is meant by phase angle and volume ratio. The representative low-ΔT

machine is of 'beta/gamma' configuration with displacer many-times the diameter of the work piston. There is an illusion of a thermodynamic volume ratio κ of similar order. In fact, amplitudes of 'hot' and cold space variations are almost entirely controlled by the displacer, signifying $\kappa \approx$ unity (while *kinematic* volume ratio λ can, indeed be of the order of 1/30–1/50). For the same reason, thermodynamic phase angle α cannot be far from 180° whatever the value of *kinematic* phase angle β!

Precedents in optimisation studies are in terms of κ and α, while both categories of machine of present concern (parallel-coaxial Vari-Engine and low-ΔT) are characterised in terms of λ and β. Possibly with this potential problem in mind, Finkelstein offers conversion formulae between κ, λ, α and β. The original context gives the misleading impression that application is confined to the Schmidt analysis. In fact, they facilitate comparison between two-piston and coaxial machine regardless of the sophistication of the cycle analysis. The universality justifies reproduction here:

$$\lambda = \sqrt{(1+\kappa^2 + 2\kappa \cos\alpha)}\ (\lambda \text{ also} = V_{sw}/V_E), \tag{22.9}$$

$$\beta = \operatorname{atan}\frac{\kappa \sin\alpha}{\kappa \cos\alpha + 1}, \tag{22.10}$$

$$\kappa = \sqrt{(1+\lambda^2 - 2\lambda \cos\beta)}, \tag{22.11}$$

$$\alpha = \operatorname{atan}\frac{\lambda \sin\beta}{\lambda \cos\beta - 1}. \tag{22.12}$$

Finkelstein also offered a chart for quick inter-conversion. The original, however, was devised before the low-ΔT era, and does not highlight the range of λ and β of present concern. Figure 22.7(a) now goes some way to offering the necessary resolution. Companion Fig. 22.7(b) is for use when conversion is in reverse from λ and β to κ and α. So far so good! But back to optimisation — via an outline of salient studies to date.

The early 1960s saw two influential developments: (1) rapidly expanding access to main-frame computing power and (2) Finkelstein's virtuoso re-working of Schmidt's analysis of 1871. An aspect of Finkelstein's contribution had been reduction of the number of dependent variables through combination into dimensionless parameters having

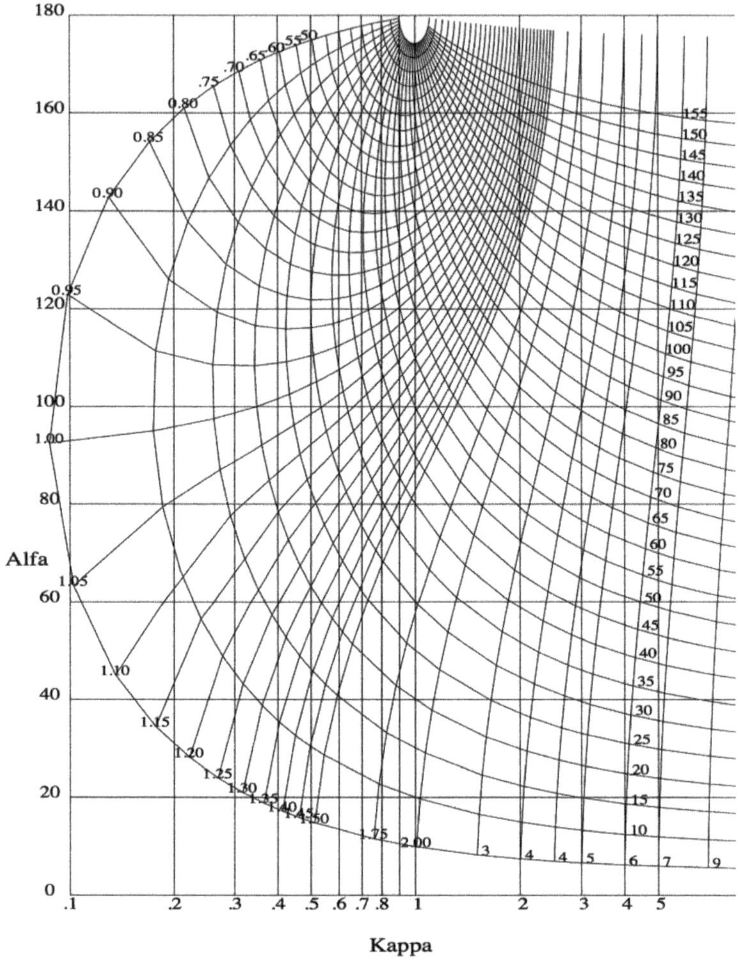

(a)

Fig. 22.7. (a) Chart for conversion from κ and α to λ and β retaining the same volume-crank-angle relationship. (b) Counterpart chart to (a) for reverse conversion from λ and β to the κ and α retaining the original volume-crank-angle history.

obvious physical significance. This made for a manageable challenge in formal optimisation, attracting the attention of academics with their penchant for differential calculus and numerical analysis.

The fact that two influential parameters, volume ratio and phase angle, could be 'optimised' may have contributed to the view that Schmidt alone could guide Stirling engine development. Indeed, analytical work

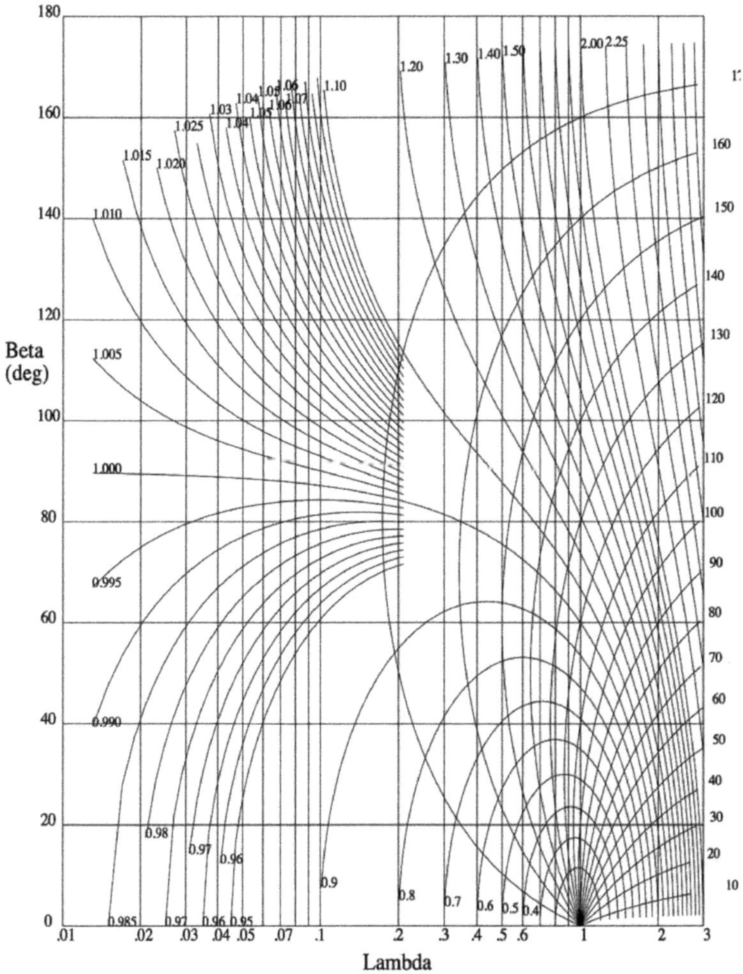

Fig. 22.7. (*Continued*).

disclosed by Philips went little further than Schmidt and its derivatives. This justifies returning to the original studies to see whether they even amounted to being exhaustive.

Optimisation as a strategy takes an 'index of performance' (or 'figure of merit' or 'object function' — preferably dimensionless) and monitors its value as the parameters are subject to 'perturbation'. A ruthlessly simple example would take ideal efficiency $\eta(N_T) = 1 - 1/N_T$

(recall Eq. (22.7)). Perturbing N_T immediately reveals that efficiency is maximum when N_T is maximum. Finkelstein's reworking yielded specific work $Z_{MRT} = W_{cyc}/MRT_C$ — or Beale number by another name. Z_{MRT} has four parameters, N_T, κ, α, υ. Finkelstein differentiated analytically with respect to two of them, κ and α and set equal to zero, thereby finding the κ and α which maximised Z_{MRT} as a function of temperature ratio N_T for various choices of dead-space parameter υ.

Walker (1964) took Finkelstein's alternative definition of specific cycle work $Zp_{ref} = W_{cyc}/p_{ref}V_{ref}$ — also a function of the same four parameters. There are various V_{ref} to choose from: Finkelstein had considered (and rejected) V_E, which drops naturally out of his analysis. His objection was that the outcome would be the machine of smallest V_E. Walker opted for $V_{ref} = V_E + V_C$ — a measure of machine volume envelope.

Kirkley (1964) reworked the Schmidt algebra in terms of the kinematics of the parallel, coaxial configuration, i.e., directly in terms of λ and β. The stroke overlap in this 'beta' layout means that envelope volume can be less than the sum of V_E and V_C. The reformulation ought to yield directly the λ and β giving the most compact coaxial Schmidt machine for given choice of p_{ref}. It has not been possible to reproduce Kirkley's graphs of λ_{opt} and β_{opt}, so the follow-up work of Takase (1972) may be more reliable.

Finkelstein identified three options for p_{ref}, viz, p_{min}, p_{max} and p_{mean} — giving three distinct pairs of κ and α which maximise Zp_{ref} for a combination of N_T and υ. Walker's computerised search for the optima chose p_{max}, a pair of values of N_T and υ and plotted the three-dimensional surface of Zp_{rmax} over respective ranges $0 \leq \kappa \leq 5$ and $0 \leq \alpha \leq \pi$.[1]

The performance of other reciprocating engines is in terms of swept volume, V_{sw}. At first sight, Walker's use of $V_E + V_C$ as V_{ref} suggests that he missed an opportunity for a useful comparison. On the other hand, optimisation is mindless in the sense of not distinguishing between maximised work and minimised V_{sw}. The illustrations which follow continue with $V_{tot} = V_E + V_C$ as V_{ref} and with $T_E = 100°C$, $T_C = 25°C$ and using υ estimated for the Vari-Engine.

Figure 22.8 is the result of repeating Walker's exercise. It yields $\kappa = 1.3$ and $\alpha = 98°$ — evidently a long way from confirming the λ and β at which exceptional low-temperature performance was recorded.

---

[1]His axis for α is labelled from 0.0 to 1.0, but 0.0 to 1.0π is evidently intended. Moreover, the 'power' axis deserved the label 'specific cycle work'.

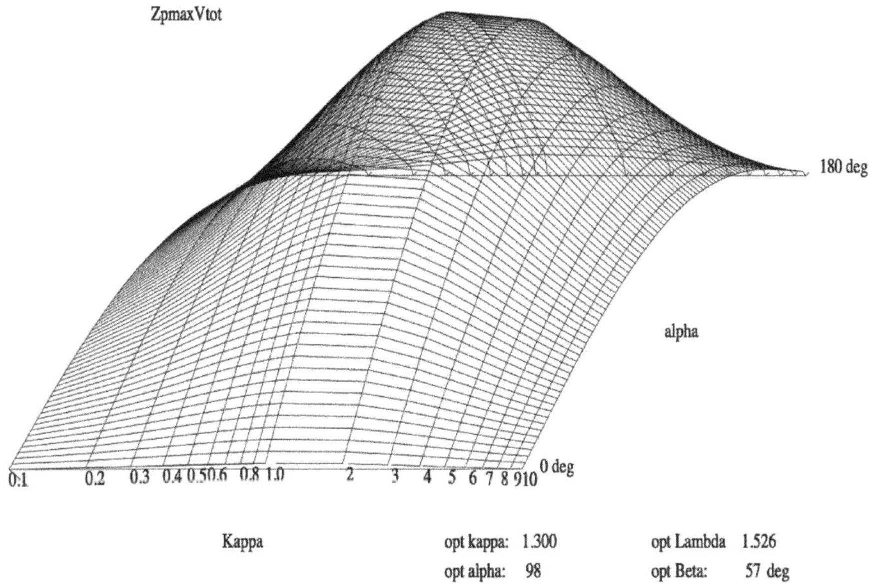

Fig. 22.8. Surface of $Zp_{max}V_{tot}(\kappa, \alpha)$ for $T_E = 100°C$, $T_C = 25°C$. Values of κ and α which maximise Z do not convert the λ_{opt} and β_{opt} of the Vari-Engine.

Use of a different criterion generally points to different optima. Finkelstein's expression for cycle work per unit mass of working fluid is

$$Z_{MRT} = \frac{W}{MRT_C} = \frac{2\pi(1 - 1/N_T)\sin\alpha}{(1/N_T + \kappa + \upsilon)^2} f_m(\rho_{TF}), \qquad (22.14)$$

$$f_m(\rho_{TF}) = \frac{1}{\sqrt{(1-\rho_{TF}^2)}[1+\sqrt{(1-\rho_{TF}^2)}]}. \qquad (22.15a)$$

The peak value of Eq. (22.14) should indicate optimum use of limited working fluid mass, M. Figure 22.9 is the surface of Z_{MRT} corresponding to the earlier exercise.

The result sets theory and experience at odds, and enquiry evidently has to proceed along different lines.

Anyone who has experimented with pressurising a running Stirling engine will know that the best way to stop it dead in its tracks is to unbalance buffer pressure and cycle pressure. The behaviour of the Karabulut

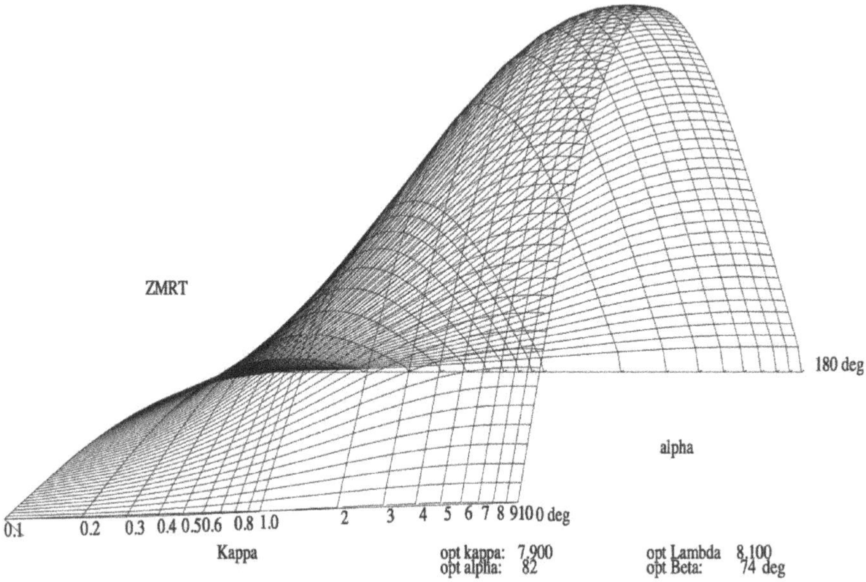

Fig. 22.9. Result of optimising per limited working fluid mass, viz wrt MRT_c. κ and α which maximise Z_{MRT} remain remote from corresponding λ_{opt} and β_{opt} of the Vari-Engine.

and Larque engines suggest that mean cycle pressure remained close to buffer pressure. (In the low-ΔT type, buffer pressure is atmospheric, and it is not difficult to see how mean cycle pressure can settle to that value. What is more, it could turn out that λ_{opt} and β_{opt} are the very values which, for reasons so far unexplored, give rise to just that balance.) Finkelstein's justification for the use of fp_{max} in his pioneer optimisation may not make it the logical choice in the present context (low temperature difference). The alternative based on mean pressure is

$$Z_{\underline{p}} = \frac{W}{\underline{p}(V_E + V_C)} = \frac{\pi(1 - 1/N_T)\sin\alpha}{1/N_T + \kappa + \upsilon} \frac{f_{\underline{p}}(\rho_{TF})}{V_E(1+\kappa)}, \tag{22.15b}$$

$$f_{\underline{p}}(\rho_{TF}) = \frac{1}{1 + \sqrt{(1 - \rho_{TF}^2)}}.$$

The surface of at Fig. 22.10 suggests that $Z_{\underline{p}}$ is not the relevant choice either!

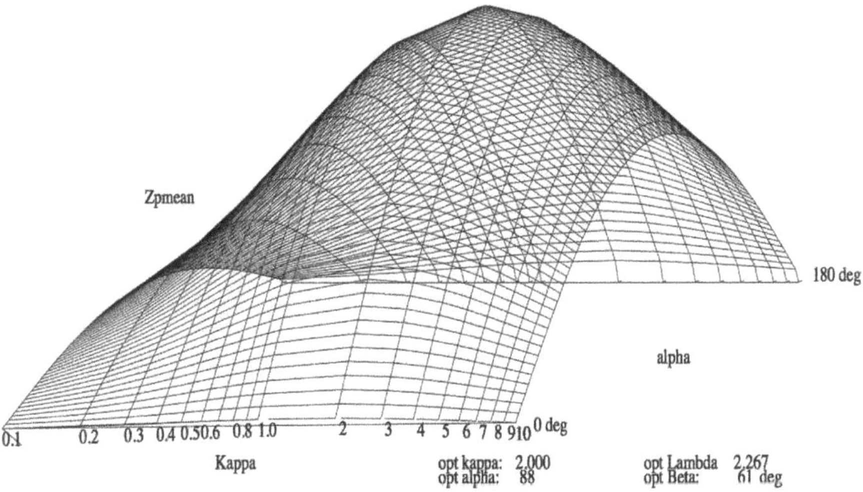

Fig. 22.10. Result of search for κ and α which maximise cycle work per $\underline{p}(V_E + V_C)$, leaving some way to go before the enigma is explained!

22.6. Optimisation per Unit Temperature Difference

What is in short supply is not pressure or working-fluid mass — but *temperature differential*. This suggests a design goal of *maximum work per cycle per given* $T_E - T_C$.

Finkelstein's expression for cycle work per MRT_C lends itself to suitable modification. Numerator and denominator of Eq. (22.14) are multiplied throughout by N_T. Noting that the term $1 - 1/N_T$ can be multiplied by T_E to give temperature difference leads directly to:

$$Z_{MR\Delta T} = \frac{W}{MR\Delta T} = \frac{2\pi (N_T\kappa)\sin\alpha}{(1+(N_T\kappa)+N_T\upsilon)^2} f_m(\rho_{TF}). \qquad (22.16)$$

Both ρ_{TF} and $f_m(\rho_{TF})$ can be evaluated in terms of N_T and κ independently, or in terms of product $N_T\kappa$:

$$\rho_{TF} = \frac{\sqrt{(1+(N_T\kappa)^2+2(N_T\kappa)\cos\alpha)}}{1+(N_T\kappa)+N_T\upsilon}.$$

The outcome is the same, but the latter is preferable because the notation is consistent with that of Eq. (22.16).

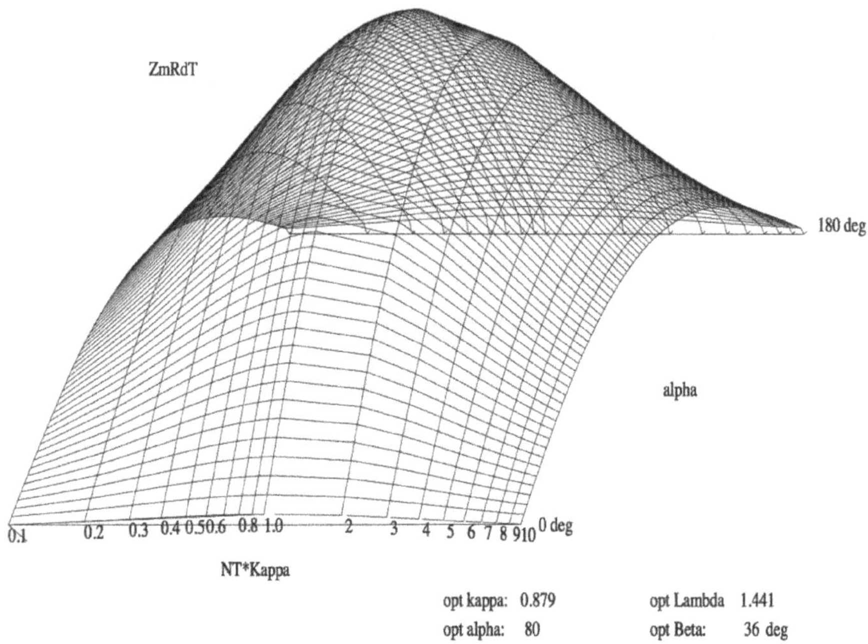

Fig. 22.11. Result of search for α and compound parameter ($N_T\kappa$) which maximise cycle work per limited temperature difference, $MR\Delta T$.

The number of parameters has been reduced by one throughout. If the exercise is promising the possibility arises of displaying the optima of $N_T\kappa$ and α over a range of the compound parameter $N_T\upsilon$ in a single chart.

Figure 22.11 shows the result for a single value (that used for previous plots). The 'optima' obviously have little to do with the experimental counterparts — those of the Vari-Engine or of the low-ΔT *genre*. This may be the point at which to abandon expectation of reconciling practice and elementary theory in the case of the latter — but not before the last ounce of value has been squeezed from its behaviour.

There is not one temperature difference but two: (a) the overall $T_E - T_C$, whose constant numerical value is the ΔT by which performance is characterised and (b) the local instantaneous differences which drive the convective heat exchange in the working spaces. A Contax™ model examined by the writer has $V_{sw} \approx 0.627$ cc and total enclosed volume $\approx$ 28.84 cc. The latter being essentially the volume swept by the styrofoam displacer on the assumption that it is not porous. Provisionally assuming adiabatic conditions, and noting the small numerical values of the ratios involved, $\Delta p/p \approx \gamma V_{sw}/V_{enc} = 0.03$ and $\Delta T/T \approx (\gamma-1)V_{sw}/V_{enc} = 2.61$. At 300 K the latter amounts to a temperature swing of 2.61°.

This not only exceeds the overall ΔT of the best-performing machines — it promises to swamp the temperature differences which determine the direction of heat exchange. That no-load *rpm* are achieved nevertheless suggests that compression and expansion are closer to isothermal than to adiabatic. Aiming for this condition would call for hydraulic radius (large ratio of wetted area to wetted volume) to compensate the small temperature difference. The same consideration applies to the mechanism of heat transfer at the exterior surfaces.

The small values of λ which characterise low-ΔT table-top models cannot be reconciled with an 'isothermal' cycle model, so there can be no expectation that the latter can be a guide to the optimum. From this point, enquires will concentrate on power-producing machines — the Vari-Engine and the Karabulut.

22.7. Optimisation per $\Delta p = p_{max} - p_{min}$

It is a feature of regenerative thermal machines in general (including the regenerative gas turbine) that adiabatic temperature swing restricts operation to characteristically-low pressure ratios.

This suggests a further, alternative picture of the optimum as the combination of κ and α which minimises pressure excursion $p_{max} - p_{min}$.

Ideal pressure ratio p_{max}/p_{min} is retrieved in terms of ρ_{TF}:

$$p_{min}/p_{max} = (1 - \rho_{TF})/(1 + \rho_{TF}).$$

Setting $p_{max} - p_{min} = p_{max}\,[1 - (1 - \rho_{TF})/(1 + \rho_{TF})]$ and simplifying leads to

$$p_{max} - p_{min} = 2p_{max}\rho_{TF}/(1 + \rho_{TF}), \tag{22.17}$$

$$\frac{W}{\Delta p(V_E + V_C)} = \frac{\pi\kappa(1-1/N_T)N_T\sin\alpha}{(1/N_T + \kappa + \upsilon)}\,\frac{(1+\rho_{TF})/\rho_{TF}}{(V_E + V_C)}\,f_{max}(\rho_{TF}), \tag{22.18}$$

$$f_{max}(\rho_{TF}) = \frac{\sqrt{(1-\rho_{TF})}}{\sqrt{(1+\rho_{TF})}[1+\sqrt{(1-\rho_{TF}{}^2)}]}.$$

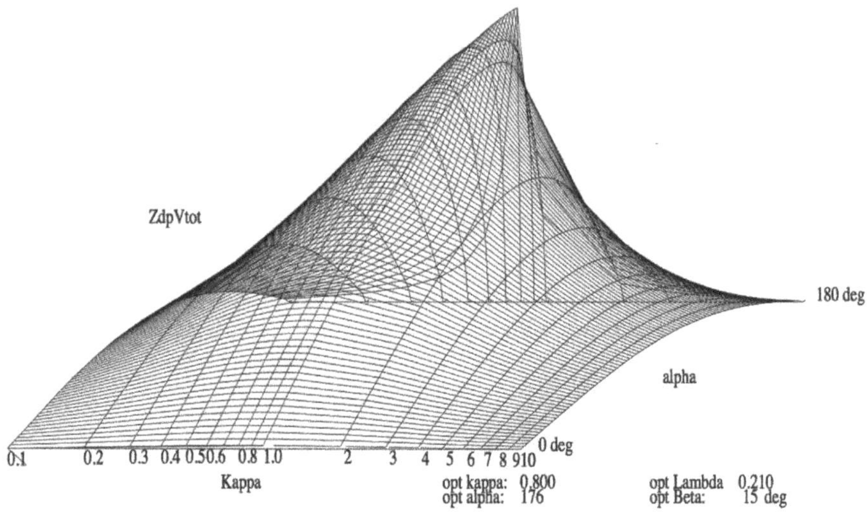

Fig. 22.12. Surface of $Z_{\Delta pVT}$ showing the optima suggested by limiting pressure difference. They still fail to convert to the λ_{opt} and β_{opt} of the Vari-Engine.

The surface of $Z_{\Delta pVT}$ generated by the search is shown at Fig. 22.12. Contrary to expectation, the κ and α giving the maximum fail to convert to anything resembling the λ_{opt} and β_{opt} of the Vari-Engine.

Two conditions for operation at low temperature difference have been suggested — and probed independently. If both are relevant they must be applicable simultaneously. At first sight it is not possible to restrict both differences — Δp and ΔT — without upsetting the dimensional balance of the algebra. On the other hand, temperature *difference* can be limited together with (dimensionless) pressure *ratio*. Likewise, pressure *difference* can feature in combination with temperature *ratio*.

22.8. Optimisation per Δp and per ΔT

The former pairing is achieved on dividing Eq. (22.16) by Eq. (22.17), the latter on dividing Eq. (22.18) by $N_{\text{T}} - 1$. Figure 22.13 is the result corresponding to the former choice.

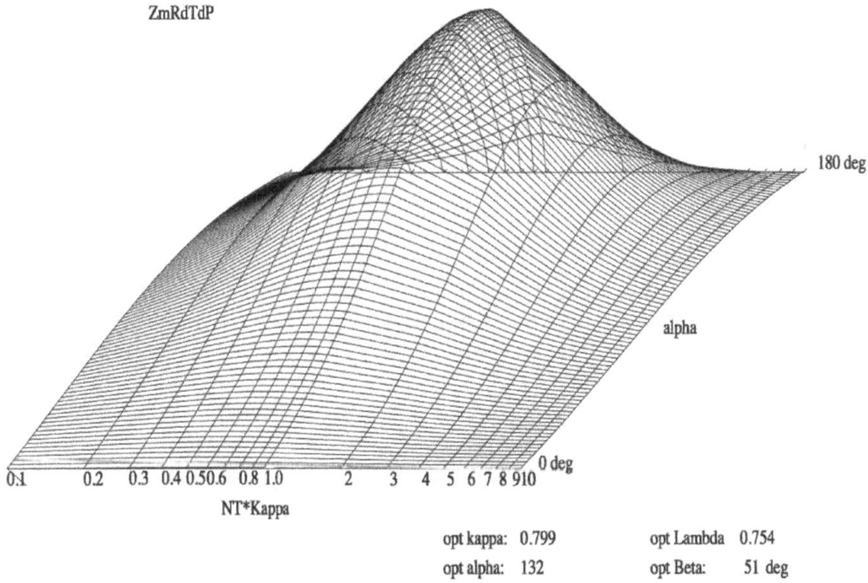

Fig. 22.13. At last! λ_{opt} and β_{opt} back-calculated from $Z_{dpdT}(N_T\kappa,\ \alpha)$ constrained by limiting both differences — pressure and temperature — close to experimental optima.

At last! Predicted optima correspond closely to the experimentally-determined counterparts.

22.9. Non-Sinusoidal Volume Variations

Given the similarities in start-up and run-down behaviour, the Karabulut and Larque engines will make for interesting comparison in terms of their respective volume-crank-angle characteristics.

Figure 22.14 annotates the kinematics of the mechanism of Karabulut *et al.* Much of the notation of the 'max flush' mechanism applies (recall Fig. 10.7).

The algebra arrives at the vertical and horizontal motions of piston and displacer gudgeon pins, starting with the piston:

$$\theta_p = \text{asin}\,[(\varepsilon - X_p)/l_p],$$
$$y_p = y_{pin} + l_p \cos\theta_p. \tag{22.19}$$

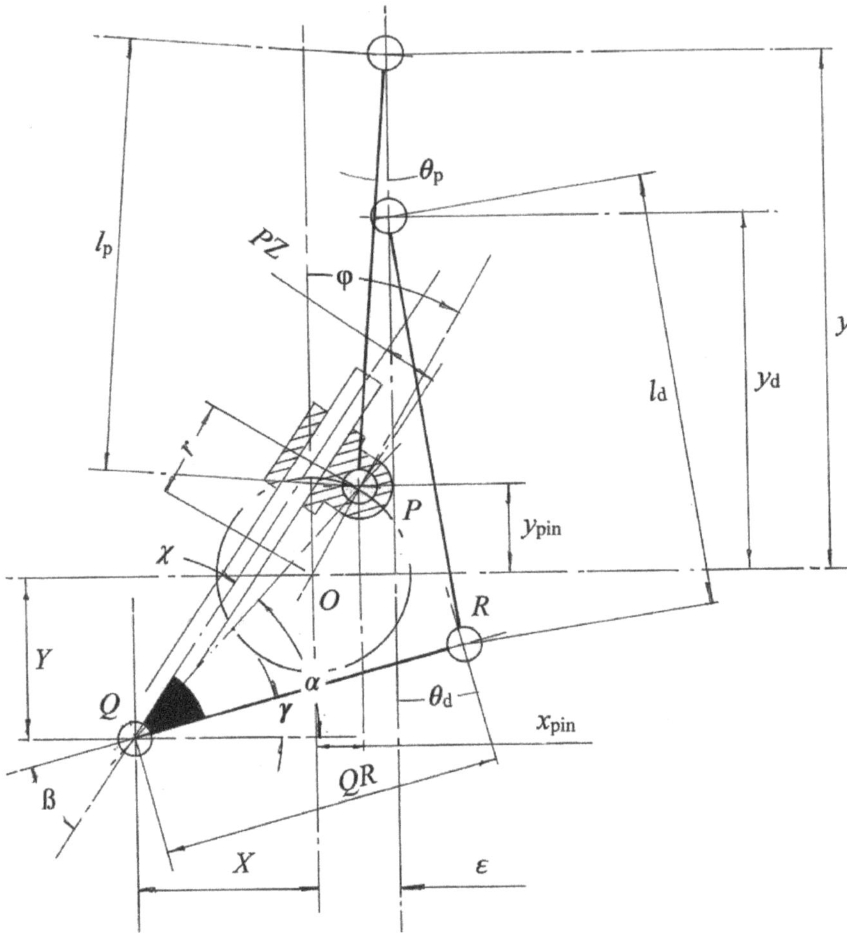

Fig. 22.14. Notation for analysis of the Karabulut slider/crank displacer drive.

Variable angle α follows in terms of the instantaneous position of the crank-pin. Noting that, with origin 0 at the crank-shaft centre-line as shown, distances X and Y negative:

$$\alpha = \operatorname{atan}\frac{y_{pin} - Y}{x_{pin} - X},$$

$$QP = \sqrt{[(x_{\text{pin}} - X)^2 + (y_{\text{pin}} - Y)^2]},$$

$$\chi = \text{a sin } (PZ/QP).$$

Angle β is set by the designer:

$$\gamma = \alpha + \chi - \beta.$$

The instantaneous coordinates of point R are as follows:

$$y_R = Y + QR \sin \gamma,$$

$$x_R = X + QR \cos \gamma.$$

Angularity of the displacer connecting rod θ_d is

$$\theta_d = \text{asin } [(x_R - \varepsilon)/l_d].$$

Finally,

$$y_d = y_{\text{pin}} + l_d \cos \theta_d.$$

Figure 22.15 (upper) is the skeleton motion diagram. The traces at Fig. 22.15 (lower) show relative vertical position of piston and displacer as it varies with crank angle. Also visible are five stages of a mechanised optimisation search. Exact linear dimensions and angles of the Karabulut machine are not known, but values scaled from the original schematic diagram suggest volume-crank-angle history more in common with that of the 'max-flush' mechanism (Fig. 12.2) than with that of the Vari-Engine (Fig. 22.4). Particularly noteworthy is the displacer's rapid return to upper dead-centre following expansion, expelling all but a small fraction of the working fluid from the expansion space in preparation for the compression phase.

Compression-space volume (indicated by the vertical distance between traces) is also minimal over much of the expansion stroke — consistent with the intuitive criterion for optimum working-fluid utilisation.

The ideal indicator diagrams of Fig. 22.16 are constructed for T_E = 75°C, T_C = 25°C. (Dead space ratio is an estimate.)

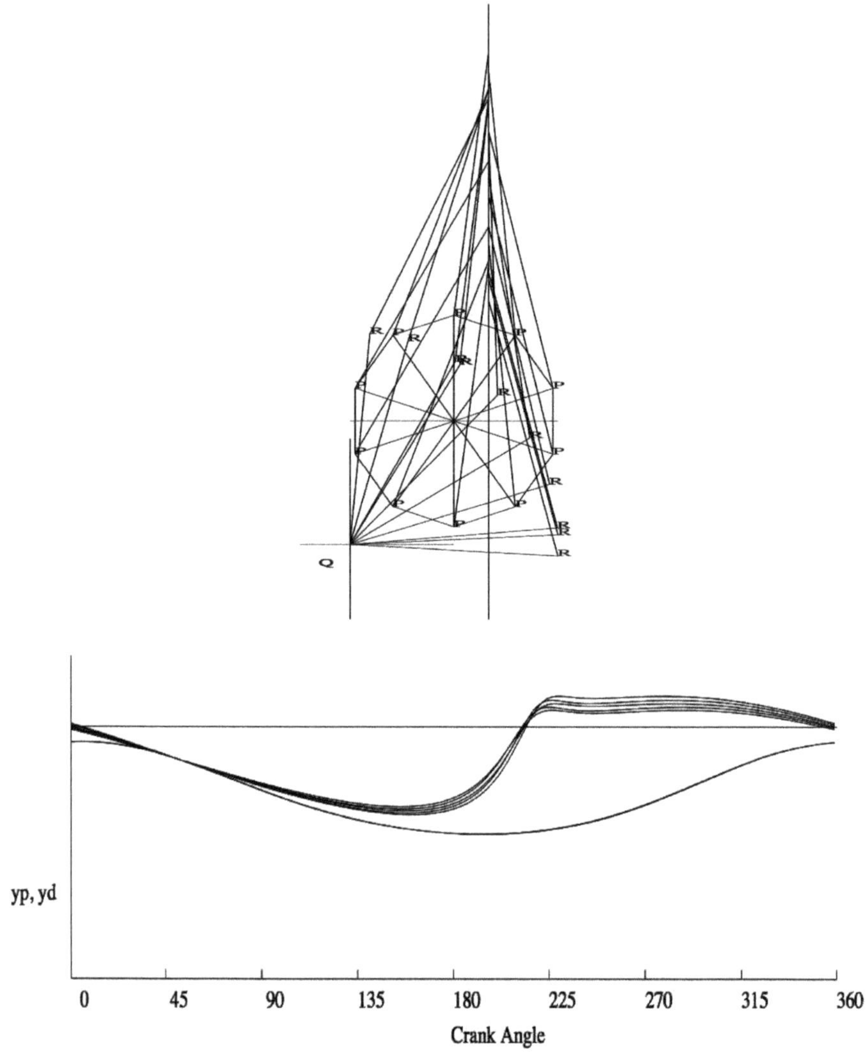

Fig. 22.15. Skeleton kinematic diagram for the Karabulut mechanism (upper) and traces of displacer and piston motion (lower).

22.10. Summing-Up

At first sight the enquiry has an inconclusive outcome. Under consideration is the exceptional — if not unique — low-temperature performance of two independently designed and built Stirling engines. Each embodies

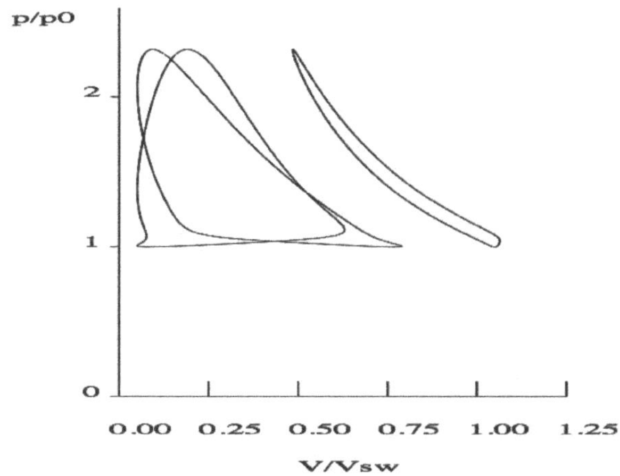

Fig. 22.16. Ideal indicator diagrams for Karabulut engine computed for $T_E = 75°C$ and $T_C = 25°C$.

a novel drive mechanism affording considerable kinematic flexibility, each allowing experimental determination of the volume variations giving best low-temperature performance.

Those observations have not, however, been consistently reconciled with any theoretical performance criterion: volume variations calculated from the kinematics of the Karabulut mechanism go some way to satisfying the intuitive requirement for effective working-fluid utilisation. This is a characteristic they share with the 'max-flush' mechanism designed for that specific purpose.

Fed with the data notified by Larque ($\lambda_{opt} = 0.75$ and $\beta_{opt} = 50°$.) the kinematic model of the Vari-Engine generates volume variations which show no distinguishing features.

Before giving up, consider Fig. 22.17, showing the motion diagram re-calculated with direction of rotation reversed and with λ reset to 0.85 and β to $-5°$. (That's right, *minus* 5!) The traces now have features in common with those of both the 'max-flush' and Karabulut kinematics: All but an insignificant fraction of working fluid is duly excluded from the compression space during 100 degrees-worth of the expansion phase.

This is followed by accelerated return of the displacer to upper dead-centre ahead of the compression stroke. The corresponding ideal indicator diagrams of Fig. 22.18 calculated for $T_E = 75°C$ and $T_C = 25°C$ suggest useful net indicated ideal work.

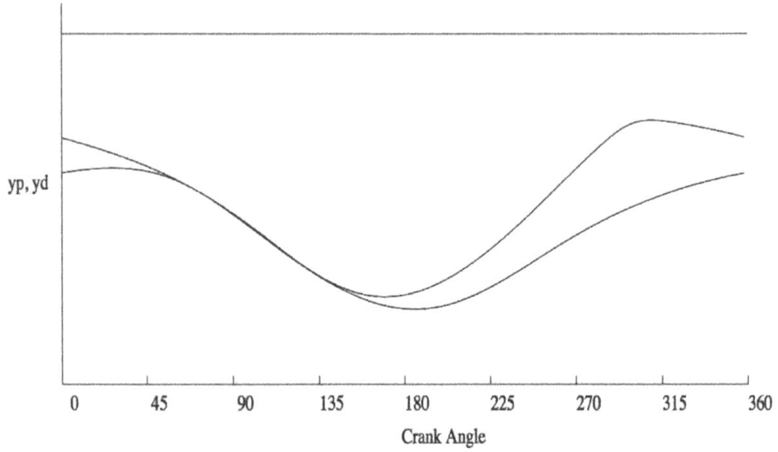

Fig. 22.17. Traces of piston and displacer motions of Vari-Engine re-computed for $\lambda = 0.85$ and $\beta = -5°C$ suggesting improved working fluid utilisation relative to the optima ($\lambda_{opt} = 0.75$ and $\beta_{opt} = 50°C$) notified by Larque.

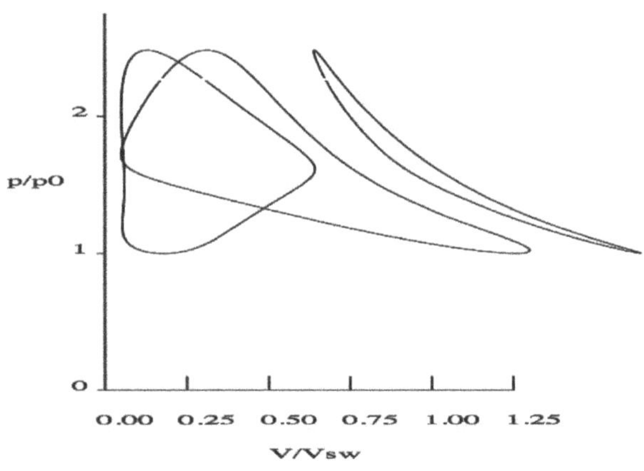

Fig. 22.18. Ideal indicator diagrams for Vari-Engine re-computed for $\lambda = 0.85$ and $\beta = -5°C$ $T_E = 75°C$ and $T_C = 25°C$.

It may be that the only certainty is in terms of the pitfalls which arise when theoretical work is applied to retrospectively to accounts of experimental work — and from a distance: it is no longer possible to liaise with Larque over the direction of rotation of the Vari-Engine.

A facetious claim has it that, if engineering equals physics plus common sense, then physics must equal engineering minus common sense. On a similar basis, Stirling Engine Science minus common sense is Schmidt — and a treacherous and potentially expensive way to do business.

Any cycle analysis (other than the 'ideal' — 2 – constant-volume, 2 – isothermal) has heat-flux reversing once per cycle in each exchanger. Any model other than the pure-adiabatic sees likewise in each cylinder. This may be why it is so difficult to anticipate by intuition alone the *net* effect over a cycle — or, indeed, by intuition 'aided' by Schmidt.

But a common sense item which can be added to Schmidt is the view that anything less than infinite heat-transfer coefficient adds an element of adiabatic temperature swing (attributable in turn to pressure swing). All other things being equal, that is an extra load on a real exchanger. Common sense says that a configuration in which pressure-swing is minimal is capable of operation under conditions more nearly isothermal (or less adiabatic!). The context of the optimisation exercise in question was low-ΔT operation — which common sense suggests is most vulnerable to temperature excursions beyond T_E and T_C. The way I see things, optimising per ΔT and per pressure-swing builds into Schmidt a bit of the common sense missing from its idealised assumptions.

23

A Cautionary Tale*

23.1. Background

The temperature limits T_E and T_C routinely used for performance predictions are those of the heat exchangers. Intentionally or otherwise, the result flatters an engine whose source of heat, like that of the internal combustion engine, is at 2000°C or thereabouts. This focuses attention on *overall exchanger temperature drop* ΔT, and the fact that, at some 1400°C, ΔT_E vastly exceeds ΔT_C (30–50°C).

The *NTU* concept ($NTU = StL/r_h$) identifies the *exterior* of the expansion exchanger as the thermal bottleneck. For the multi-tube exchanger of the GPU-3, where combustion gas flow is perpendicular to a single bank of tubes, exterior L/r_h is about 10, whereas internally it is about 160.

A project was undertaken which, had it merited a name, would have been called *the search for a balanced design*. A cycle simulation in use at the time used Availability Theory to calculate net performance by subtracting *lost available work* from an ideal lossless cycle. For the tentative engine specification, this suggested an additional imbalance: pumping

---

*The chapter title is borrowed from a valedictory, wistful account by Professor S. A. Tobias. He had committed the overwhelming majority of his research facilities and £millions of ministry of Technology funding to the development of high-speed metal-working (or High Energy-rate Forming — HERF). The British manufacturing industry was already in decline under pressure from abroad, the effect being felt particularly in the machine-tool industry. Tobias could have been kinder to himself: he was not to know that his ambitious and enterprising effort would be insufficient to turn the situation around.

loss component (both exchangers) was a small fraction of loss[1] due to heat transfer restrictions.

Hardware design was approached 'from the inside out', i.e., by specifying *NTU* and d*p*/*p* requirements and calculating back to exchanger geometry.

The process led to an expansion exchanger comprising a bank of intercalated, parallel, spiral-wound tubes which promised to relax a number of limitations of the more conventional arrangement. Internally, intense secondary flows would be generated — a phenomenon which had attracted the writer's attention decades earlier in dealings with colleague Ewart Brundrett (Brundrett and Baines, 1964) at the University of Toronto. There would be an accompanying improvement in heat transfer intensity to the external surfaces achieved by forcing the combustion gases along the axis of the tube bank rather than across as in the majority of designs. (Hargreaves reveals that the usual single or double pass perpendicular to the tube bank is barely adequate — even with tubes closely-spaced and with combustion products forced through under pressure.) Finally, the helical configuration promised to absorb the thermal expansion which otherwise threatened bending stresses in the V-structure of the engine.

The heat transfer provision at the compression end involved tapered fins generated by wire erosion integral with the piston crown. At inner dead-centre these engaged with matching fins on the under-side of the water-cooled heat exchanger, all but eliminating unswept volume at this crucial point in the cycle.

Details of the regenerator design would be particularly pertinent here. Unfortunately, all relevant records and accompanying cycle simulations were lost when both hard-drives of two, independent, low-mileage Mac computers crashed irretrievably in quick succession. (The incidents taught an invaluable lesson: do not use two machines for mutual back-up!)

An exercise in the application of Dynamic Similarity principles suggested that achieving the objective was not dependent on choice of working fluid. On paper the concept benefited from the compression ratio given by a volume ratio of 2½:1 and air/N2 as working fluid. A net brake

---

[1]NB: This was *internal* loss: had it taken account of the temperature gradient over the *external* surface, the loss in the expansion exchanger in particular would have been orders-of-magnitude times pumping loss.

output of 1 kW at 1500 rpm appeared within reach. The project was accordingly labelled the *mRT*-1k.

There was no immediate prospect of proceeding to construction, but a fortuitous situation encouraged completion of a set of dimensioned, toleranced, orthographic detail drawings. The catalyst was an interest shown by colleague Duncan James[2] in using the design to exercise his skills in the use of proprietary 3D CAD packages.

The concept attracted the attention of Sustainable Energy Systems (SES) Ltd which had a history of involvement in Stirling engines, and which was providing consultancy services to an Italian organisation, REET (Renewable Energies and Environmental Technologies). It was agreed that SES would purchase the *mRT*-1k in the form of the original orthographic drawings for supply to REET on terms agreed between them.

The responsibility of supplying a viable design focused attention on a matter still outstanding: verifying the ability of the expansion exchanger to counteract the tendency for increased adiabatic temperature swing called for flow-friction and heat transfer data specific to helical-wound tubes. When it became clear that the search was unlikely to succeed, the writer got cold feet! The drawings were supplemented by Design Notes recommending a return to a volume ratio V_C/V_E of unity. REET proposed to take over by bypassing the files generated by CADStudios and to use their in-house CAD facilities to generate a new set of machining files directly from the writers orthographic drawings — ideal timing (one might have thought) for implementing reversion to the revised volume ratio. The recommendation to return to κ = unity was reinforced twice via SES before REET began work.

It was therefore disconcerting to learn somewhat later that CAD modelling had been completed retaining the 2½:1 volume ratio. Later still came news that the helical-tube expansion exchanger had been abandoned — together with any hope of exploring the original concept.[3]

---

[2]Now Director of a consultancy flourishing under the name CADStudios http://www.cadstudios.co.uk/.

[3]CADStudios continued to supply invaluable input to the original paper study. Applying a finite-element stress-analysis package revealed worrying stress concentrations at the transitions between the connecting rod shanks and the big-end eyes. REET were notified via SES, and it is understood that the recommended change of material for the components was implemented.

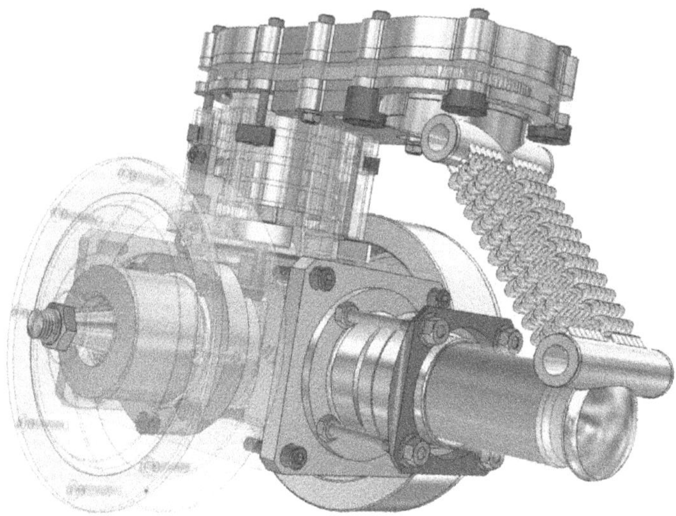

Fig. 23.1. AutoDesk model of *mRT*-1k generated by Duncan James with selected features in 'ghost' view. Reproduced by permission of CADStudios.

Figure 23.1 is a 'ghost' view generated by CADStudios with their blessing to reproduce. The design is somewhat dominated by the twin flywheels specified to cope with the anticipated cycle pressure swing at the relatively low rpm of operation on air. Figure 23.2 is a photograph of the machine as modified and assembled at REET. Machining and fabrication appear to be of exemplary standard.

The last information to reach the writer was to the effect that the *mRT*-1k had functioned as a refrigerator (as many otherwise unsuccessful projects do!) and done so with impressively little vibration but that, in prime-mover mode, it had yet to demonstrate that it was capable of producing power.

23.2. A Lesson Learned

Despite the fact that machining files had been generated contrary to contractual agreements, the outcome of the *mRT*-1k project will no doubt be attributed to a flawed concept. So near and yet so far! There is (to the knowledge of this writer) no 'honest' figure for the efficiency the Stirling engine based on *actual* source temperature. Had the *mRT*-1k been

Fig. 23.2. *mRT*-1k as assembled by REET showing fabrication of unknown flow-passage geometry substituting the helical-tube expansion exchanger of the original concept.

constructed to specification (incorporating either of the volume ratios) and produced sufficient power to have merited instrumenting, the overall ΔT_E of the helical expansion exchanger could have offered unique insight into potential.

The writer hopes to attract interest in constructing apparatus to vindicate claims made in this book regarding working fluid distribution — but will do so on this occasion bearing in mind that a Stirling engine is a Stirling engine! It is characteristic of — and possibly unique to — the type that events in the expansion space instantaneously affect those in the compression space — indeed, every location in the gas path. Achieving any potential benefit from drastically-altered volume variations calls for a number of digressions from the norm. These *interact*.

Venturing from the mainstream a second time means going out on a limb: To paraphrase an observation by former UK politician 'Rab' Butler: '*Engineering is the art of the possible.*' There is no such thing as a promising variant of the Stirling principle until it is delivering useful power (think of the ultimate in thermal efficiency — the hopelessly impotent Carnot 'engine').

24

Towards the Definitive Gas Process Model

24.1. For and Against

A cycle model can be formulated so as to serve one of two purposes:

(a) to characterise the *genre*;
(b) to model a specific machine in fine detail.

The better the formulation serves *b*, the less likely it is to fulfil the purpose of a generic model — and *vice versa*.

At first sight, requirement (a) ought to call for no more than a one-dimensional ('slab'-flow) formulation. However, such simulations rely on heat-transfer and flow-friction 'correlations' which, as earlier chapters have amply demonstrated, are treacherous at the modest *Ma* arising at the performance boundaries of air- and N2-charged machines. If the correlations do not characterise the exchange processes, then a simulation embodying them can hardly characterise the *genre*. A rethink is called for.

Achieving more accurate representation amounts to a choice between two contrasting approaches: (a) solution of the full Navier–Stokes equations in three dimensions, or (b) modelling in terms of the kinetic theory of gases. In his salutary *Think before you compute*, Hinch (2020) recommends leaving the coding of Navier–Stokes to the professionals, who already offer software packages such as Fluent™, marketed as extensively-tested and ready-to-go. Hinch justifies in terms of the challenges and pitfalls encountered in discretisation. But these confront not only the individual researcher, but also the software house in equal measure. A fixed-grid scheme imposed on a turbulent fluid involves a measure of compromise. CPU time to cyclic equilibrium for a detailed model of gas-path geometry is measured in hours rather than seconds. Errors accumulated over that time will probably never be known. So, while proprietary packages generate impressive, colour coded animated images, the independent-minded researcher may still prefer a formulation over which he/she has a measure of control.

The Lagrange approach, which came to the rescue in earlier chapters does not work in three dimensions (or, indeed, in two). This refocuses attention on a 'first-principles' cycle model based on the kinetic theory of gases introduced in an earlier account (Organ, 2014). This also admitted how far there is to go in terms of storage and processor speed before such a model becomes usable. However, there have been promising developments since:

24.2. Quantum Computing

What follows reproduces the essence of an announcement by the Bloomberg Media Group:

Chinese scientists claim to have built a quantum computer that is able to perform certain computations nearly 100 trillion times faster than the world's most advanced super-computer. That's exponentially (sic) *faster than existing supercomputers. While still in its infancy, quantum computing is seen as the key to radically improving the processing speed and power of computers, enabling them to simulate large systems and drive advances in physics, chemistry and other fields.*

Chinese researchers are competing against major U.S. corporations from Alphabet Inc.'s Google to Amazon.com Inc. and Microsoft Corp. for a lead in the technology. Google said last year it has built a computer that could perform a computation in 200 seconds that would take the fastest supercomputers about 10,000 years, reaching quantum supremacy. The Chinese researchers claim their new prototype is able to process 10 billion times faster than Google's prototype. (Bloomberg 06 December 2020)

Progress of this magnitude is difficult to grasp. Indeed, a sceptic might draw attention to the caveat '*certain* computations'. Moreover, the Bloomberg piece makes no explicit mention of storage capacity nor of speed of access: for the orders-of-magnitude increase in processor speed to be harnessable, these aspects would have to keep in step.

This writer first encountered the digital computer (the laboratory-sized IBM 7090) in 1962, and has made uninterrupted use of it since. Over that time there has been the impression that advances achieved have surpassed, rather than fallen short of predictions. With this as justification, the earlier gas-kinetic model is pursued in terms of an improvement to the earlier picture of molecule/molecule collision.

24.3. Recapitulation

One cubic centimetre of gas at 0°C and atmospheric pressure holds about 2.685×10^{19} molecules (Jeans, 1967). The number is way beyond the memory capacity of current computers — but this is not the point: every phenomenon inherent in the model may be modelled, developed and tested using a far smaller number — collision with stationary or moving containment, molecular interaction, etc. Then, if/when quantum computing becomes reality, all that will be necessary is to change the scope of array declaration(s) and the input data read by the program. Moreover, if the model behaves convincingly when applied to a ruthlessly-basic gas-path (e.g., that of the coaxial 'hot-air' engine) then extending to a more sophisticated configuration means simply elaborating the data defining the gas-path outline.

From the gas-process modelling perspective there is no way that the 'hot-air' engine can be called a 'simplified' Stirling engine. Gas flowing to and from the variable-volume spaces via the annular gap is subject to a component of cyclically-reversing shear between stationary enclosure and moving displacer. Except at two instants per cycle there is a temperature difference across the flow. This changes sign and magnitude twice per cycle.

There is no relevant heat-transfer or flow 'correlation' — which in any case defies description in terms of one-dimensional or 'slab' flow. Under the assumption of laminar flow, an able analyst might come up with a mathematical model, although the general case promises to be pretty daunting.

An attraction of the gas-kinetic treatment is that it dispenses with the need to think laminar vs turbulent: molecules collide with each other and/or with the enclosure — and that's it!

It is also an attractive challenge in coding: it presents at two, separate levels which can be tackled independently, neither of which is particularly daunting when dealt with in isolation. One level is bookkeeping — keeping tabs on a manageable number of specimen particles — instantaneous *x, y* and *z* coordinates and corresponding velocity components, *u*, *v* and *w*. Crude collision models serve to get things up-and-running, one model for contact with the enclosure, one for molecule-to-molecule impact. The latter also calls for a 'hit-or-miss' criterion.

A decision is necessary as to whether to formulate in two dimensions or in three. Jeans (1967) offers a manageable algebra for collision in three-dimensional space between a pair of 'rigid, elastic' spheres (the 'crude' collision model). Problems loom when upgrading this to, say, an attraction/repulsion model such as the Lennard-Jones.

The complication evaporates somewhat if the containment induces axisymmetric flow. Figure 24.1 shows a coaxial 'hot-air' engine schematically in cross-section — a configuration which promises just this.

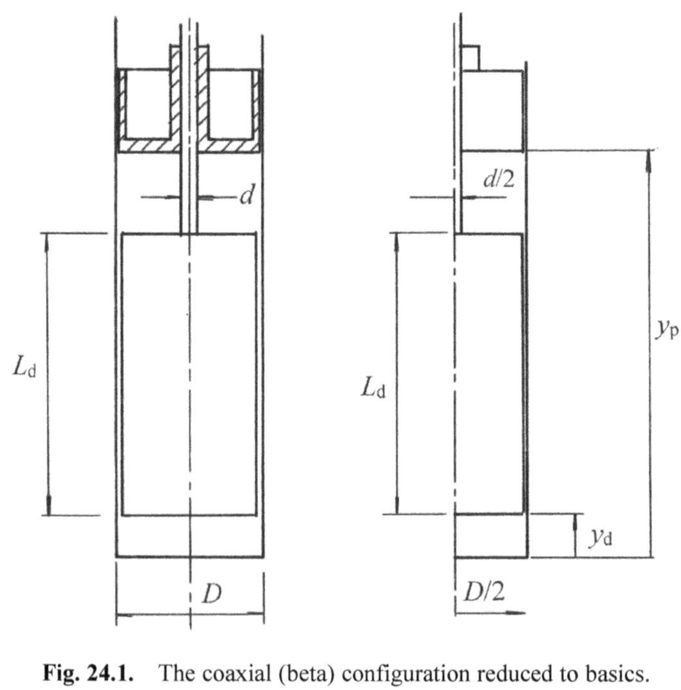

Fig. 24.1. The coaxial (beta) configuration reduced to basics.

With no net flow circumferentially it can be argued that a slice through the vertical axis characterises the three-dimensional case. There is then no loss of generality in considering one symmetrical half (image on right-hand side).

The earlier formulation (this author, 2014) took this case to exercise the bookkeeping (track-and-trace) aspect of coding. The combination of hardware and software available then available allowed on-screen animated display in real time as solution proceeded (the real time of the solution — not that of any 'real' assembly of molecules). A display which is almost as informative is in terms of streaks representing the paths of molecules over the (scaled) time of the integration step. Many of the assumptions of the earlier model are reused. All are over-simplifications, justified at this stage on the basis that, if they allow the nuts-and-bolts of the simulation to be got up-and-running and tested, then they can be relaxed one-by-one at leisure. Rigid-body impact will be discarded first in favor of a more realistic model — that of attraction-repulsion. The updated list follows.

24.4. Assumptions

A molecule striking a wall at angle of incidence rebounds at incidence angle $\pi - \theta$. (Jeans offers an improved model derived from viscosity considerations.)

Rebound speed is determined in terms of local wall temperature (and of the instantaneous velocity of the enclosure if the contact point is in motion).

Collision mechanics is in terms of Lennard-Jones' attraction/repulsion model (1924).

A collision, together with preceding and subsequent events, is confined to two molecules: the 'three-body problem' does not arise.

A molecule which collides with the enclosure does not also collide with another particle during the same integration step.

Integration time-step can be chosen in relation to molecular speed so that a molecule experiences a maximum of a single collision during the interval.

24.5. Boundary and Initial Conditions

These are not the variable volumes of the traditional cycle simulation, but instantaneous locations and speeds of various faces of the enclosure (Fig. 24.1). There is no restriction on the variation with time (or crank angle ϕ). Simple-harmonic motion offers a realistic case. Instantaneous vertical location of the lowermost face of the piston can be written $y_p = \frac{1}{2}S_d[1 + \cos(\phi - \beta)] + L_d + overlap$, in which S_d is the stroke of the displacer and L_d its axial length. Instantaneous velocity is evidently $-\frac{1}{2}S_d\sin(\phi - \beta)$. Likewise, instantaneous location of upper and lower faces of the displacer can be written in terms of $y_d = \frac{1}{2}S_d(1 + \cos\phi) + L_d$ or +0 plus a constant to set required piston/displacer clearance.

Enclosure temperature varies from T_E at the lowermost extent to T_C at the upper limit of piston travel, and between the same limits over L_d of the displacer shell. Provisionally the variation is linear in both cases, but the restriction is readily relaxed with no change to the nuts-and-bolts of the solution.

Functioning of the model should be independent of number n of molecules, allowing it to be explored and verified with n limited to a few thousands. Vacant spaces are populated with molecules set on a uniform rectangular grid. Initial particle speed $|\underline{V}|$ is set according to absolute temperature using the Maxwell–Boltzmann distribution:

$$|\underline{V}| = \sqrt{(8V^2/3\pi)}, \tag{24.1}$$

in which

$\underline{V}^2 = 3\ kT/m$,
T = absolute temperature [K],
k = Boltzmann's constant ($\approx 1.38 \times 10^{-23}$ J/K),
m = molecular mass.

For each molecule, an initial x-component u of V is allocated randomly such that $\sqrt{(u^2 + v^2)} = |V|$, where v is the corresponding initial y component. The latter is back-calculated as $v = \sqrt{(|V|^2 - v^2)}$. Molecules j thus head off at respective speeds $|\underline{V}|_j$ but each in a different direction.

24.6. Lennard-Jones Potential V_{L-J}

A familiar form is

$$V_{L-J} = 4\varepsilon\,[(\sigma/r)^{12} - (\sigma/r)^{6})]. \tag{24.2a}$$

In Eq. (24.2), r is inter-molecular distance [m], σ is the 'size' of the particle [m] (distance at which the particle-particle energy V is zero) and ε is the depth of the potential well [J] (also called the dispersion energy). For purposes of repetitive computation, first evaluate $x = (\sigma/r)^6$, then

$$V_{L-J} = 4\varepsilon x[x - 1]. \tag{24.2b}$$

By definition, force F_{L-J} along an axis connecting the isolated pair of particles is given by $F_{L-J} = \partial V/\partial r = \partial V/\partial x.\partial x/\partial r$:

$$F_{L-J} = 24\varepsilon/r[-2(\sigma/r)^{12} + (\sigma/r)^{6})]. \tag{24.3}$$

The upper curve of Fig. 24.2 is the conventional display of Eq. (24.2), i.e., normalised to show V_{L-J}/ε vs r/σ. The lower curve is the corresponding relationship between F_{L-J} and r/σ (Eq. (24.3)).

24.7. Specimen Collisions

The earlier collision model envisaged perfectly elastic collisions in the 'billiard-ball' tradition, with change of direction at the instant of impact. Under the Lennard-Jones model, molecules on conflicting courses might be thought of as avoiding collision by 'swerving' to varying degrees rather than impacting.

The three illustrations immediately following accompany a pair of atoms (or molecules), j and k in sequences of 'stills' at uniform increments of time. In all three examples molecule j has horizontal velocity component u_j positive, i.e., it approaches from left-to-right. The corresponding component u_k of molecule k is opposite in sign, i.e., it always sets off right-to-left. In all cases, vertical components v_j and v_k are zero before proximity induces the 'swerve'.

For Fig. 24.3(a), the particles approach with a small vertical displacement between respective horizontal trajectories. Approach speed of particle j is ten times higher than that of particle k, which it pushes it back

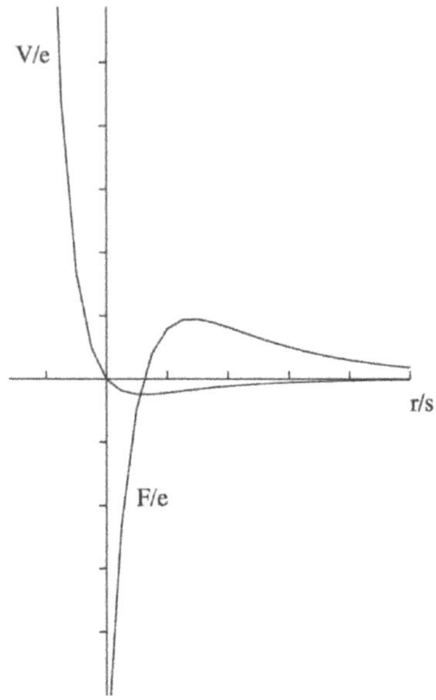

Fig. 24.2. Lennard-Jones potential V_{L-J}/ε as a function of r/σ (upper curve) and corresponding Lennard-Jones force F_{L-J}/ε (lower).

on its tracks while suffering only slight downwards deflection from its track rightward.

In Fig. 24.3(b) approach velocities are equal and opposite. Vertical displacement between horizontal tracks has been increased somewhat — but remains less than notional molecular diameter. The encounter reverses both velocity components u_j and u_k and the initial vertical offset results in equal and opposite components v_j and v_k of vertical velocity.

In Fig. 24.3(c) initial vertical offset is increased to a value greater than particle radius. With initial u_j and u_k equal and opposite, impact is avoided by swerving. Equal and opposite components v_j and v_k of vertical velocity are acquired as before, but there is no reversal of either x-component.

24.8. Overall Computational Economy

Each integration step has three tasks (a) to calculate the distance and direction travelled by each molecule assuming no obstruction and (b) to inspect for a potential collision (i) with the enclosure or (ii) with another

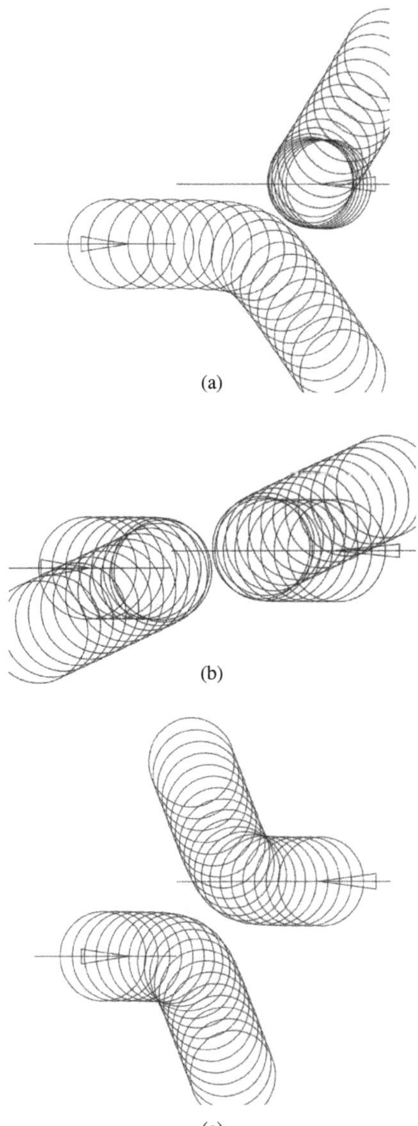

Fig. 24.3. Specimen encounters under the Lennard-Jones attraction/repulsion model. (a) Speed of molecule j much greater than that of molecule k. Latter repulsed backwards by the encounter. (b) Initial speed of molecule j and that of molecule k identical. Vertical offset of initial tracks increased. Mutual repulsion of both molecules into reverse x-direction. (c) Approach velocities u_j and u_k equal and opposite. Molecules deflected off course by equal and opposite amounts.

molecule. Were it not for the possibility of (ii), there would be n calculations per step. The number is actually far greater, since each molecule has to be checked for proximity to *every one of the other n – 1 molecules individually*!

This reveals a fundamental distinction between visual observation and a mechanised scan: it is obvious at a glance that, if molecule j currently occupies the expansion space it is not a candidate for collision with molecule k currently in the compression space. The mechanised scan has to arrive at the judgement via a calculation of the distance. At first sight the number of calculations is thus $n(n - 1)$. On this basis, possibility (ii) is responsible for the heaviest computational overhead.

It is probable that a sharper intellect than that of this writer could devise — or has already devised — an algorithm which short-circuits the tedium. Options so far explored have generally been at the cost of additional bookkeeping. All except one, that is as follows:

Assume that the initial scan of the integration step focuses on molecule 1 and calculates distances to molecules 2, 3,..., n. The following scan pivots on molecule 2 and calculates distances to molecules 1, 3, 4,..., n. However, distance between 2 and 1 has already been inspected as reciprocal distance from 1 to 2, eliminating the need for a repeat. By the time molecule 3 is the pivot, distances 3-1 and 3-2 have already been checked. It follows that the scan gets shorter as it proceeds, and can be coded as:

```
do i = 1, n
    do j = i, n
            dist = sqrt((x(i) – x(j))**2 + (y(i) – y(j))**2)
            if(dist < test) then
                    (implement collision model)
            endif
    end do
end do
```

(It goes without saying that coding is in double-precision, calling for DSQRT to be specified explicitly in FORTRAN.)

The number of inspections has been reduced to $n(n - 1) - \sum n - j$ where the sum is over values of j from 1 to $n - 1$. This is closer to a 50% reduction than to the much-needed order-of-magnitude.

24.9. Specimen Particle Tracks

Figure 24.4 is the computer-generated equivalent of the symmetrical half of the schematic 'hot-air' engine of Fig. 24.1. A streak joins the locations of a particle at the start and end of a time step in the integration process. Display and formulation relate as follows:

(1) The speed of a particle — and thus streak length to scale — is a function of absolute temperature. In the orientation of Figs. 24.1 and 24.4, the upper part of the machine is at T_C, and streak length short. Fluid temperature increases downwards — as does streak length.
(2) A pair of tracks which cross suggests a collision. For Fig. 24.4 slightly fewer than 5000 molecules are tracked — but not a single collision

Fig. 24.4. Symmetrical right-hand half of hot-air engine corresponding to earlier Fig. 24.1 (right-hand figure) showing (linear) particle tracks over a representative integration interval. Note short track length in low-temperature (upper) region relative to that of higher-energy particles in heated region.

Fig. 24.5. 'Over-driven' case (unrealistically-high rpm), captured at the instant with displacer moving downwards at peak cyclic speed. Molecular 'cloud' in compression (upper) space complies with two intuitive expectations (a) pressure and rarefaction waves marked by heavy and light particle concentrations respectively, and (b) 'jetting' of flow entering upwards from the annulus, unable to adjust to the speed of descent of the displacer and outlining a region of low pressure (sparsely-populated).

detected. This is because none of the points of intersection displayed was reached *simultaneously*[1] by both particles.

Figure 24.5 may be familiar from the 2014 account, having been generated using different software and hardware. It is for an 'over-driven' case (unrealistically-high rpm), captured at the instant when the displacer was moving downwards at the peak of its cyclic speed. The molecular 'cloud' in the compression (upper) space complies with two intuitive expectations (a) pressure and rarefaction waves marked by heavy and

[1]That is, within attraction/repulsion range.

light particle concentrations respectively, and (b) 'jetting' of flow entering upwards from the annulus, unable to adjust to the speed of descent of the displacer and outlining a region of low pressure (sparsely-populated).

24.10. Further Computational Considerations

Coding the earlier collision model had involved 10 calls to trigonometric functions and 3 calls to the square-root function per integration step. The subroutine which implements the Lennard-Jones model requires only a single call to each of square root, cos and sin, and the number of lines of code is halved. The initial expectation was that improved realism would be achieved at improved computational efficiency. This failed to recognise that a great deal happens in a very short distance over which the attraction force has any effect on a pair of fast-moving particles. The 'swerves' illustrated in Fig. 24.3 are the result of numerical integration *within* the basic time-step dt of the main algorithm. That a large number of sub-steps d(dt) are required might have been anticipated from the shape of Fig. 24.2. Indeed, the number is so large that the displays of Fig. 24.3 were achieved by suppressing all but one plot in 50.

Two factors make it possible to display the 'swerve' in Fig. 24.3:

(a) the force field extends by only a few molecular diameters and
(b) inter-molecular distance r has been 'normalised' by characteristic molecular size σ. This means that the arithmetic is carried out in numerical values not far from unity.

A way of seeing the outstanding coding challenge is in terms of simultaneous, coordinate systems, one macroscopic, the other microscopic. Each grid element of the former is further subdivided by the latter. The collision criterion for r calculated by the algorithm of Section 24.8 does not resolve down to the sub-molecular scale of the microscopic grid, on which the 'swerve' integration process relies. It can, however, be thought of as the first of a succession of screening tests of increasing resolution — a priority being to avoid elaborate and time-consuming testing of every pair of molecules. The cheap-and-cheerful test (specimen code under Section 24.8) serves to dismiss the majority of cases. An instance which survives against the odds is pursued to see whether separation distance is decreasing or increasing: if the latter, the check sequence is aborted. If the former, inspection jumps to a criterion of finer resolution.

25

Pushing the Boundaries

25.1. Raising Rated rpm

The preceding chapters suggested a list of criteria to be met by an engine (Stirling or thermal-lag) if it is to be sufficiently simple, sufficiently low/medium-technology and of sufficient power-to-weight to make a commercially-viable contribution to energy-generation in a post-hydrocarbon economy. Summarising:

(1) working-fluid utilisation to be significantly improved;
(2) to be optimised for unsophisticated working fluid (air/N2);
(3) to be self-pressurising — or viable without pressurisation. Preferably the latter;
(4) peak specific cycle work (e.g., Beale number N_B) to be achieved at rpm 3–4 times the current norm.

The final criterion has received least attention so far. This chapter attempts to remedy.

If there were any possibility that trial-and-error development of the Philips' MP1002CA air engine could have led to meeting criterion 4, the modifications would surely have been identified and implemented long ago. The costs of cut-and-try development can be justified only if changes have been reliably evaluated in advance. That makes modelling by computer essential — and, in the present context, not by any old model.

Anything already 'out there' — and worthy of being called a computer simulation — models flow resistance with the aid of 'correlations' — friction factor C_f correlated to Reynolds number *Re* viz, C_f vs *Re*. As Chapters 7–9 have discussed at length, correlations 'of choice' were acquired under conditions of steady, incompressible flow, *Re* coverage being achieved by sleight-of-hand which obscures compressibility effects. A simulation embodying a C_f vs *Re* 'correlation' *is not equipped* to reflect the choking to which modest values of approach *Ma* (~0.05 — Chapter 8) make the wire-gauze regenerator susceptible. *A simulation which does not predict cannot design.*

But the present context demands more that correlations sensitive to *Ma.* If attention is confined to engines of any consequence, the rpm proposed have never previously been achieved except by change of working fluid from air/N2 to lighter gases — and then only at high pressure. Getting around this while retaining air/N2 calls for taking realistic account of compressibility — pressure waves and of their interactions. And modelling needs to do more than simply flag performance penalty: it must indicate where compressibility effects can be harnessed to advantage.

Shapiro's lucid exposition laid the ground-work in 1954, but has remained largely unexploited — and largely unappreciated. The neglect may reflect the fact that the study pre-dated the digital computer. Illustrating the power of the method through a worked example of any consequence would have been a prohibitive challenge in manual arithmetic and drawing-board construction.

The advantages of presenting again in full here will be best appreciated as development proceeds. (There are minor errors to be corrected. More importantly, the original treatment undermines its own defining merit: it resorts unnecessarily to the 'smearing' effect of interpolation.)

25.2. Derivation According to Shapiro

His depiction of the control surface (Fig. 25.1) allows for gradual variation of free-flow area *A* (his notation) with distance *x*. The law of mass conservation becomes:

$$\partial/\partial x(\rho u A)\mathrm{d}x = -\,\partial/\partial t(\rho A \mathrm{d}x). \tag{25.1}$$

The corresponding momentum theorem requires a term for equivalent frictional resistance. Shapiro defines a 'wall-friction term' *F*:

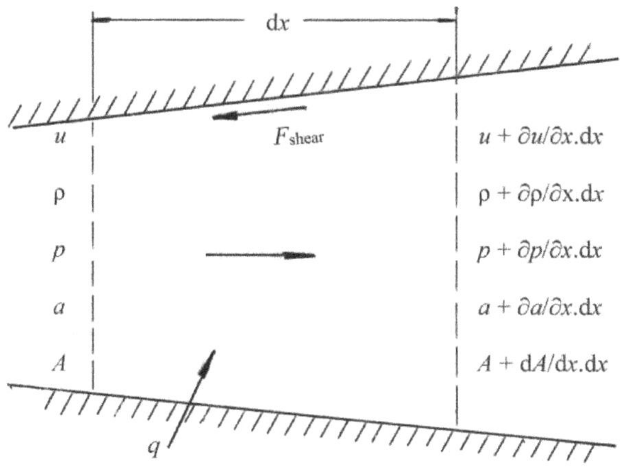

Fig. 25.1. Shapiro's control volume notation.

$$F = -\frac{1}{2} C_f u^2 / r_h \times sign(u). \tag{25.2}$$

Sign(u) is introduced to ensure that friction acts opposite to the direction of motion. The momentum law becomes:

$$\begin{aligned}&-\partial/\partial x(pA)\,\mathrm{d}x + p(\mathrm{d}A/\mathrm{d}x)\,\mathrm{d}x - \frac{1}{2} C_{\mathrm{f}} \rho u^2 p_{\mathrm{w}} \mathrm{d}x \\ &= \partial/\partial t(\rho A u \mathrm{d}x) + \partial/\partial x(\rho A u^2)\mathrm{d}x.\end{aligned} \tag{25.3}$$

Shapiro expands and subtracts Eq. (25.1)

$$\partial u/\partial t + u\partial u/\partial x + \rho^{-1}\partial p/\partial x + F = 0. \tag{25.4}$$

In terms of his q' for heat-rate per unit mass of working fluid, the First Law of Thermodynamics becomes:

$$\begin{aligned}&q' \rho A \mathrm{d}x \\ &= \partial/\partial t\left[(\rho A \mathrm{d}x)\left(c_v T + \frac{1}{2}u^2\right)\right] + \partial/\partial x\left[(\rho u A)\left(c_v T + p/\rho + \frac{1}{2}u^2\right)\right]\mathrm{d}x.\end{aligned} \tag{25.5}$$

(In the original 1954 account at p. 973 the final dx is missing.) Shapiro expands Eq. (25.5) and subtracts the earlier continuity equation from the result:

$$\mathrm{D}/\mathrm{d}t\left(c_{\mathrm{v}}T+\frac{1}{2}u^{2}\right)=q'-\rho^{-1}(u\partial p/\partial x+pA^{-1}\partial Au/\partial x). \qquad (25.6)$$

In Eq. (25.6) D/dt is the total or substantial derivative $\partial/\partial t + u\partial/\partial x$. Using the identity $c_{\mathrm{v}}T = (\gamma - 1)^{-1}p/\rho$, the left-hand side is expanded:

$$\mathrm{D}(c_{\mathrm{v}}T)/\mathrm{d}t = [\rho(\gamma-1)]^{-1}[\partial p/\partial t + u\partial p/\partial x - p/\rho(\partial\rho/\partial t + u\partial\rho/\partial x)].$$

He evaluates the term $\partial(Au)/\partial x$ using the continuity equation:

$$\partial(Au)/\partial x = -A\rho^{-1}(\partial\rho/\partial t + u\partial\rho/\partial x). \qquad (25.7)$$

Further expansion promises a term $\partial p/\partial x$. Shapiro heads off its appearance by substituting from his momentum equation. This is an unexpected step, since $\partial p/\partial x$ (and $\partial p/\partial t$) are required in the expressions he is aiming for.

The energy equation becomes:

$$(\gamma-1)\rho(q' + uF) = \partial p/\partial t + u\partial p/\partial x - a^{2}(\partial\rho/\partial t + u\partial\rho/\partial x). \qquad (25.8)$$

The development calls for a path-line equation to relate properties while following a fluid particle. Shapiro makes uncharacteristically heavy-weather of this — but in any case, present purposes are served by reusing the path equation already developed here for the earlier Lagrange formulations. Converting to Shapiro's notation:

$$\mathrm{D}p/p|_{\mathrm{path}} = [\gamma/(\gamma-1)]p\mathrm{D}T/T - \rho(uF/R + q')\mathrm{d}t|_{\mathrm{path}}. \qquad (25.9)$$

There is now available a set of six simultaneous equations linear in the unknown partial differential coefficients $\partial u/\partial x$, $\partial u/\partial t$, $\partial p/\partial x$, $\partial p/\partial t$, $\partial\rho/\partial x$ and $\partial\rho/\partial t$. These may be laid out systematically:

$\partial u/\partial x$	$\partial u/\partial t$	$\partial p/\partial x$	$\partial p/\partial t$	$\partial\rho/\partial x$	$\partial\rho/\partial t$	RHS
$\mathrm{d}x\partial u/\partial x$	$+\mathrm{d}t\partial u/\partial t$					$= \mathrm{d}u$
		$+\mathrm{d}x\partial p/\partial x$	$+\mathrm{d}t\partial p/\partial t$			$= \mathrm{d}p$
				$+\mathrm{d}x\partial\rho/\partial x$	$+\mathrm{d}t\partial\rho/\partial t$	$= \mathrm{d}\rho$
$\rho\partial u/\partial x$				$+u\partial\rho/\partial x$	$+\partial\rho/\partial t$	$= -\rho uA^{-1}\mathrm{d}A/\mathrm{d}x$
$u\partial u/\partial x$	$+\partial u/\partial t$	$+\rho^{-1}\partial p/\partial x$				$= -F$
		$+u\partial p/\partial x$		$+\partial p/\partial t - ua^{2}\partial\rho/\partial x$	$-a^{2}\partial\rho/\partial t$	$=(\gamma-1)\rho(uF+q')$

The first three rows express mathematically the conditions that the properties themselves (i.e., u, p, ρ) are continuous.

The equation set lends itself to solution by the methods for determinants: the coefficients of the partial differentials are laid out in determinant notation. Shapiro takes the example of the solution for $\partial u/\partial x$

$$\partial u/\partial x = \frac{\begin{vmatrix} \mathrm{d}u & \mathrm{d}t & 0 & 0 & 0 & 0 \\ \mathrm{d}p & 0 & \mathrm{d}x & \mathrm{d}t & 0 & 0 \\ \mathrm{d}\rho & 0 & 0 & 0 & \mathrm{d}x & \mathrm{d}t \\ -\rho u A^{-1}\mathrm{d}A/\mathrm{d}x & 0 & 0 & 0 & u & 1 \\ -F & 1 & 1/\rho & 0 & 0 & 0 \\ (\gamma-1)\rho(uF+q') & 0 & u & 1 & -ua^2 & -a^2 \end{vmatrix}}{\begin{vmatrix} \mathrm{d}x & \mathrm{d}t & 0 & 0 & 0 & 0 \\ 0 & 0 & \mathrm{d}x & \mathrm{d}t & 0 & 0 \\ 0 & 0 & 0 & 0 & \mathrm{d}x & \mathrm{d}t \\ \rho & 0 & 0 & 0 & u & 1 \\ u & 1 & 1/\rho & 0 & 0 & 0 \\ 0 & 0 & u & 1 & -ua^2 & -a^2 \end{vmatrix}}$$

The denominator is the determinant of the coefficients, sometimes denoted Δ. The numerator is the very same array of coefficients, but with the column corresponding to the specific unknown ($\partial u/\partial x$ in this example) substituted by the column from the RHS.

There are two things one can do with equations: (a) solve them or (b) *not* solve them, but inspect instead. Shapiro pursues the latter course with a view to exploiting a property of the u, p, ρ surface in the $x - t$ plane: *If both numerator and denominator are zero, then the result of evaluating the determinant is indeterminate,* i.e., *is 0/0*. The physical equivalent of the indeterminacy is a locus in $x - y$ coordinates along which the gradient (in the fluid property of interest) may undergo arbitrary discontinuities. These loci have been called characteristic directions (or Characteristic curves — or *Characteristics*).

Evaluating the denominator by the rules for determinants and setting equal to zero results in eight terms. There is a single term in $(\mathrm{d}x/\mathrm{d}t)^3$, the rest being of the form $u(\mathrm{d}x/\mathrm{d}t)^2$, $u^2\mathrm{d}x/\mathrm{d}t$, $a^2\mathrm{d}x/\mathrm{d}t$, etc., each evidently cubic in units of length/time, viz (length/time)^3 and suggesting three roots.

On substituting $\mathrm{d}x/\mathrm{d}t = u$ the eight terms boil down to $u^2 - a^2 = 0$, revealing the three characteristic directions:

$$\mathrm{d}x/\mathrm{d}t|_{\text{path}} = u, \tag{25.10}$$

$$\mathrm{d}x/\mathrm{d}t|_{\text{I, II}} = u +/- a, \quad \text{Mach lines.} \tag{25.11}$$

Expanding the numerator by the same rules and equating to zero yields two relationships, this time between (total) changes Du and Dp in state variables u and p:

$$\mathrm{D}u\Big|_{\text{I,II}} = -/+ \mathrm{D}p\Big|_{\text{I,II}} / \rho a -/+ auA^{-1}\mathrm{d}A/\mathrm{d}x.\mathrm{d}t\,|_{\text{I,II}} \tag{25.12a}$$

$$+/-(\gamma-1)(q'a)\mathrm{d}t\,|_{\text{I,II}}\,\mathrm{d}t - F[1-/+(\gamma-1)u/a]\mathrm{d}t\,|_{\text{I,II.}} \tag{25.12b}$$

(Shapiro proceeds to say that '*it has been found more convenient to work with the a, u characteristics*.' Convenient maybe — but not the way to go, and for two reasons (a) in the Stirling engine context, interest is in the pressure–distance–time (p–x–t) history rather than the surface of a–x–t. More importantly (b) Shapiro's conversion leads to a term in the gradient of the lengthwise distribution of entropy $a^2\gamma^{-1}\partial/\partial x[ln(p^{-1}a^{\gamma/(\gamma-1)})]$. Anyone happy with entropy gradients is welcome to them — but the objection ought to be that interpolation is required in evaluating the term, whereas the previous $p - u$ system was free from the compromise of numerical 'smearing'.)

Shapiro's formulation provides for the cross-sectional area of the flow passage to vary — but gradually. Until more is known, it is assumed that raising the rpm at which peak N_B is achieved will require eliminating, or minimising, interruptions in free-flow area — bends and abrupt expansions/contractions.

25.3. Preparation for Implementation

Two steps are involved in applying the Method of Characteristics to the Stirling engine: (a) linking dummy terms for heat transfer and friction, q' and F with something approaching reality and (b) mechanising the process of numerical integration to coincide with the characteristic directions.

Shapiro's q' is heat rate per unit mass of working fluid. In terms of coefficient h of convective heat transfer $q' = hp_w dx(T_w - T_g)/\rho A dx$. The ratio of free-flow area A to wetted perimeter p_w is hydraulic radius r_h. Multiplying numerator and denominator by $c_p|u|$ and recalling the definition of Stanton number St as $h/\rho c_p|u|$ allows re-writing as

$$q' = St(T_w - T_g)c_p|u|/r_h. \tag{25.13}$$

The extent of meaningful rpm coverage depends on whether St is evaluated from Re alone (i.e., from the usual $StPr^{2/3} - Re$ steady-flow correlation) or with due account of Mach number Ma. In the meantime, using Eq. (25.13) as a 'placeholder' allows the nuts and bolts of the method to be coded and tested. Likewise, Shapiro's F (Eq. (25.2)) is provisionally evaluated from local, instantaneous Re pending availability of Cf of the appropriate form $Cf(Re, Ma)$.

25.4. Integration Step: The Unit Process

In the more common rectangular cartesian coordinate frame, a step in numerical integration of the function $\partial y/\partial x = f(x,y)$ from x_0 and y_0 to new values x and y could proceed by tentatively evaluating as $y \approx y_0 + \partial f(x_0, y_0)/\partial x \times \Delta x$ with iteration to the required accuracy. The computation then steps ahead by Δx from the new values of dependent variable y.

The characteristic directions are not (necessarily) orthogonal, and in general, angularity varies throughout the $x - t$ plane. A single integration step involves working (iterating) between two planes, the Mach plane (or physical $x - t$) plane and a state ($a - u$ or $p - u$) plane. The Method of Characteristics evolved in the context of homentropic gas dynamics with the state plane displaying Δa vs Δu, as in Figs. 25.2(a) and 25.2(b). This is the easiest case to illustrate before taking particle paths into account:

The t and x coordinates at points 1 and 2 are supposed known. Each step builds on the previous one. As indicated in the diagram, states a and u are assumed known at intersections 1 and 2. The coordinates t and x of point 3 in the physical plane are first located approximately by finding the intersection of a I-line element projected from point 1 at gradient $dt/dx_I = 1/(u_1 + a_1)$ with a II-line element projected from point 2 with gradient $dt/dx_{II} = 1/(u_2 - a_2)$. The construction yields approximate values for time

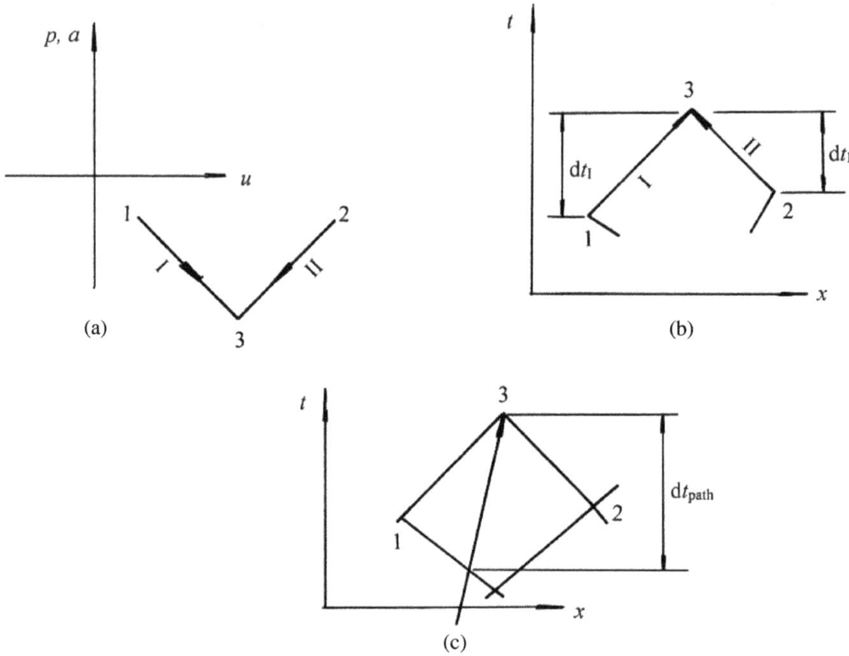

Fig. 25.2. Physical (Mach) plane and state plane. Solution alternates between the two.

increments dt_I and dt_{II}. It typifies every such step to which has to be taken, with the exception of those at left- and right-hand boundaries.

(For the homentropic case, a at a given intersection defines p at that point.)

Where friction and heat transfer are involved (as in Shapiro's re-formulation) the unit process calls for an additional step to allow the information on fluid states which propagates along path lines to be absorbed into the construction. Figure 25.3 shows the physical ($t - x$) plane modified to suit. State equations (25.9), (25.12a) and (25.12b) are simultaneous and linear in the unknown differences dp, da and du, which may respectively be written $p_3 - p_1$ (or $p_3 - p_2$), $a_3 - a_1$ (or $a_3 - a_2$) as appropriate. The path line reaching point 3 will have passed close to point 4, in the process cutting either the I-characteristic between 4 and 2, or the II-characteristic between 4 and 1. Intersection is at auxiliary point t^*, x^*, located by projecting a path line backwards from point 3 at provisional gradient $dt/dx_{path} = 1/u_4$. This yields provisional values of p^*, a^*, u^* and of dt_{path}. Substituting into the three state equations and solving (e.g., in

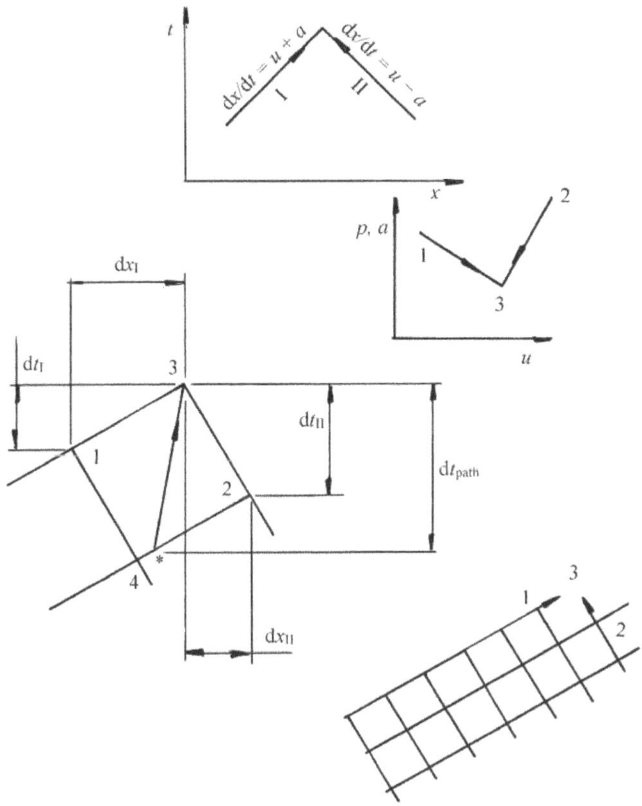

Fig. 25.3. Unit process in the Mach plane taking information from the particle path.

3 × 3 determinant form) yields provisional values of p_3, a_3 and u_3. Tentative values of the respective time increments dt_I, dt_{II} and dt_{path} may now be picked off in the physical plane.

The $t - x$ coordinates of point 3 are now improved by use of mean values of u and a between 1 and 3, and between 2 and 3. The unit process is iterated to convergence.

The construction step applies equally to the special case of a stationary fluid. The calculation at this point is trivial, since local I and II directions are simply $dx/dt_I = +a$, $dx/dt_{II} = -a$, suggesting starting construction with the gas stationary. The first step then consists of a I-wave running left-to-right at gradient $dx/dt_I = +a$, where a is given by the temperature of the immediately adjacent containing surface. The waves at this stage are of 'zero strength', i.e., the disturbances in pressure, temperature (or acoustic speed) and particle are 'virtual'. The initial, zero-strength wave is divided into a convenient number of points from which the set of II waves emanates.

In anticipation of mechanised solution, state variables are normalised with respect to reference acoustic speed a_0, thus u/a_0, a/a_0. Distance coordinate x is normalised by flow-passage length L, viz x/L. Time itself is normalised by nominal time of end-to-end wave passage, as $t/(L/a_0)$. Construction of the Mach plane in terms of normalised distance and normalised time ensures the same degree of resolution regardless of absolute values of a, u. It also ensures that, for all but extreme values of $u + a$, $u - a$, Mach lines traverse at close to +/–45 degrees.

25.5. Specimen Solutions

One of several features distinguishing a solution by the Method of Characteristics is that it has to be appropriately 'started up'.

The real engine starts from rest and accelerates to rated rpm as the gas process interactions reach cyclic equilibrium. The computer model can be forced to do the same — but it that is not a good use of CPU time. Unsurprisingly, starting the simulation instantaneously to rated speed simulates phenomena which *would* arise in practice if such an impulsive start were possible: compression and/or rarefaction waves.

The problem with compression waves is that they can steepen into shocks, which the numerical method has difficulty handling (*although see later heading*!). There is a choice of ways forward, illustrated here by reference to the opposed-piston layout assumed for Fig. 25.4: In the first method, the left-hand (expansion-space) piston is held stationary at inner or outer dead-centre (u = zero) while the right-hand (compression) piston is started from outer dead-centre (at which precise point its velocity is zero — regardless of rpm). The left-hand piston is allowed to move instantaneously to rated rpm when the phase lag between the two pistons reaches the value of α prescribed for the design. Subsequent motions remain out-of-phase by constant phase angle α. Alternatively, and based on expectation of α close to 90°, the left-hand piston is started from inner or outer dead-centre and with the right-hand piston at or close to *mid-expansion stroke.* Starting instantaneously to rate rpm centre generates a rarefaction fan. In contrast with a compression, a fan offers no computational problems.

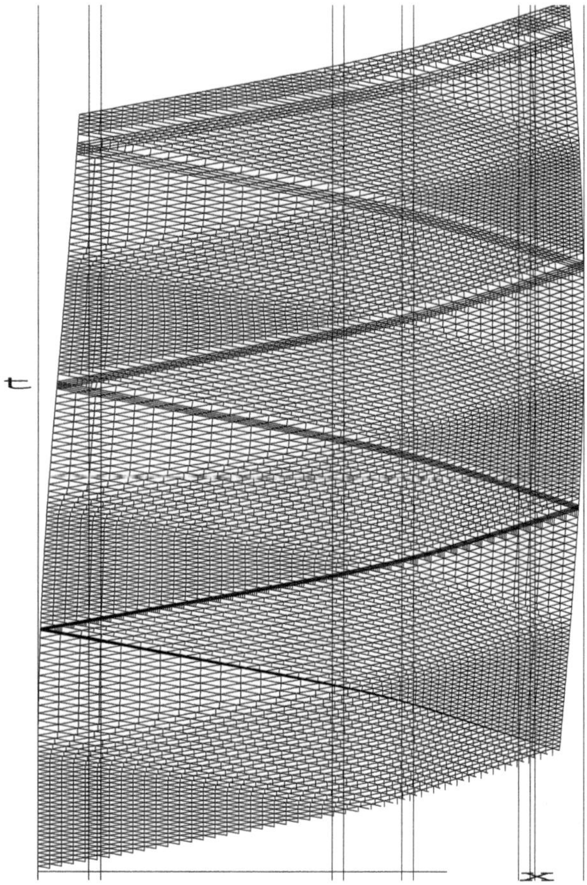

Fig. 25.4. Specimen $t - x$ (Mach) plane construction. May be thought of as plan view of the pressure (or velocity — or temperature — or acoustic speed) surface.

An earlier illustration of the Mach plane construction (the authors' 2007 text at p 186) used the former option, thus accounting for the absence of expansion fan. Fig. 25.4 is the Mach plane construction for a gas path owing much to Philips' MP1002CA air engine. It uses the latter start-up option, with rpm instantaneously reaching 1500.

Progress of the expansion fan is well-marked, but raises an interesting reflection on the solution method: supposing that some analyst with a Brunsviga hand-cranked stepped-gear calculator had summoned the patience to carry the construction this far, it is highly improbable that he/she would have had the staying-power to proceed to the corresponding pressure relief of Fig. 25.5, leaving a woefully incomplete picture.

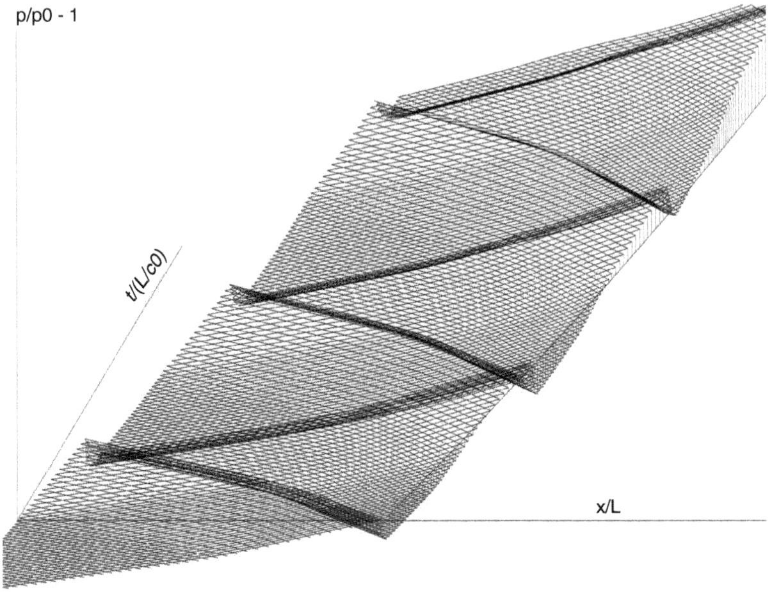

Fig. 25.5. Pressure relief corresponding to Mach plane solution of Fig. 25.4.

It has already been agreed that there is no laboratory counterpart to the impulsive start-up, so neither is there a real counterpart to the stepped pressure surface. The value of the example is in illustrating the inherent robustness of the construction. It will also serve to illustrate the challenge of re-designing the MP1002CA for operation at elevated rpm. A figure of 4500 was proposed earlier. Figure 25.6 is the Mach plane for the same elapsed time from start-up — obviously covering a proportionately greater number of complete cycles.

Like the earlier Mach plane, there is almost no hint as to the underlying pressure excursions. Figure 25.7 leaves no doubt! Evidently, the display has everything to do with the vertical scale chosen for $p/p_0 - 1$ — the pressure excess. However, the point here is that the scale is unaltered from Fig. 25.5: (a) the Method of Characteristics can cope and (b) wave effects in an unmodified MP1002CA (if it could be driven at 4500 rpm) pose daunting problems.

If there are challenges ahead, one might as well anticipate as many as possible.

It is frequently remarked of the Method of Characteristics that it comes unstuck when shock waves threaten. Possibly the best analogy for a shock wave in a gas is an ocean wave at the instant of breaking. This is

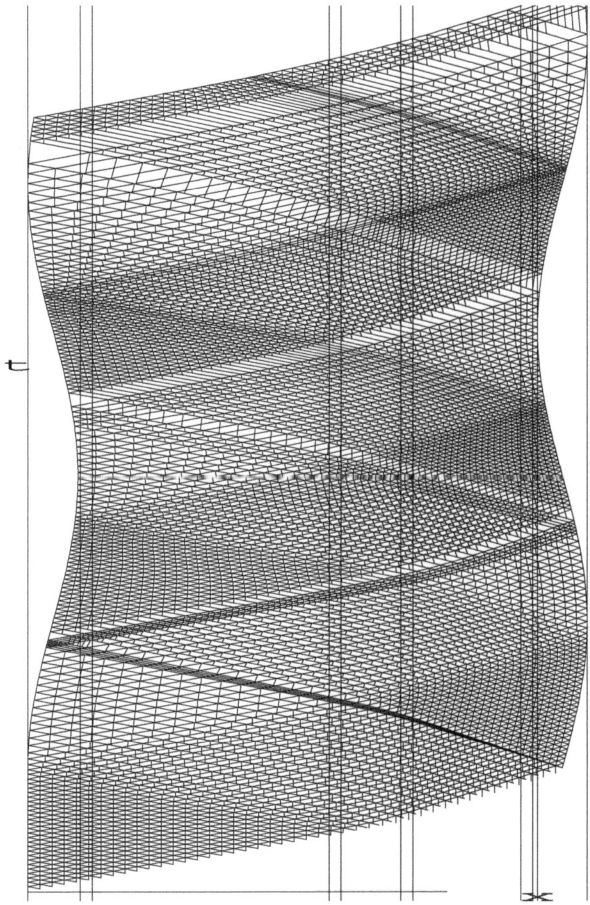

Fig. 25.6. Mach plane construction for the data of Fig. 25.4, but with rpm raised to 4500.

the point at which the crest begins to overtake, calling for two values of height to define the wave-front. The incipient shock in the gas might be more appropriately thought of as a vertical wall corresponding to an arbitrary number of pressure values at its leading face.

But helpful insight is at hand — and from an unlikely source: Elon Musk's HyperLoop concept!

25.6. Shock Waves

Millions have been spent pursuing Musk's brain-child, which envisages population centres connected by partially-evacuated tubes carrying

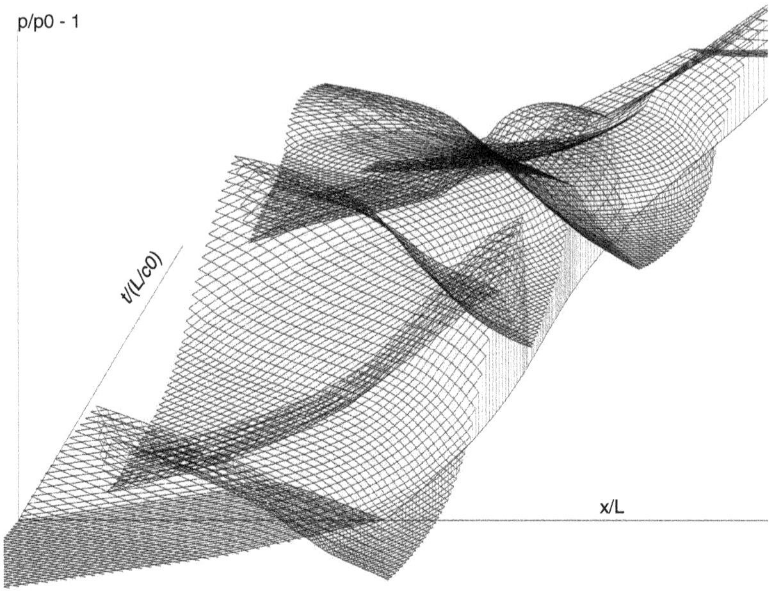

Fig. 25.7. Pressure relief corresponding to Mach plane solution of Fig. 25.6.

passenger 'pods' at speeds close to the speed of sound. Travel at close to Mach-one guarantees shock waves — steady shock-waves — whether in free-air or a duct, whether at atmospheric pressure or in partial vacuum. Acceleration up to those speeds means *unsteady* pressure waves which can coalesce into shocks, complicating the picture.

Given the proposed frequency of services, realising short journey times promised by peak speeds calls for acceleration and deceleration to occupy the least possible fraction of stage length. '*The same g-forces as rail*' of the Virgin HyperLoop website hardly fits the bill. The harsher the acceleration (and deceleration) the more intense is the unsteady pressure difference acting between front and rear faces of the vehicle. That difference, acting on some 4 m^2 of cross-sectional area, even at rarefied pressure, means periodic disturbing forces. A pod suspended by magnetic levitation and travelling at 1270 km/hr is a dynamic system. Such a system can be driven to destructive oscillations by an almost insignificant disturbing force *if it acts at the natural frequency of the system.*

Test sections exist, and in late 2020 a pod, with passengers, was accelerated to 43 m/s or 172 km/hr. (https://virginhyperloop.com/pegasus).

This is 13% of target (sonic) speed of 1270 km/hr., At $M = 0.13$ it is below the criterion for onset of compressibility effects in steady flow ($M = 0.3$). It might therefore be wondered whether the design team has been alerted to potential problems of accelerations to *eight times that speed.*

To explore further, the Method of Characteristics has been applied to a HyperLoop journey cycle specified in terms of acceleration, cruise and deceleration stages — and has confirmed the intuitive concerns. (A report tendered — unsolicited — to Musk's team and to Sir Richard Branson has drawn no reaction.) The report is reproduced here as Appendix J because it called for extension of the Method of Characteristics (MoC) to deal with shock formation. This extends simulation coverage to the physical boundaries of the Stirling gas-cycle envelope — and in a way in which *the MoC is uniquely capable of handling.*

To avoid collision, two HyperLoop pods travelling in a common duct will likely do so in synchronism — possibly achieved electromagnetically. Such a system presents the same challenge in unsteady gas-dynamics as the opposed-piston Stirling engine with $T_E = T_C$, with uniform gas-path cross-section, with phase angle α of zero degrees (but with somewhat greater piston blow-by!). This section pursues the HyperLoop example with no loss of relevance to the Stirling case. (To the extent that a HyperLoop journey and its counterpart modelled by the MoC both start from rest there is slight but advantageous simplification in implementation.)

A shock is the indirect outcome of the temperature increase which results from pressure rise when the latter is adiabatic — or substantially so. The corresponding rise in local acoustic speed causes a wave (as defined in terms of Mach lines) to overtake the one which preceded it. In the Mach plane, increased speed manifests as reduced gradient $\mathrm{d}t/\mathrm{d}x$. At first sight this thwarts application of the unit process (earlier Figs. 25.2 and 25.3).

Figure 25.8 shows a I-wave overtaking its predecessor as the former approaches the right-hand moving boundary. Crossing point p does not have a counterpart in the state plane It therefore does not call for application of the unit-process — which would otherwise generate an unwanted additional Mach line. From this point it is necessary only to keep in mind that the integration process steps forward *not* (necessarily) in terms of elapsing time, but along I and II directions — whither they may lead!

One thing the computer-coded MoC does not welcome is an increase in the inventory of Mach lines. This need not be a problem: appropriate

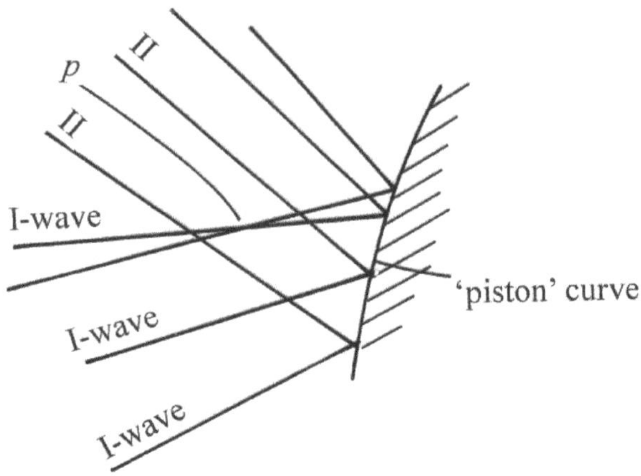

Fig. 25.8. Mach plane showing, at point p, a I-wave overtaking its predecessor in the construction. In applying the unit process past this point it is necessary to distinguish 'real' intersections of I and II waves from crossing-points such as p which have no counterpart in the state plane.

bookkeeping maintains the necessary distinction between 'real' intersections and points (such as p in Fig. 25.8) which superimpose in the Mach plane view, but which do so without actually intersecting.

Figure 25.9 is the Mach plane for the acceleration-cruise-deceleration cycle modelled in the study (Appendix J). The left-hand boundary is the leading face of the trailing pod, the right-hand boundary being the trailing face of the leading pod. The compression wave (converging fan) steepens to a shock by the mid-journey point (half-way along x/L-axis between start-point of both pods). Traces thereafter appear darker because there are actually two layers of Mach lines superposed (*c.f.* the overhang of the ocean wave).

Figure 25.10 is a carpet plot of the surface of pressure excess, $p/p_0 - 1$, corresponding to the previous figure. The comparison highlights the difficulty in inferring the latter from the former.

There is no denying the idealisations behind these solutions (and their disconcerting implications for HyperLoop). At first sight, the build-up of lengthwise pressure difference across the pod is countered by flow around the periphery. Conversely, around a streamlined pod, flow is nozzle flow which can choke at upstream speeds one third of rated coast speed. There again, this is no ordinary nozzle, since one face is stationary while the other travels at 340 m/s — quite some shear gradient $\partial u/\partial y$!

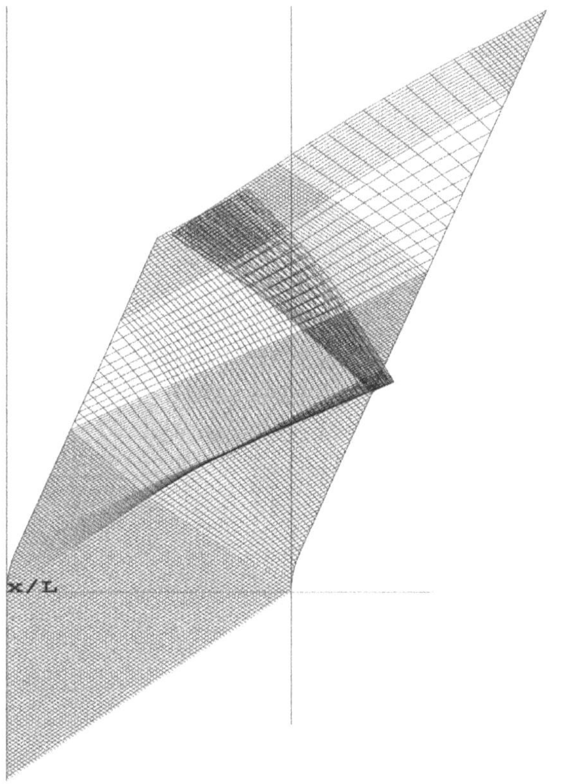

Fig. 25.9. Mach plane construction for the start of the HyperLoop journey modelled in Appendix J. Darker regions are where two Mach planes superimpose, one for the slower wave, the other for the wave which has overtaken (shock).

Fortunately, this is a rare instance in which the Stirling case is relatively straightforward — or will be when the matter of flow-area discontinuity has been dealt with.

25.7. Abrupt Discontinuities in Free-Flow Area[1]

The idealised case just illustrated (parallel gas-path with no interruptions in free-flow area) offers two wave-reflection sites, one at each opposing

[1] Selected text introducing the figures is reproduced from the author's 1997 by permission of Messrs John Wiley & Sons.

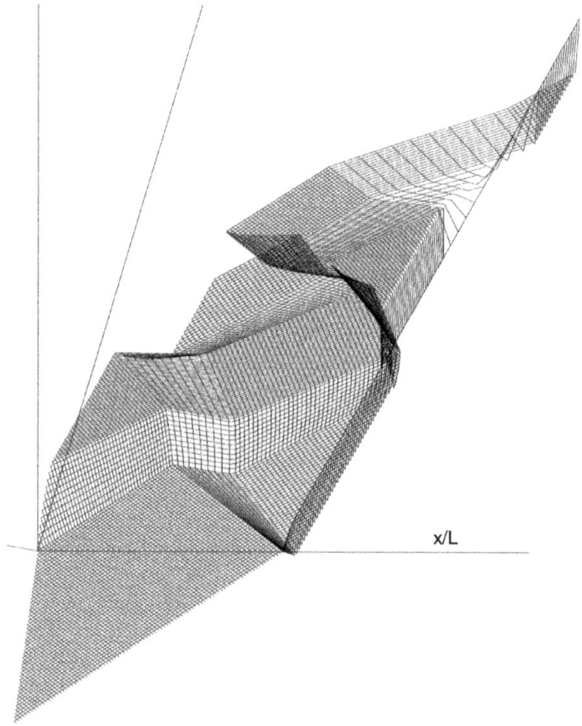

Fig. 25.10. Carpet plot of pressure surface corresponding to Mach plane of previous figure.

piston face. Since the waves are reflected in full (without any transmitted component) integration can proceed through many reflections without adding to the original inventory of Mach lines.

In contrast, a Mach line encountering a discontinuity in the form of a step-change in free-flow area resolves into both a reflected and a transmitted component. Carrying the latter forward with the solution promises an unmanageable number of lines after only a fraction of a cycle. Keeping storage (and coding) within reasonable bounds calls for a method whereby 'new' waves generated by reflection are absorbed into the lattice of original waves *without sacrificing thermodynamic information*.

Figure 25.11 is in terms of a pair Mach lines outlining a single wavelet in the lattice depiction. At a discontinuity they divide. States immediately to left and right of the discontinuity are labelled 4 and 5, respectively. The reflected component — a left-running 'additional' II-wave — must

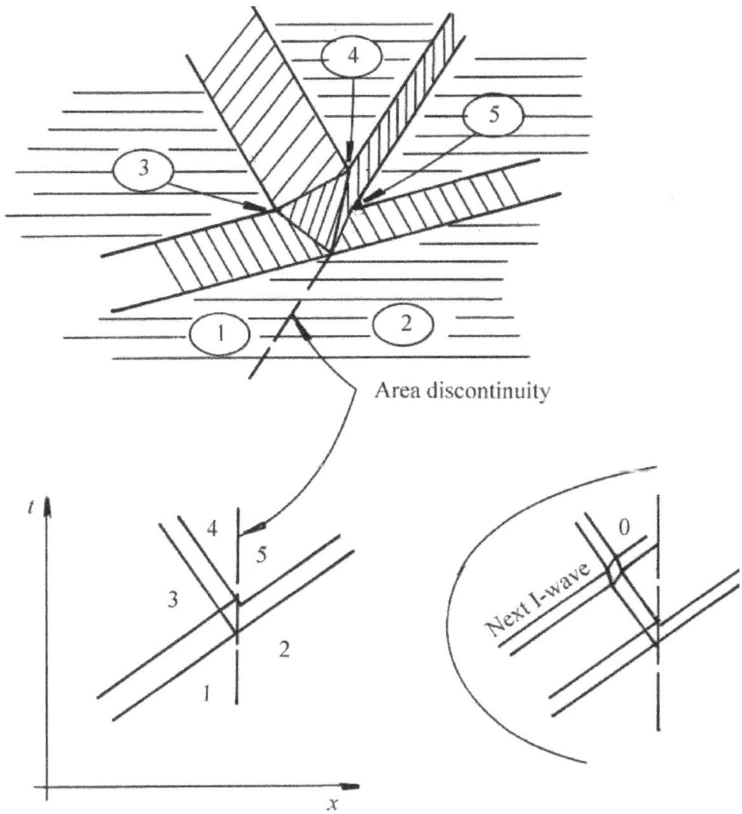

Fig. 25.11. A wavelet of the I-family (outlined by the pair of Mach lines) encountering an abrupt area-change divides into a reflected (II-family) part and a transmitted part.

eventually be intercepted by some future right-running (type I - wave suggested as point O on inset 'plan' view of the physical plane) which had formed part of the original inventory. In the lattice system, fluid states are resolved where I and II Mach lines intersect. Applying this process at this 'extra' point O ensures that both I and II elements proceed with up-to-date state information. To this extent, retaining the reflected component leftward past point O is superfluous: its presence offers finer subdivision of the flow field — but the field is assumed to be appropriately sub-divided by the original set of Mach lines. Discarding the leftward element *after incorporation into the I-wave* sacrifices no information.

25.8. Defining Equations

Sudden contraction or expansion in a real flow promises irreversibilities over and above those due to the heat transfer essential to this context. However, the present focus is on the wave phenomenon *per se*. This can be illustrated and implemented using the state equations for the classic homentropic case.

Figure 25.12 shows the modified unit process in the Mach plane. The x and t coordinates of points 1 and 2 are assumed known, as are states a_1, u_1 and a_2, u_2. If states a_3, u_3 can be determined, coordinates x_3 and t_3 follow.

A wavelet of the I-family (outlined by the pair of Mach lines) encountering a discontinuity in free-flow area divides into a reflected (II-family)

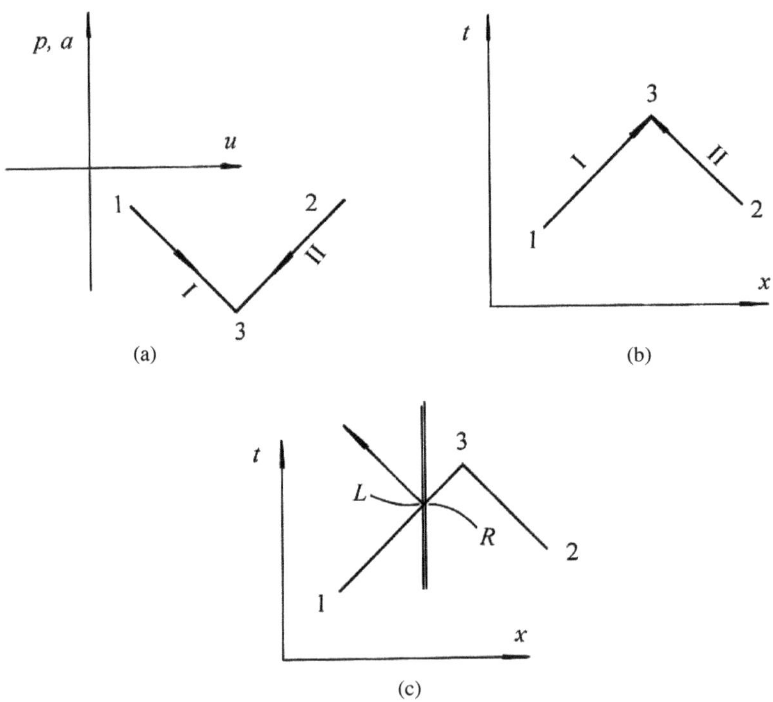

Fig. 25.12. (a) and (b) Traditional 'unit process' in state and physical planes. (c) Unit process spanning flow area discontinuity in Mach plane.

part and a transmitted part. Without the discontinuity, states a_3, u_3 follow from applying the Riemann relationships between points 1 and 3 and between 2 and 3. The discontinuity, represented by the double line in Fig. 25.12(c), introduces additional unknowns a_L, u_L a_R, u_R and, indeed, t_{LR} — time of arrival at the discontinuity. There is a further unknown in the sense that it is not known at this stage whether the discontinuity will be encountered by a wave of type I, or a wave of type II.

It is assumed that the original Mach-line net has been chosen sufficiently fine that the unit process anticipated can span one area discontinuity at most.

Wherever the discontinuity is encountered, it is always possible to define an auxiliary wave of the opposite family such that both meet precisely at the discontinuity. Along the I-wave between points 1 and L the appropriate Riemann relationship applies: $da = -1/2(\gamma - 1)du$. In terms of finite differences:

$$a_L - a_1 + \frac{1}{2}(\gamma - 1)(u_L - u_1) = 0. \qquad (25.14)$$

Points R and 2 are linked by a II-wave, for which the Riemann relationship is $da = +1/2(\gamma - 1)du$:

$$a_R - a_2 - \frac{1}{2}(\gamma - 1)(u_R - u_2) = 0. \qquad (25.15)$$

Equations (25.14) and (25.14) embody four unknowns: $a_L - u_L$, $a_R - u_R$, calling for two further relationships between left and right corresponding to the total of six unknowns. By convention these are provided by the steady-flow energy equation and, in the present (provisional) formulation, by the homentropic condition. The former derives from the energy conservation statement:

$$c_p T_L + \frac{1}{2} u_L^2 = c_p T_R + \frac{1}{2} u_R^2. \qquad (25.16)$$

Recalling that $c_p - c_v = R$, that $c_p/c_v = \gamma$ and that acoustic speed $a = \sqrt{\gamma RT}$ allows re-writing:

$$a_L^2 + \frac{1}{2}(\gamma - 1)u_L^2 = a_R^2 + \frac{1}{2}(\gamma - 1)u_R^2. \qquad (25.17)$$

Equation (25.17) is nonlinear, but linearises on expansion later in terms of du and da.

The final equation expresses the isentropic condition across the discontinuity. It derives from the combination of steady-flow mass conservation, $\rho_L u_L A_L = \rho_R u_R A_R$. Making the substitution and writing out in standard form:

$$A_R/A_L - (u_L/u_R)(a_L/a_R)^{2/(\gamma-1)} = 0.$$

Particle velocity u takes positive and negative values, and thus may pass though zero. To head off difficulties with division by zero, the expression is multiplied through:

$$A_R/A_L(u_R a_R^{2/(\gamma-1)}) - (u_L a_L^{2/(\gamma-1)}) = 0.$$

Once linearised, the set of equations may be solved for the four unknowns given the states a and u at points 1 and 2.

The resulting 4 × 4 determinant is an elementary proposition as determinants go, but has to be seen in the context of the ultimate aim: modelling the large number of discontinuities of the representative Stirling regenerator calls for the arithmetic per unit process to be kept to a minimum. Thus it is noted that the linear equations (25.14) and (25.15) are sufficiently straightforward to permit a_L and a_R to be expressed in terms of u_L and u_R respectively and thus to be eliminated from the later equations. Finally there result two simultaneous, linear equations in u_L and u_R of the form:

$$a_{11}u_L + a_{12}u_R = b_1, \qquad (25.18)$$

$$a_{21}u_L + a2_{22}u_R = b_2.$$

The solutions are explicit:

$$u_L = \frac{b_1 a_{22} - b_2 a_{12}}{a_{11}a_{22} - a_{21}a_{12}}, \qquad (25.19a)$$

$$u_R = \frac{b_2 a_{11} - b_1 a_{21}}{a_{11}a_{22} - a_{21}a_{12}}. \qquad (25.19b)$$

a_L and a_R are determined by back-substitution. Each evaluation of the state variables calls for iteration to convergence, but the number of arithmetic operations is now minimal.

25.9. Generalised Unit Process

In Fig. 25.13 the integration sequence (which marches from lower left to upper right) is illustrated in the Mach plane. Integration has reached a stage where the current unit process spans a discontinuity. Four possibilities are illustrated, perhaps better categorised as two alternatives, each with two variants:

(1) discontinuity cut by the I-wave of the previous integration sweep;
(2) discontinuity cut by the II-wave of the previous sweep.

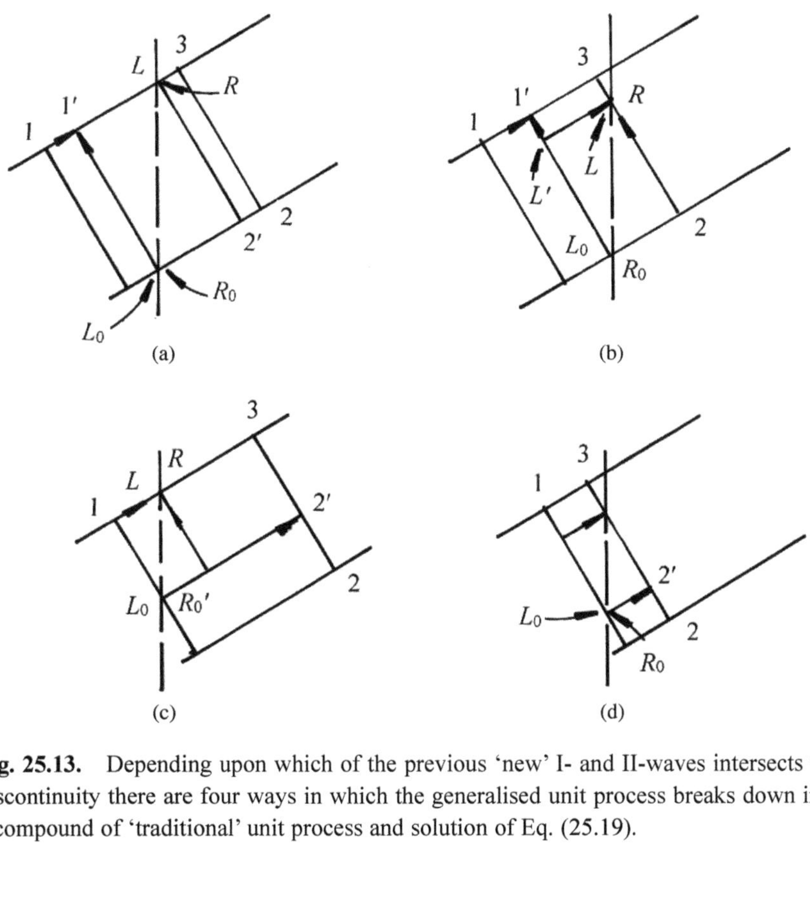

Fig. 25.13. Depending upon which of the previous 'new' I- and II-waves intersects the discontinuity there are four ways in which the generalised unit process breaks down into a compound of 'traditional' unit process and solution of Eq. (25.19).

For either (1) or (2) either (a) the discontinuity is spanned by new I-wave (between points 1 and 3); or (b) it is spanned by the new II-wave (point 2 to point 3).

Other groupings of the alternatives are possible: the only criterion is minimisation of the number of alternative sequences of switches in the code for the mechanised solution.

Each of the four cases is seen to comprise:

- one or more 'conventional' unit processes, whereby a new point is defined as an intersection of I- and II-waves which proceed from respective start points to intersection without interfering with the area discontinuity;
- an intersection point coinciding precisely with the discontinuity.

The unit process of type 1a provides an illustration:

(1) An approximation to point 3 is established by initially proceeding as though the discontinuity did not exist. This is achieved by applying the 'traditional' unit process based on known points 1 and 2.
(2) States 1′ are determined explicitly — and accurately — in terms of states 1 (along I-wave and states L_0 (along II-wave), on the basis that the states at L_0 are known from the previous integration sweep.
(3) States at 2′ are tentatively pro-rated between states at R_0 (determined in previous sweep) and states at 2 in the ratio of the (tentative) distances 1′ to L and R to 3.
(4) Equations (25.19a) and (25.19b) are evaluated for states L and R, respectively, left and right of the discontinuity spanned by the new I-wave.
(5) Point 3 is recalculated via a 'traditional' unit process based on points R and 2.
(6) The sequence, with some modification of step 3, iterates to convergence: instead of pro-rating a linear distance to find states 2', these are now determined authentically by establishing point 2′ on the earlier I-wave on projecting a II-wave backwards from point R at a gradient duly calculated in terms of the means of states at R and 2′.

In marginal cases, point 3 may flip to the opposite side of the discontinuity from its position at first estimate. In such cases, computation moves wholesale to the appropriate alternative unit process.

At first sight the method is compromised by the requirement for an interpolation. On the contrary: in the lattice method the gradation of states between successive lattice points is, by definition, linear. To this extent, states at an interpolated point are exact.

25.10. Solution in the Mach Plane

Between points 1 and 3 the gradient of the Mach line is expressed in terms of mean states between 1 and 3: $dx/dt_I = \frac{1}{2}[(u_1 + u_3) + (a_1 + a_3)]$. In discrete form:

$$(t_3 - t_1)(u_3 + u_1 - a_3 - a_1) = 2(x_3 - x_1). \tag{25.20}$$

The step in the physical plane between points 2 and 3 is again of second-order accuracy, being carried out in terms of the means of states between 2 and 3:

$$(t_3 - t_2)(u_3 + u_2 - a_3 - a_2) = 2(x_3 - x_2). \tag{25.21}$$

25.11. Simulated Experiment

Figure 25.14 represents a tube closed initially at both ends and filled with a gas under pressure in excess of that of the surroundings. It is an elementary matter (Shapiro, 1954) to calculate the pressure which will give rise to specified Mach number $Ma = u/a_0$ at exit if one end of the tube is opened suddenly, as by bursting of a diaphragm in the counterpart laboratory demonstration. Thus, initial conditions may be chosen such that Ma_0 is, say, 0.1. (This, incidentally, is one half of the value arising in Philips' MP1002CA air engine at rated rpm — and far below the 0.3 normally referred (White, 1979) to as distinguishing compressible from incompressible flow under steady conditions.)

The disturbance resulting from the sudden opening propagates leftward as an expansion fan. Immediately to the right of the fan, gas velocity $u = -[2(\gamma-1)]da$, where da is the change of acoustic speed induced by expansion. The fan is reflected from the left-hand enclosure as a continuing expansion. Behind the reflected expansion wave, pressure falls to uniform value until return of the wave reflected from the open end with reversed sign — i.e., as a compression. A two-dimensional version (Mach-line net) will have been constructed by all students of unsteady gas dynamics in the course of deriving the 'square-wave' pressure variation at

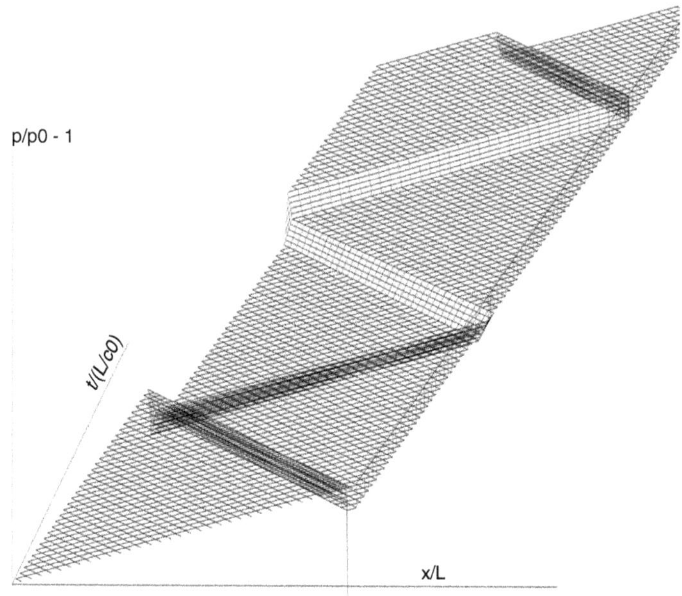

Fig. 25.14. Carpet plot of pressure excess $p/p_0 - 1$ resulting from sudden opening of right-hand end of pressurised uniform, parallel duct.

the closed end. Construction when area-change is involved becomes a matter for the computer.

Figure 25.15 shows the effect on the pressure wave-form of a single step-change in area (by a factor of 2). The pressure surface reveals the increased complexity of the *p*–*t*–*x* history — showing above all that this has been modelled without expanding the inventory of Mach lines with which construction started.

Figure 25.16 extends the construction to two successive discontinuities, calling for a finer grid. In the presence of multiple reflection sites a propagating expansion fan is modified beyond recognition at Ma = 0.01 — i.e., at some 3 percent of the value normally considered to differentiate compressible from incompressible flow in the steady case.

25.12. Discussion

The power of computer simulation has never been harnessed to explore the full potential performance envelope of the Stirling engine, or that of its close relative, the thermal-lag engine.

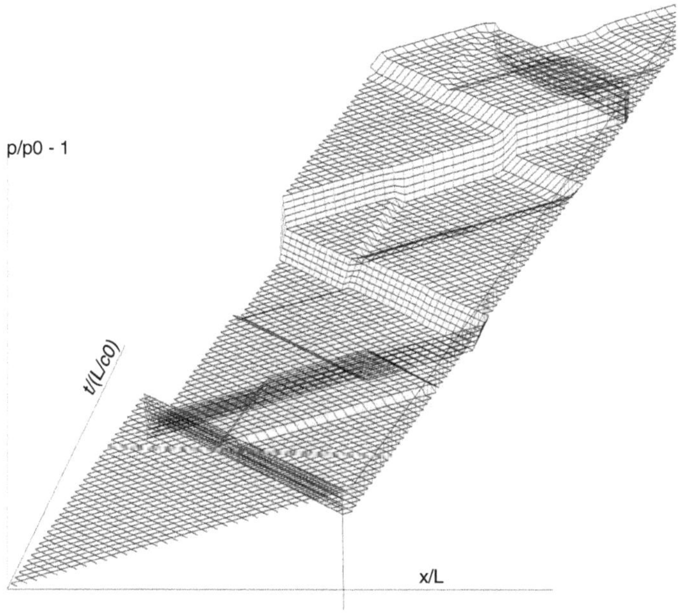

Fig. 25.15. As for previous figure, but for duct having abrupt change in cross-sectional area.

This can be claimed in confidence because a computer simulation needs data — data correlating flow conditions with heat transfer intensity and with flow resistance (the penalty of that intensity) and data essential to the task have yet to be acquired.

There is a subsidiary reason: the tradition in modelling the gas process interactions of the Stirling engine is to work in Euler coordinates. This necessitates compromise which glosses over the phenomena which this chapter has highlighted. It can be — and probably has been — argued that routine operation does not give rise to features such as expansion fans. Quite so. But the point is that, if it did, the Euler formulation does not detect, and having failed to detect, does not display.

The matter of shock waves raises a similar point: it appears highly unlikely that an engine could drive itself so hard as to generate shock-waves. But the MoC takes the *possibility* in its stride, offering a graphical depiction while other algorithms plough ahead unaware.

The MoC is the authentic and mathematically respectable algorithm for solution of the one-dimensional compressible flow-field. It is promoted here because, when coded for computer implementation, it does not

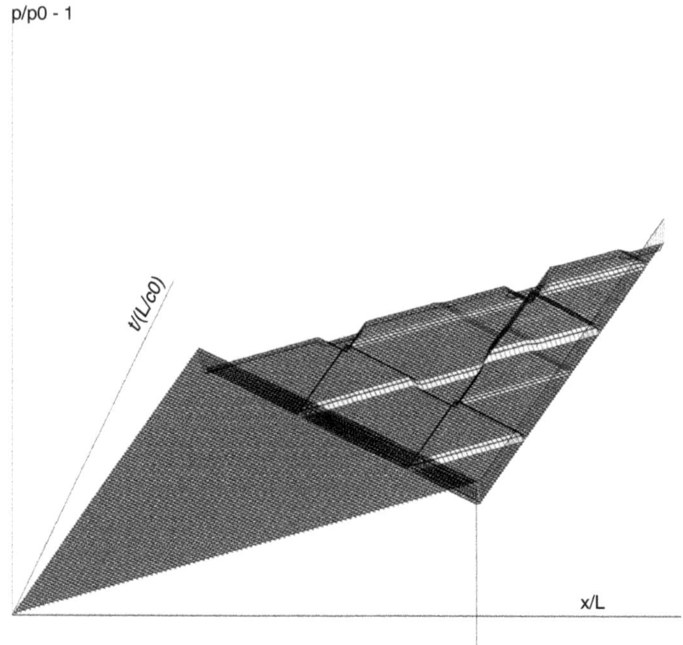

Fig. 25.16. Initial stages in construction of solution as for Fig. 25.14, but for duct having a succession of two abrupt changes in cross-sectional area.

rely on *a single compromise.* It has been shown to cope robustly with systems of interacting waves from multiple reflection sites.

An earlier treatment (the author's 1997 text) explored the response of the wire-mesh regenerator to unsteady flow in terms of linear-wave analysis (LWA). It concluded that this was better characterised as an impedance (i.e., having a reactive component) than as a pure resistance. The power of LWA, particularly as expounded by Lighthill (1952), takes in its stride reflections and re-reflections at an indefinite number of successive sites, and so makes light work of the 300–500 or so sites of the representative wire-gauze stack.

Implementations of LWA to date, however, characterise the individual site in terms of geometry only — specifically in terms of area ratio. The effect of that ratio on the propensity for choking is not considered. This is no longer good enough — as earlier chapters have been at pains to demonstrate. It may well be possible to modify LWA for sensitivity to local upstream *Ma*. This evidently varies between a maximum and zero over each flow reversal, suggesting a theoretical model that could be dismissed as esoteric in the extreme.

Conversely, there is no doubting the result of laboratory experiment advised by, and designed around, a searching analysis.

Performance prediction of the Stirling engine based on currently accessible design data has reached the end of the road. If there is to be a new generation of affordable (i.e., air/N2-charged) machines, design has to be on the basis not only of *awareness* of the physical phenomena which arise and interact at elevated rpm, but *exploitation*! After all, a pressure wave shifts mass, either opposing or complementing the efforts of the displacer.

Modular coding of a cycle model based n the MoC provides for relevant heat transfer and flow correlations to be simply 'plugged-in' if/as/when they become available.

26

The Stirling Engine and the Environment

26.1. Reprise

The Stirling engine is an energy converter. Its commercial future — if there is to be one — can be discussed in terms of energy.

Our primary source of controllable energy is fossil fuel — hydrocarbons — in the form of coal, oil and gas. The entire material well-being of the so-called 'developed' world can be traced to negotiable energy: a gallon of diesel fuel does the day's work of ten fit labourers, allowing one to drive the tractor while the other nine get on with being neuro-surgeons, hedge-fund managers, tarot-card readers, candle-stick makers, baggage-handlers, designers of nuclear-powered submarines and full-time, professional celebrities.

And it is ludicrously cheap: after prospecting, drilling, extraction, transport, refinement, storage, distribution, (in the UK) double-taxation and a profit for the retailer, cost per litre is comparable with that of Coca-Cola.

Energy 'needs' per individual can be back-calculated from figures for CO_2 emissions, for which the UK average is 10 tonnes per year (MacKay 2009). In the USA it's 25 tonnes! The UK figure converts to about 6 kW/person — or 30-times the energy-rate involved in merely staying alive (~200 W). It amounts to 'leverage' over our native muscle-power and means, amongst other things, that babies can now fly before they can crawl — and that wars can be fought with tanks (~1 gallon/mile) submarines and drones.

There is a growing body of scientific evidence that human activity (i.e., energy consumption) is contributing to climate change and that the change is for the worse. Before his untimely death, David MacKay was an eminent scientist — Chief Scientific Adviser to the UK Department of Energy and Climate Change. His conclusion was that the only way of reining in climate change to the time-scale remaining is by drastic reduction in the use of fossil fuels.

Thus arises a problem: no alternative has been lined-up. To respond promptly to the exhortations of the remarkable young Greta Thunberg would mean a return to travel by donkey, cultivation by the horse-drawn plough and doing laundry in the creek. Worse still from the point of view of those who control matters, it would mean letting go of the most prolific cash-cow ever milked by mankind.

The lack of preparedness, given recent images of bush-fires raging, glaciers retreating, perma-frost thawing and polar ice-caps shrinking, might be thought remarkable — even irresponsible. Remarkable, that is, until it is recalled that readiness to the timescale and extent now required has to be initiated from the very top — the top of government, the top of the fossil-fuel industries and the top of the investment funds who thrive off them.

So why no action? Indeed, why the very the reverse — energetic denial that problem even exists? Simple: acknowledging a threat of the magnitude of climate change creates an obligation to respond. And responding meaningfully means coming up with a solution. Those in power do not have a solution. This leaves the well-worn political escape-route: deny the existence of the problem.

I hear howls of indignation: has not the UK Government acknowledged concerns by providing for a Citizens' Panel to debate possible solutions? Yes. But if the agenda is anything to go by (Guardian 24 January 2020) with tongue in cheek: taxation is the laziest and most intellectually-impoverished of all tools: to suggest increasing the tax on aviation in the context of the declared target of 60% increase in air transport capacity by 2050 is nothing short of cynical.

And whoever put biomass on agenda has not done the arithmetic: overall life-cycle return (cultivation to portable fuel) under UK conditions is 1 Watt per square metre — or 0.1% of peak insolation of 1 kW over the same area. As David MacKay comments '*scarcely worth talking about*' (2009 p. 44).

The problem to be addressed (preferably not by a panel of well-intentioned estate agents, personal trainers, calligraphers, dentists, uncle

Tom Cobley *et al.*) is how to make an orderly transition from spending irreplaceable capital to spending out of income. The capital in question is solar energy, concentrated into fossil fuels over millions of years. According to Rampen (2015), we are consuming at a million-times the rate of original formation.

MacKay at p. 178 gives world-wide rate of 'capital' consumption as 15,000 GW (Pro-rating the earlier 6 kW per head of UK population would give 40,000 GW — and we owe it to the rest of the world to plan with the larger figure in mind.) Rate of 'income' in terms of incident solar energy based on 1 kW per m^2 of projected area of the globe is about $12{,}700 \times 10^3$ GW — in other words, of the order of 1000-times the of rate of consumption.

Not all of this can be harnessed, and of that which can be captured at peak intensity, a maximum of 50% can be converted to electricity. Focused solar energy is nevertheless the solution which, in MacKay's words, adds up. He calculates that solar collectors covering a square 600 km × 600 km in the North-African desert would supply the electricity needs[1] of all of Europe and North Africa.

'*Technology has put more shirts on more backs than the charity of all the monarchs in Christendom*' (LeMoyne, n.d.) — technology traceable to James Watt and Robert Stephenson, no less! It has given us the lifestyle we have come to regard as a right. Technology can get us out of it — but only if mobilised before it is too late.

This is not to suggest a crash programme in mass-manufacture of internally-focusing Stirling engines. If the Stirling engine has a track-record, it is in terms of false starts and unfulfilled expectations — although this is not the stumbling block. The latter is the overwhelming inertia and myopia of the institutions in whose interests it lies to appease the One True God of Economic Growth[2]. A technological solution which threatens to interrupt the flow of dividends to investors and shareholders faces the interminable uphill struggle of the least winnable of all wars — a religious war…but for a crucial difference: too much ground held by the Economic Growth side (EG) threatens annihilation of *both* sides.

---

[1] Insufficient to cover wholesale conversion to electric transport.

[2] US Treasury Secretary Stephen Mnuchin recommends Greta Thunberg take a course in economics before offering advice — evidently blind to the facts (a) that there are no courses on the economics of a dying ecosystem from which Greta could choose and (b) that it is conventional economics which are the source of the problem.

26.2. A Case for Pessimism

The EG side has a head-start with representatives in the media trained in sound-bite tactics: Piers Morgan (ITV's *Good Morning Britain*) reassures viewers that *the world is not going to end tomorrow* — neglecting to mention that it is on course to end in *somebody's* tomorrow. BBC *Top Gear*'s Jeremy Clarkson (who emits more than his fair share of CO_2 written, spoken and when in *any* gear) rages that the 'idiot' Greta Thunberg *has killed the car show*. (*Greta Thunberg ... you spoilt brat.' The Sun* newspaper (online) 27 September 2019).

The Economic Growth camp has the whole-hearted support of the UK government. Rather than phasing out coal production, as it earlier committed to doing, Westminster has recently (December 2019) given the go-ahead for opening a new deep coal-mine — the first in 30 years — in Whitehaven, Cumbria, to be called The Woodhouse. Over its lifetime Woodhouse is expected to be responsible for 420,000,000 tonnes of CO_2. Try off-setting that by planting saplings!

In April 2019 the Scottish government declared a 'climate emergency' on the strength of which it claimed to lead the UK in measures to cut CO_2 emissions. A date was set for achieving 'zero net carbon emissions'. Simultaneously appeared announcements of 'investments' in North-Sea oil drilling for the period 2019-30: Rosebank-Lochnager: £4.4bn; Captain: £700; Pilot: 450m; Avalon: £450m; Liberator: 400m; Capercaille: £400m. Total: £6.8bn. In order to cope with increased output there is to be modernisation and expansion of at least two refineries, Grangemouth and Hunterston.[3] The same government is also considering plans to reduce airport tax to encourage air travel, while Scotland's airline, Loganair, has announced a significant expansion of its routes — both consistent with predictions that UK airport capacity is set to increase 59% by 2050.

In the USA the White House administration is busy undoing all measures to protect the environment put in place by President Obama.

Vladimir Putin's position is clear from his attempted rubbishing of the remarkable young Greta Thunberg, which has been translated as:

'No one has explained to Greta that the modern world is complex and different and ... people in Africa or in many Asian countries want to live at

---

[3] *Scotland's new fossil fuel plans will help fry the planet.* National, 15 September 2019, p. 14.

the same wealth level as in Sweden.' (*World News*, Reuters, 02 October 2019). In other words, keep drilling!

Apparently, no one has explained to Vladimir that temperatures in some African and Asian countries are already quite high and that, at a few degrees higher, wealth-level becomes the least of the problems.

President Bolsonaro of Brazil has reiterated his commitment to accelerating de-forestation of the Amazon — and '*... Europe can mind its own business*'.

Or consider the object lesson in cynicism from online source (https://oilprice.com/Energy) under the headline *How Much Oil Is Up For Grabs In Syria?*

> *Though it's as yet uncertain whether the Pentagon is actually going to execute the plan, Trump is mulling keeping a small US troop contingency in Syria in order to "secure the oil". The president said at a cabinet meeting Monday: "I always said if you're going in, keep the oil. We'll work something out with the Kurds so that they have some money so that they have some cash flow. Maybe we'll get one of our big oil companies to go in and do it properly."*

The Economic Growth side is up against an opponent which has no generals. Were I the monarch I would summon the Prime Minister with instructions to bring a list of effective measures and a timetable for remedial action. Regrettably — given that thermal run-aways are not renowned for giving second chances — the royal focus is on conferring further symbolic honours[4] upon those already weighed-down with them.

26.3. A Case for Optimism

This is largely in terms of straws-in-the-wind.

- (Posted by Associated Press, 10 October 2019.) The Brazilian government has put up for auction licences for exploration of oil fields close

[4] *Kate Middleton has been awarded a significant royal honour by her grandmother-in-law* news.com.au December 2019.

The Queen has made Prince William Lord High Commissioner to the General Assembly of the Church of Scotland. The Cambridge, 26 January 2020.

to Abrolhos Marine National Park in the state of Bhaia. The auction, due to be held Rio de Janeiro on October 10th, 2019 was boycotted by all 17 companies scheduled to bid '*...under pressure from environmental organizations*'.

- Under the heading '*How the military will spark the death of fossil fuels*', Zaremba (2019) draws attention to a report by the news site Quartz that *U.S. defense leaders recognize the threat of climate change, even if their commander-in-chief doesn't.* The report continues: *American military strategists have long viewed climate change as a 'threat multiplier.'* In its Quadrennial Defense Review the Pentagon cited changing weather as *aggravating stressors abroad such as poverty, environmental degradation, political instability, and social tensions.*

On another front, the potential for compelling and massive side-benefits cannot be overlooked indefinitely:

Reduced international tensions: The aggressive demand for mineral resources in general is associated with war, territorial dispute and land dispossession. Look no further than Trump's attempt to buy Greenland or Bolsonaro's treatment of the indigenous peoples of the Amazon. And note Cyprus' investment in drones[5] to monitor Turkey's incursions in search of drilling sites in the Mediterranean.

A lessening of the blight of coastal and marine contamination: *Brazil is dealing with oil deposits over 1500 miles of its north-east coastline, with 130 previously pristine beaches off the coast of Bahia reportedly devastated* (*Bloomberg*, 10 October 2019).

Emergency teams are fighting to protect some of the world's most unique and fragile ecosystems after 600 gallons of oil spilled into the waters around the Galapagos Islands when a cargo vessel overturned (*The Telegraph*, 23 December 2019).

26.4. A Case for Despair

On 28 January 2020, The Whittle Laboratory at the University of Cambridge UK received a visit by the Prince of Wales.

---

[5]The National, 08 October 2019.

According a report by Harriet Brewis for the Evening Standard: *He warned that we 'haven't got time to waste' if we want to save the planet from climate change: 'We have run out of time now to rescue this poor old planet from man-made emissions and all the complications we're now facing... And the future has to happen much quicker than it used to, that's the problem, so we have to find the best ways to bring everybody together for once so we don't have endless argy-bargies about it but actually get down together... If you mount a military campaign, because you're up against it, you have to produce the answers, overnight practically.'*

His Royal Highness returned from Cambridge to Highgrove as he had arrived — by Royal helicopter. Independently estimated cost: at least £12,000 and $2\frac{1}{2}$ tonnes of CO_2.

26.5. But What About Action?

The crisis is not without parallel — albeit a loose one: The 1930s faced a different threat to existence — the Second World War. 'Appeasement', '*Peace in our time*' sum up the official approach. Fortunately, private industry saw things differently: The de Havilland aircraft company recognised the threat and reasoned that UK metal-working facilities would be stretched to capacity in armament manufacture, while the woodworking and cabinet-making skills of the substantial wooden-furniture industry would be facing redundancy.

The company's resourceful response, in the face of official indifference, was to design and develop an all-wood aircraft, the legendary Mosquito. When the Air Ministry woke up to the unprecedented performance and versatility of the world's first multi-role combat aircraft they could not order in sufficient numbers. The Mosquito went on to inflict greater damage on the enemy than the Spitfire of more romantic association.[6]

Given the dire consequences of continued procrastination, a low-technology prime-mover, uniquely capable of functioning from sunlight independently of water supply, must surely merit serious consideration. Not for routine operation in the British climate, obviously, but as a manufacturing and export opportunity as well as a pilot study in making the transition from fossil fuels to 'renewables'.

[6] de Havilland Mosquito bomber DVD 96069 www.deltaleisuregroup.co.uk.

This book is about the Stirling engine rather than the environment, but where the two aspects can come together to mutual advantage is the academic environment, as illustrated anecdotally below.

26.6. A Role in Academia?

This sophisticated concept has the potential to exercise every topic[7] in the mechanical engineering undergraduate syllabus — kinematics (velocity and acceleration diagrams), dynamics (balancing), tribology, thermodynamics, gas dynamics, fluid mechanics, combustion (stoichiometry) materials science, tribology, stress-analysis, structural analysis, bearing selection, heat transfer (convection, conduction, radiation) gear design, computer-coding, use of proprietary software (Fluent™, ProEngineer™), traditional engineering drawing — and more. In a faculty of engineering where I had the privilege of holding a post, the conjugate heat exchange problem was taught and illustrated in terms of the every-day phenomenon of 'turning on the hot tap'. This is the regenerator problem by another name, and the opportunity arises of finally doing a professional job of a problem of which the majority of texts make a pig's breakfast.

In that same faculty, a newly-appointed head proposed a revolution in teaching. He saw a choice between two extremes. One was 'design-project-centred', with each relevant 'chalk-and-talk' class directing its emphasis and worked examples at the challenges of the core design project. Laboratory experiments, drawing-office exercises and take-home problems would be revamped accordingly. One candidate project for the mechanical track was the Shell Mileage Marathon vehicle (a competition which was attracting entries surpassing 10,000 miles/gallon. A worthy challenge).

The other extreme put applied mathematics at the centre, with its syllabus and worked examples selected from, and at the service of the specialist topics.

The prospect appealed at a number of levels. Imagine the benefits of generations of graduating engineering students wise to the fact that fluid flow is not always steady — and not always from left to right — as it invariably is on the blackboard.

[7] An exception: turbo-machinery.

Tragically, the enterprising professor died prematurely, and before the experiment could be put to the test. His successor evidently subscribed to the view that '*if it's not broken, don't fix it*' and we reverted to business as usual.

With hindsight, the workload throughout the four-year transition period would have been crippling. To a lecturer appointed from outside, the quality and scope of the established teaching support was eye-opening: two weeks following issue of a set of homework problems, model solutions were available as hard-copy and on micro-fiche (and subsequently online). Each college library kept multiple bound copies of past examination papers with model solutions. Timetabled 'examples classes' were available for optional attendance over and above the corresponding lecture and college supervision. It would have been necessary to discard a well-oiled, tried-and-tested system, and quite arguable that, in context (a long-established university) an inspired concept was a non-starter.

But that context is not typical: university courses have since been set up from scratch. A least one — NMiTE[8] (University of Hereford, 2019) — has a strong project element ('*We want to bring the passion back to engineering.*') The University of the Highlands and Islands (UHI) has recently opened a campus (Lews Castle, Stornaway). It publishes a flexible, enlightened syllabus for its four-year honours BEng in Energy Engineering (University of the Highlands and Islands, 2019) which places reassuring emphasis on renewable energy and 'future energy systems'.

It is too late for me to influence events, but opportunities to implement the earlier aborted experiment appear to be on the increase rather than the reverse. The long-term impact of generations of engineering graduates already competent in the technology of regenerative thermal machines can only be imagined.

26.7. Back to the Present

Prospects have to be weighed bearing in mind that, regardless of external appearances, the global economy ticks in the capitalist way. How else can it be that the system of carbon credits, promoted (unregulated) as a step towards saving the planet, allows trading for profit on the stock

[8]NMiTE — New Method in Technology and Engineering.

exchange — and has become a lucrative revenue stream for the burgeoning enterprise of scamming? (Financial Conduct Authority, 2018).

The prototype reality check was articulated as early as the 1960s by the chairman of the once-mighty ICI, Paul Chambers: '*ICI is in business not to make chemicals, but to make money*.' The wider world feigned shock, but the cards were on the table.

An orderly transition to a renewable energy economy depends on convincing capitalism that it can flourish — i.e., that investors and shareholders can continue to take a profit uninterrupted. The world of investment is blinkered — but not naïve. Unorthodox technology will not be embraced before being demonstrated, proven and fully costed.

That process needs to start now. And it would help to have the media on-side: rather than bullying Miss Thunberg from the safe distance of a newspaper column, Mr Clarkson might consider emerging from the 20th century and using his undisputed communication skills to put across the message crucial for the 21st and beyond.

The final arbiter will not be logic, but Mother Nature. When sea-level rise coincides with a high tide, the Thames barrier is overwhelmed and the wine cellars of the Palace of Westminster take in water, those who don't 'get'[9] climate change may finally do so. Initially, action will follow the routine pattern: blame — blame for the scientists who failed to couch their warnings in the form of sufficiently convincing sound-bites.

But then what? Difficult to predict. A lot could depend on how much preparation has been taking place under private initiative while officialdom has been conducting business-as-usual. The transition itself (or more likely the rear-guard action) will consume energy. Fossil sources must be conserved for that purpose in preference to being used for setting new land speed records.

Those interested in the outcome may not have long to wait.

[9]Claire Perry O'Neill former president of this year's UN Climate Change Conference, says Boris Johnson 'doesn't get' climate change. *The Telegraph*, 4 February 2020.

Appendices

Appendix A

A Guide to *In Situ* Regenerator Testing

A.1. Context

There seems little chance of a major breakthrough in regenerator performance prediction before the matter of correlations is resolved (Chapters 7, and 8). In the meantime, it may be justifiable to press ahead using existing resources — but with due regard for Dynamic Similarity implications.

The scenario is a Stirling engine already constructed and delivering power. The designer requires a percentage change (increase or decrease) in regenerator heat transfer capability specified in terms of NTU (Number of Transfer Units) = StL_r/r_{hr} and, independently, a different percentage change in fractional pressure drop dp/p.

The regenerator is a close-packed stack of square-weave gauzes. The treatment extends to matrices of other geometries for which flow friction and heat transfer correlations are available in conventional form, i.e., C_f vs Re and St vs Re. Gauzes have wire diameter d_w and mesh number m_w giving hydraulic radius r_{hr}. Under the close-packing assumption, stack length L_r in the flow direction is $2n_gd_w$.

Subject to a condition, the L_r and wire diameter d_w needed to effect the required changes can be reduced to a pair of design charts. The condition is that the revised L_r and d_w be achieved without changing regenerator dead volume V_{dr}.

The underlying theory is elementary and is embodied in design charts in nomogram form.

A.2. Introduction

A substantial literature addresses the 'regenerator problem'. Relevant work by mathematicians dating from the 1920s can be found under the heading 'conjugate heat exchange problem'. On the other hand, anyone who has attempted to reconcile the operating characteristics of a particular Stirling engine with the parameters of a real regenerator — hydraulic radius r_{hr}, overall flow-passage length L_r, volume porosity $¶_v$, etc. will have found even the masterful distillation of the subject by Jakob (1957) of little direct help.

While by no means remedying the situation, this appendix offers a tool having potential for guiding experimental development of a regenerator for a specific application, i.e., for an existing engine at given operating conditions — rpm, charge pressure p_{ref}, charge gas, etc.

It is taken for granted that an aim is maximum cycle efficiency. This calls for a high value of thermal recovery ratio, η_T and, in turn, a high Number of Transfer Units, $NTU = StL_r/r_{hr}$. However, in the context of forced convective heat transfer, the aim is not the raw maximum: classic regenerator thinking and corresponding temperature solutions sidestep the fact that heat transfer comes at a cost in terms of fractional pressure drop dp/p with its associated parasitic pumping power. It is noteworthy that regenerators of 'bench-mark' engines — GPU-3, V-160, P-40, etc. — do not make use of the 'ideal' gauze, which would otherwise be that offering the highest available ratio of surface area to volume, i.e., the smallest hydraulic radius r_{hr}. Design is accordingly a compromise between the benefit of enhanced temperature recovery and the accompanying penalty of increased pumping power.

To appreciate the proposed method, note that square weave gauzes of common volume porosity $¶_v$ share geometric similarity and therefore the same flow correlation regardless of the parameters of that correlation — *Re* and/or *Ma* — or of absolute size.

Volume porosity may be expressed in terms of d_w and mesh number m_w [wire/m]:

$$¶_v \approx 1 - \tfrac{1}{4}\pi d_w m_w \sqrt{[1 + (d_w m_w)^2]}. \tag{A.1}$$

According to Eq. (A.1), any combination wire diameter and mesh spacing which retains the initial value of the product $d_w m_w$ leaves $¶_v$, and thus flow-passage geometry, unchanged.

Hydraulic radius r_{hr} may be expressed in terms of $¶_v$ and d_w:

$$r_{hr} \approx \frac{1}{4} d_w ¶_v / (1 - ¶_v). \tag{A.2}$$

Equation (A.2) says that r_{hr} may be varied *ad lib* without changing $¶_v$ — as long as the change in m_w is in inverse proportion to that in d_w.

Now note that a given set of operating conditions (rpm, p_{ref}, etc.) will, over a cycle, generate a particular distribution of mass m and mass rate $m' = dm/dt$. With the aid of the identity $|m'| = \rho|u|A_{ff}$, Reynolds number $Re = 4\rho|u|r_h/\mu$ can be expressed as $4|m'|r_h L_r/(\mu V_{dr})$. (The conversion makes use of $V_{dr} = A_{ff}L_r$.) Modest changes to r_{hr} and/or L_r have relatively little effect on dm/dt and hence of Re *provided they leave dead volume* V_{dr} *unchanged* (see sensitivity analysis at the end of this chapter). Changes in Re are then overwhelmingly a function of changes to r_{hr} and/or L_r.

The vehicle for illustrating the method will be the GPU-3. An estimate of peak Re is 67 — well below the critical value read from Fig. 4.3 with charge pressure of 69 bar and r_h of 0.0302 mm. Consistent with this, the high acoustic speed of the charge gas (H2) gives peak Ma of 0.007 — safely below the criterion of $Ma < 0.1$ used to construct Fig. 4.3. This justifies fitting algebraic expressions of the form $StPr^{2/3} = a/Re^b$ and $Cf = c/Re + d$ to correlations complying with Kays and London format — although other instances of the method, particularly those involving air or N2, require to be treated on their respective merits.

Interest accordingly lies in the ratio of Re modified by the changes to installed Re, viz, in Re/Re_0, It is unnecessary to calculate either Re or Re_0, since, *under present assumptions*, the ratio reduces for all locations and crank angles to $r_{hr}L_r/r_{hr0}L_{r0}$.

The relationship between Re and C_f, and between Re and $StPr^{2/3}$ is the link between the proposed changes to dp/p and to NTU, and the L_r, d_w and mesh number m_w required to effect those changes.

A.3. Assumptions

An increase in calculated NTU results in an improvement in temperature recovery ratio η_T, and likewise a decrease leads to deterioration.

An increase in calculated C_fL_r/r_h (where C_f is friction factor) leads to increased dp/p and thus to increased pumping loss and likewise a decrease leads to a reduction.

A.4. Rationale

Suppose that recovery ratio η_T, estimated by some rational method (i.e., in terms of Reynolds number Re and known values of d_w, m_w and L_r) is 97.5% and that calculated fractional pressure drop dp/p is 0.005. The numerical values are of no absolute significance because an alternative — but equally rational — formula will give a different numerical result.

Suppose, however, that d_w and/or m_w and/or L_r is/are changed and *the original method of calculation is reapplied.* Let the original values of recovery ratio and pressure-drop ratio be distinguished from the new values by subscript $_0$, viz, as NTU_0 and dp/p_0. Unlike respective absolute values, the *ratios* NTU/NTU_0 and $(\mathrm{d}p/p)/(\mathrm{d}p/p)_0$ have significance to the extent that there is justifiable expectation that a figure in excess of unity signifies an increase.

A.5. Algebraic Development

The development infers operation at some fixed condition — rpm, charge pressure p_{ref}, working fluid and cycle temperature limits T_E and T_C.

By any definition of recovery ratio η_T, it is a strong function of NTU:

$$NTU = StL_r/r_h, \tag{A.3}$$

where St is Stanton number $h/(\rho|u|c_p)$. If the correlations of Kays and London (1964) are to be relied upon, St relates to Re in steady, incompressible flow through the wire-screen stack by:

$$StPr^{2/3} = a/Re^b. \tag{A.4}$$

(Some of Kays' and London's tests involved a small number of layers of crossed rods — *not* hundreds of fine wire screens! And the results *do not scale* to all flow conditions arising in all Stirling engines (Organ 2007, 2014).)

In terms of mass rate m', Re becomes:

$$Re = 4|m'|r_{hr}L_r/(\mu V_{dr}). \tag{A.5}$$

Thus, at fixed operating conditions, *NTU* is a function of r_{hr} and L_r (and, by extension, of d_w and m_w). Let *NTU* denote a target value to be achieved by modifying r_{hr} and L_r:

$$\frac{NTU}{NTU_0} = \frac{a/Re^b(L_r/r_{hr})}{a/Re_0^b(L_{r0}/r_{hr0})}.$$

From Eq. (A.5), $Re/Re_0 = r_{hr}L_r/r_{hr0}\,L_{r0}$. Substituting and rearranging,

$$L_r = \{(NTU/NTU_0)(r_h/r_{h0})^{(1+b)}\,L_{r0}{}^{(1+b)}\}^{1/(1-b)}. \tag{A.6}$$

From the standard definition of C_f, pressure gradient dp/dx over elemental distance dx is

$$\mathrm{d}p/\mathrm{d}x = -\frac{1}{2}\rho u^2 C_f/r_{hr}.$$

A potential inconvenience is the wide variation in ρ in flow direction x due to the intense temperature gradient. Calculation accordingly proceeds in terms of $\rho u = m'/A_{ff}$ where A_{ff} is free-flow area.

For given volume porosity (and again for steady, incompressible flow) the commonly-applied relationship between *Re* and friction factor C_f is

$$C_f = c/Re + d. \tag{A.7}$$

Recalling that $A_{ff}L_r = V_{dr}$, over some characteristic length in the flow direction (which may as well be L_r)

$$\mathrm{d}p/p = -\frac{1}{2}m'^2(c/Re + d)(p_{ref}V_{dr})^{-2}\,L_r{}^3/r_{hr}. \tag{A.8}$$

At a given operating condition dp/p is again a function of r_{hr} and L_r (and again, by extension, of d_w and m_w). With dp/p denoting target fractional pressure drop at location x and crank angle ϕ, the r_{hr} and L_r required to achieve it are related through:

$$r_{hr}{}^* = \frac{(L_r{}^3/r)(c/Re + d)}{(L_{r0}{}^3/r_{hr0})(c/Re_0 + d)}. \tag{A.9}$$

Symbol r is the ratio of target dp/p to installed (d$p/p)_0$, i.e., is dp/dp_0.

The unknown Re may be rewritten $(Re/Re_0)Re_0$ where Re_0 is the installed value. Ratio Re/Re_0 simplifies to $r_{hr}L_r/r_{hr0}L_{r0}$ as before. The equation resulting from elimination of ratio L_r/L_{r0} between Eqs. (A.6) and (A.9) has non-integer powers of r_{hr}/r_{hr0}. The complication is immaterial: display in nomogram form permits any variable of choice (dependent or independent) to be read off in terms of the other two.

A.6. Redesign Charts for GPU-3

Figure A.1 and those which follow are constructed on the basis of the rated operating point of the GM GPU-3. Given a target ratio NTU/NTU_0 and a (generally different) target dp/dp_0 they display fractional values of L_r/L_{r0} and of r_{hr}/r_{hr0} (and/or of d_w/d_{w0}) calculated to achieve both targets simultaneously. *The same pair of target values must be used with both charts.*

For square-weave wire gauze, hydraulic radius r_{hr} is proportional to wire diameter d_w. Figure A.2 is the graphical equivalent of Eq. (A.2) and may be useful for quick conversions.

For porosity ¶$_v$ unchanged between design alterations, Fig. A.3 serves to determine both r_{hr} and d_w.

A cardinal condition of the method is that dead volume V_{dr} should remain unaffected by the changes. Calculating new mesh number m_w via the definition of volume porosity ¶$_v$ in terms of d_w while retaining the original ¶$_v$ value keeps matrix thermal mass $\rho_w c_w V_{dr}(1 - ¶_v)$ unchanged. This in turn maintains the original thermal capacity ratio and thus the same nominal matrix temperature swing.

Inverting Eq. (A.4) yields the redesign value of m_w. Figure A.4 achieves this graphically.

The requirement that the redesign process should leave dead volume V_{dr} unchanged means that an increase in L_r over L_{r0} is accompanied by a reduction in free-flow area A_{ff} of the stack. All other things being equal, thermal short is reduced, and gas flow is more nearly one-dimensional — both arguably beneficial.

If, on the other hand, $L_r < L_{r0}$ the situation is reversed. The absolute value of thermal short is difficult to estimate because of the unknown thermal conductivity of the stack. However, working again in terms of ratios eliminates unknowns and offers a reassuring idea of the percentage

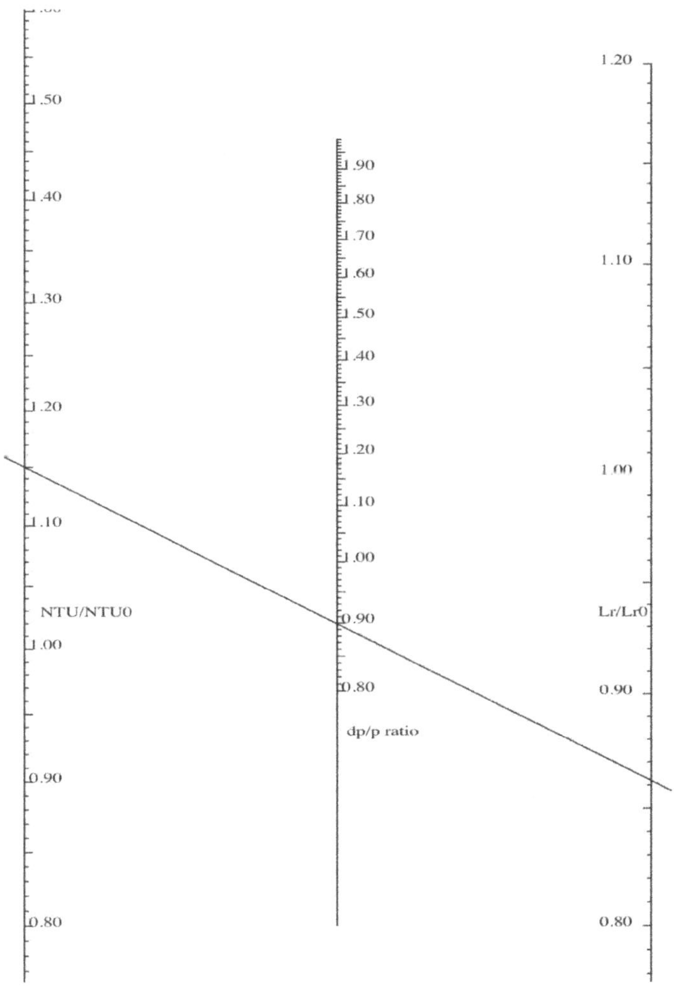

Fig. A.1. Solution in nomogram format of Eq. (A.6) jointly with Eq. (A.9) for L_r/L_{r0}.

increase (or decrease) in the combined thermal leak — stack plus housing. Appendix A.1 has the assumptions and the algebra.

A.7. What Target *NTU*?

Regenerator operation is cyclic by definition, and at no point in a cycle are conditions at steady state. There is accordingly no unique value of

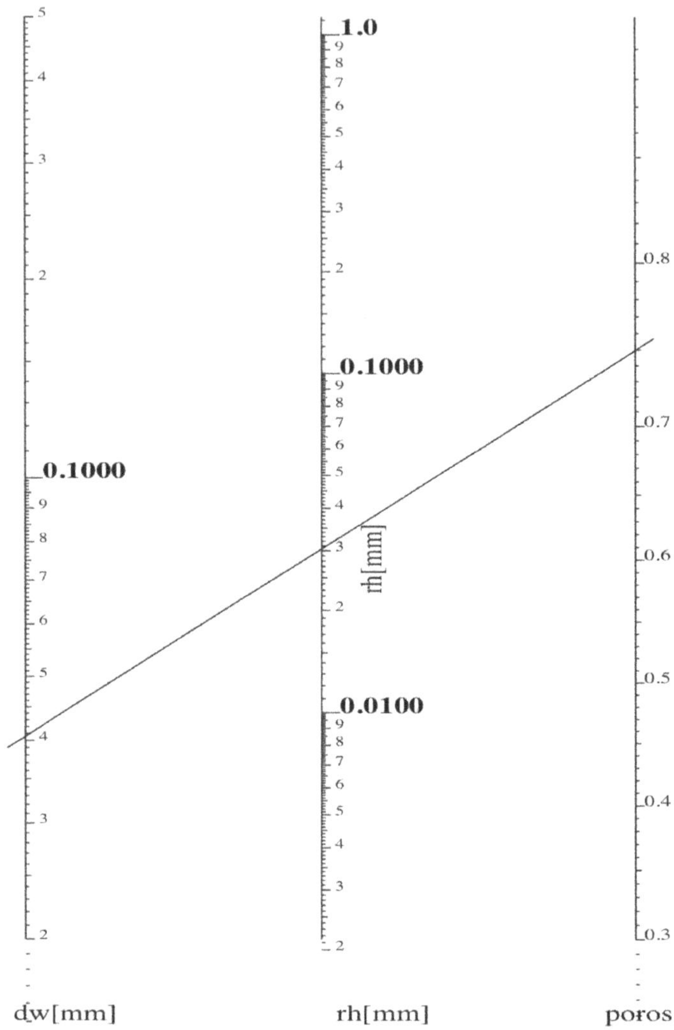

Fig. A.2. Hydraulic radius r_h in terms of d_w and $¶_v$ according to Eq. (A.2).

ratio η_T, and it could be argued that targeting a percentage *change* in η_T (rather than an absolute value) is the only approach which makes sense.

Figure A.5 nevertheless sheds light. It displays Eq. (A.9) which in turn gives the dependence of η_T on NTU for the hypothetical case of steady flow with uniform NTU:

$$\eta_T = (NTU - 1)/NTU. \tag{A.9}$$

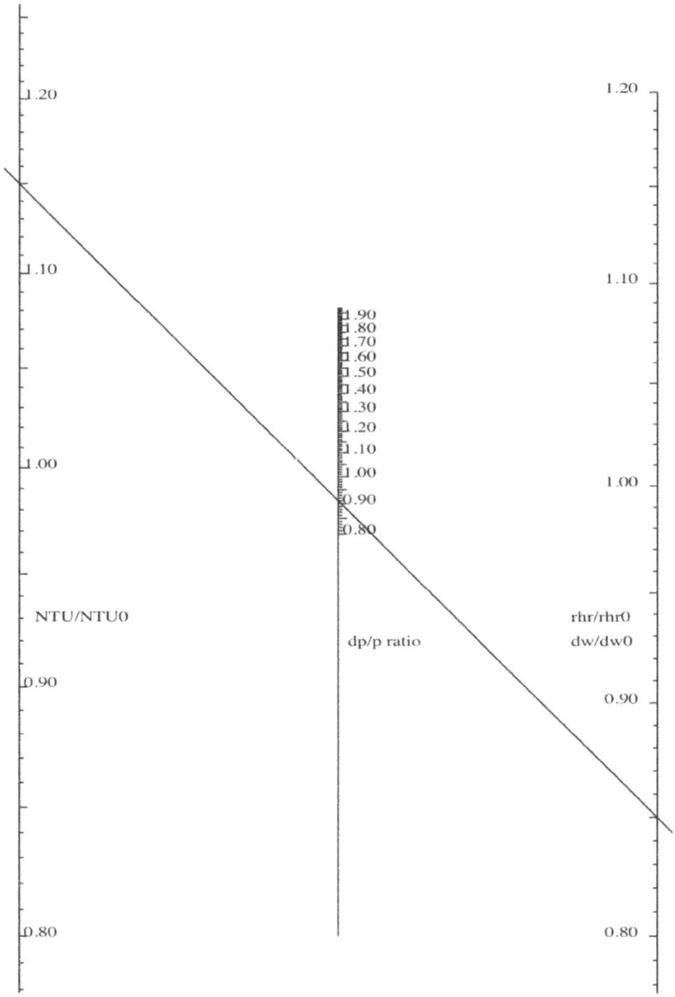

Fig. A.3. Solution in nomogram format of Eq. (A.6) with (A.9) for r_{hr}/r_{hr0} (and/or d_w/d_{w0}).

It shows rapidly-diminishing marginal returns in η_T as *NTU* value increases above 10. (There is no reason to think that a more sophisticated relationship than Eq. (A.9) would show a different *trend.*) On the other hand, early work by NV Philips of Holland asserts that a small change in regenerator efficiency has a disproportionate effect on overall engine performance. The only way forward is to try!

The example continues by reference to the GPU-3, targeting 115% of installed *NTU* with d*p*/*p* reduced to 90%. Joining these two values by a

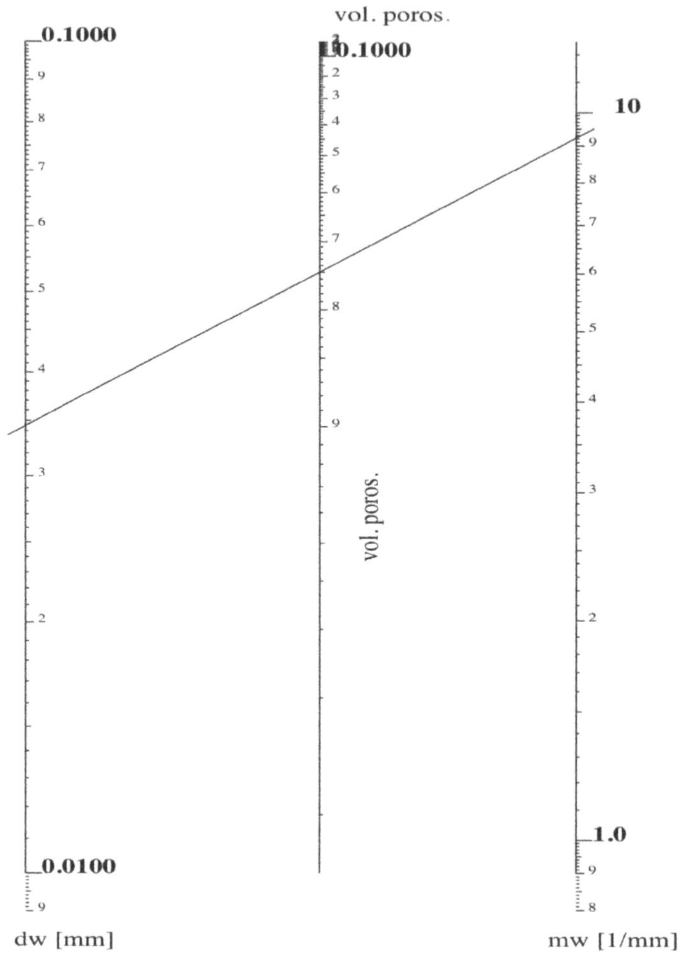

Fig. A.4. Mesh number m_w in terms of d_w and $¶_v$ according to Eq. (A.1). (NB $¶_v$ scale elapses *downwards*.)

straight line in Fig. A.1 shows stack length L_r reduced by 13.8% to 86.2% of L_{r0}.

At a fixed value of volume porosity, r_{hr} is proportional to d_w (Eq. (A.2)), so percentage values of r_{hr}/r_{hr0} and of d_w/d_{w0} are the same. Turning to Fig. A.3 and entering with the previous pair of target values reveals that the required d_w is some 85% of d_{w0}.

The latter being 0.0406 mm means that the new d_w is 0.0345 mm. The new m_w required to retain the original ¶$_v$ (Eq. (A.1) or Fig. A.4) is $m_{w0}/0.85 = 7.84/0.85$ wire/mm = 9.22 mm^{-1}.

There is little point in specifying a gauze unless supplies are commercially available. A square-weave mesh in stainless steel to AISI from the screen-printing range of G Bopp and Co (Anon, 1986) is their designation SD 75/36 and has $d_w = 0.036$ mm. with $m_w = 9.0$ mm^{-1} (230 mesh).

Haver and Boeker (Anon, 1998) advertise precisely this specification as part of their standard range.

The perfectionist will no doubt now reverse the calculation and check revised *NTU* and d*p*/*p*.

Thermal short as a fraction of installed value (Section A.9 below) is the inverse square of length ratio, viz $(L_{r0}/L_r)^2$. In this case it is $(1.0/0.862)^2 = 1.346$, or a 34% increase.

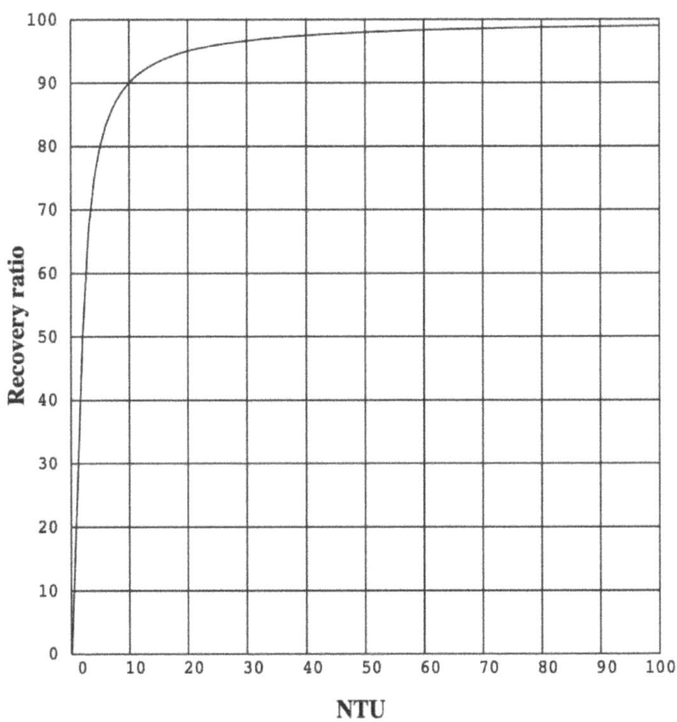

Fig. A.5. Temperature recovery ratio (%) in function of *NTU* according to Eq. (A.9).

A.8. Discussion

The scale values of the design charts reflect the assumption that Kays' and London's heat transfer and flow correlations are applicable to conditions in the Stirling engine. In general they are not — and at any number of levels (see earlier chapters). The formulation *per se* has passed numerical tests which seem pertinent.

Without further input — particularly from experiment — discussion would be largely speculative, and is accordingly withheld.

A.9. Thermal Short q'_L

The matrix stack is a close fit in a parallel, cylindrical housing of uniform radial thickness t. Housing and stack are of common axial length. With k_{eff} and k_w for thermal conductivity of stack and housing respectively:

$$q'_L = -\frac{1}{4}\pi D^2 k_{eff}(T_C - T_E)/L_r - \pi D t k_w (T_C - T_E)/L_r$$

$$= -\pi D(T_C - T_E)/L_r \left\{\frac{1}{4} D k_{eff} + t k_w\right\}. \quad \text{(A.10)}$$

Hoop stress s_h induced by pressure p in a parallel cylinder of wall thickness t (and infinite axial length L_r) is

$$s_h = pD/2t \; (t \ll D). \quad \text{(A.11)}$$

On the basis that installed and modified housings are assumed designed to the same factor of safety *fos* over hoop stress s_{hy} at yield, Eq. (A.11) is substituted into Eq. (A.10), giving:

$$q_L = -\pi D^2 (T_C - T_E)/L_r \left\{\frac{1}{4} k_{eff} + p\, fos\, k_w / 2s_h\right\}.$$

Noting that free-flow area $A_{ff} = \frac{1}{4}\pi D^2 ¶_v$ and that $A_{ff} = V_{dr}/L_r$:

$$q_L = -4V_{dr}/¶_v (T_C - T_E)/L_r^2 \left\{\frac{1}{4} k_{eff} + p\, fos\, k_w / 2s_h\right\}. \quad \text{(A.12)}$$

Under present assumptions, the only term which varies during the proposed optimisation search is L_r. However, because the algebra extends without difficulty to allow for variable porosity $\P_v$, this is retained in the final expressions:

$$q_L/q_{L0} = \P_{v0}L_{r0}{}^2/\P_v L_r{}^2.$$

A.10. Sensitivity Analysis

Pressure drop dp for the regenerator as a whole is notionally dp = $\frac{1}{2}\rho u^2 C_f L_r/r_{hr}$. Equation (A.4) is used to eliminate C_f and ρ replaced by p/RT. Setting $u^2/RT = Ma^2$ and rearranging:

$$\mathrm{d}p/p = \frac{1}{2} Ma^2 L_r/r_{hr}(c/Re + d). \tag{A.13}$$

Re has been expressed in terms of m' (Eq. (A.5)) whose possible variation with dp/p is now to be investigated. The u of *Ma* may likewise be expressed in terms of m', which can be eliminated between *Ma* and *Re*:

$$Ma = (Re\mu\sqrt{RT})/(4pr_{hr}). \tag{A.14}$$

Substituted into Eq. (A.13), Eq. (A.14) yields an equation which is quadratic in *Re*:

$$\frac{dp}{p} = \frac{1}{2}\frac{\mu^2 RT}{16p^2} L_r/r_{hr}^3\{cRe + dRe^2\}. \tag{A.15}$$

Solved by standard means, Eq. (A.15) yields two solutions for *Re* in terms of dp/p, L_r and r_{hr}. Inspection reveals the positive root to be the relevant solution, which is of the form:

$$Re = Re(\mathrm{d}p/p, L_r, r_{hr}). \tag{A.16}$$

Sensitivity of *Re* to a change in L_r (for example) is revealed by differentiation as $\partial Re/\partial L_r$. Evaluating the resulting differentials using values of R, μ etc., relevant to the present study (GM GPU-3) reveals the sensitivity of *Re* to changes in dp/p to be two orders of magnitude less than sensitivity to changes in L_r, and three orders less than sensitivity to changes in r_{hr}.

Appendix B

Reynolds Number *Re* and Mach Number *Ma* Terms of Dimensionless Parameters and Variables of Stirling Machine Operation

$$Re = 4\rho|u|r_h/\mu. \tag{B.1}$$

Speed u is eliminated with the aid of the expression for mass-rate m', viz, $\rho u = m'/A_{ff}$. For the ideal gas ($p/\rho = RT$) ρu may be written $\rho u = m'p/RTA_{ff}$. Dimensionless mass-rate is defined $\sigma' = m'/\omega M$.

Recalling the definitions of Stirling parameter N_{SG} and speed parameter N_{MA} (Notation), and noting that dead-space ratio $\delta_r = V_{dr}/V_{sw} = A_{ff}L_r/V_{sw}$ allows Re to be expressed:

$$Re = \frac{4\sigma' N_{SG} N_{MA}^2 (L_r/L_0)^2}{2\psi\delta_r (L_r/r_h)}. \tag{B.2}$$

In Eq. (B.2), ψ is Finkelstein's dimensionless pressure (see Notation).

Mach number Ma in terms of dimensionless parameters and variables.

$$Ma = u/\sqrt{(\gamma RT)}. \tag{B.3}$$

Speed u is again eliminated with the aid of the expression for mass-rate m', viz, $\rho u A_{ff} = m'$. In the case of an ideal gas ($p/\rho = RT$) u may be written $u = m'RT/pA_{ff}$. Dimensionless mass-rate is $\sigma' = m'/\omega M$.

Void volume $V_{dr} = A_{ff}L_r$. Dead-space ratio is $\delta_r = V_{dr}/V_{sw}$. Taken together with the definitions of specific speed — or speed parameter — N_{MA}, Ma becomes:

$$Ma = \frac{\sigma' N_{MA} L_r \sqrt{T/Tc}}{2\psi\delta_r L_{ref}\sqrt{\gamma}}. \tag{B.4}$$

Appendix C

A Perturbation Formulation

C.1. Introduction

A high priority in studies of regenerative thermal cycles is a tool for reliable prediction of thermodynamic performance in advance of hardware manufacture. The picture usually conjured up is a comprehensive, computer-based representation of diffusion, dissipation and convection phenomena as they interact in creating the practical cycle.

Where the object is to compare specific phenomena in rival numerical models the full simulation may not be the best way forward. A topical instance is provided by the two, contrasting approaches to local, instantaneous pressure-drop — the C_f–Re dependence built into the steady-flow correlations of Kays and London (1964) vs the dp/p–Ma–Re picture of Pinker and Herbert (1967). The outcome of either is influenced by cycle temperature distribution via the path-line equation — but in different ways. The clearest distinction is promised by a formulation which decouples the pressure-drop of interest from thermal convection effects.

In the perturbation treatment, decoupling is achieved by formulating in terms of the difference, or 'error', between the ideal, lossless cycle and a cycle embodying the candidate pressure-drop models. The 'isothermal' assumptions are the basis for the 'ideal' model.

An unexpected insight emerges as a bonus: analytical and numerical models of physical processes fall into the categories of 'initial-value' and 'boundary-value' problems. The Stirling cycle model has been described as being of both types (Urieli, 1984). On the other hand, unless the numerical 'engine' is put into motion starting from zero angular speed, realistic initial conditions (pressure, velocity and temperature distributions) are unknown. It appears to be an assumption that, after a sufficient number of cycles, the resulting 'cyclic equilibrium' of the numerical model will represent conditions in the real engine.

The absence of any way of demonstrating the validity of the expectation means that it cannot be used to check the numerical model for correct functioning.

On the other hand, this models the *conservation laws* — the laws of conservation of mass, momentum and energy. Such a model must first-and-foremost be *conservative.* There is little hope unless mass is accurately conserved, a criterion which is readily checked. If the test is satisfactory it becomes realistic to check for other features. For example, conservation — or at least persistence — of initial conditions.

A formulation along these lines might also be expected to preserve *unrealistic* initial conditions — or at least track them through a plausible number of cycles.

C.2. Formulation

It is assumed that the 'real' value of a property (pressure, velocity, etc.) at any location and instant is the sum of the 'ideal' value and an 'error', here denoted ε. On this basis, the ideal gas equation applied to volume V under independently controlled variation becomes:

$$(p + \varepsilon_\mathrm{p})V = R(m + \varepsilon_\mathrm{m})(T + \varepsilon_\mathrm{T}).$$

Multiplying out, subtracting the ideal gas relationship $\mathrm{p}\mathrm{d}V = mRT$ and omitting products of perturbation quantities leaves:

$$\varepsilon_\mathrm{p}V - R(\varepsilon_\mathrm{m}T + \varepsilon_\mathrm{T}m) \approx \varepsilon_\mathrm{m}\varepsilon_\mathrm{T}.$$

The term on the right of second order yields to iteration. If, as in the present case, reference values are the fixed values of the ideal 'isothermal' analysis, T is fixed, ε_T is zero, and

$$\varepsilon_\mathrm{p}V \approx RT\varepsilon_\mathrm{m}. \tag{C.1}$$

If V is one of the variable-volume spaces of the Stirling engine, change of mass content $\mathrm{d}m$ is equal to the rate at which mass enters or leaves:

$$\mathrm{d}m = p\mathrm{d}V/RT + V\mathrm{d}p/RT. \tag{C.2}$$

In terms of 'known' and perturbation quantities, and with T assumed pre-determined:

$$\mathrm{d}m + \mathrm{d}\varepsilon_\mathrm{m} \approx (p + \varepsilon_\mathrm{p})\mathrm{d}V/RT + V(\mathrm{d}p + \mathrm{d}\varepsilon_\mathrm{p})/RT.$$

Multiplying out this last equation and subtracting Eq. (C.2):

$$\mathrm{d}\varepsilon_\mathrm{m} \approx (\varepsilon_\mathrm{p}\mathrm{d}V + V\mathrm{d}\varepsilon_\mathrm{p})/RT. \tag{C.3}$$

The method calls for the regenerator to be divided into subvolumes ΔV_r. For convenience these are of uniform size. Taking account of the (linear) temperature gradient across the representative element between j and $j + 1$, instantaneous mass content is

$$m + \varepsilon_\mathrm{m} = (p + \varepsilon_\mathrm{p})\Delta V_\mathrm{r} \log_\mathrm{e}(T_j/T_{\mathrm{j+1}})/[RT_{\mathrm{j+1}}(T_j/T_{\mathrm{j+1}}) - 1].$$

When pressure changes by $\mathrm{d}p + \mathrm{d}\varepsilon_\mathrm{p}$ the *change* in the left-hand side $\mathrm{d}\varepsilon_\mathrm{m}$ is $\mathrm{d}\varepsilon_\mathrm{p}\Delta V_\mathrm{r} \log_\mathrm{e}(T_j/T_{\mathrm{j+1}})/[RT_{\mathrm{j+1}}(T_j/T_{\mathrm{j+1}}) - 1]$.

The variables to be solved for are $\mathrm{d}\varepsilon_\mathrm{m}$ and $\mathrm{d}\varepsilon_\mathrm{p}$. The n subdivisions of the matrix yield n+1 equations in coefficients of $\mathrm{d}\varepsilon_\mathrm{m}$ and of $\mathrm{d}\varepsilon_\mathrm{p}$ — of which a total of 2(n+1) values are to be determined. There are a further n equations to come from perturbing the momentum equation(s). The final equation making the necessary 2(n+1) total is simply the statement that the total of the $\mathrm{d}\varepsilon_\mathrm{m}$ over an integration step should be zero:

$$\mathrm{d}\varepsilon_\mathrm{m1} + \mathrm{d}\varepsilon_\mathrm{m2} + \cdots \mathrm{d}\varepsilon_\mathrm{mj} + \mathrm{d}\varepsilon_\mathrm{mj+1} + \cdots = 0. \tag{C.4}$$

The ε_mj and ε_pj accumulated as the cycle proceeds give local, instantaneous difference between the ideal, lossless pressure distribution ($\mathrm{d}p/\mathrm{d}x$ = 0) and the 'real' counterpart.

Before proceeding it is worth anticipating the potential of the approach for guiding corresponding experimental work. With reference to Fig. C.1, temperature limits T_E and T_C can be set equal ($\mathrm{d}T/\mathrm{d}x$ = zero), in which case, specifying phase angle $\alpha = \pi$ causes reference (ideal) velocity u and mass rate $\mathrm{d}m/\mathrm{d}t$ at any crank angle α (but not necessarily the corresponding perturbed values) to be uniform throughout the system.

It also permits separation of the effect of cyclic pressure swing from the uniform-pressure picture of traditional regenerator theory. This is achieved on setting any chosen T_E and T_C and adjusting crank offsets r_e and r_c to achieve $r_e/r_c = T_E/T_C$.

C.3. Specimen Momentum Equation

The overwhelming number of cycle models to date have used the incompressible expression for steady-flow:

$$\Delta p = -\frac{1}{2}\rho u^2 C_f \Delta x / r_h. \tag{C.5}$$

The reference (ideal, isothermal) cycle is free of flow friction so $\Delta p = p_{j+1} - p_j$ is zero. Corrected in terms of the 'errors' of the present approach, ε_{pj} and ε_{pj+1}, pressure drop is finite:

$$p_{j+1} + \varepsilon_{pj+1} - p_j - \varepsilon_{pj} = -\frac{1}{2}\rho u^2 C_f \Delta x / r_h. \tag{C.6}$$

The formulation has been exercised using data defining the apparatus shown in concept in Fig. C.1. Phase angle α and volume ratio $\kappa = r_c/r_e$ are continuously variable, as are charge pressure p_{ref} (not usually the case in the counterpart wind-tunnel test) and *rpm*. For any given choice of p_{ref}, *rpm* and working fluid, three specific values of α are of particular interest when volume ratio κ is unity, viz. 0, π and $\pi/2$.

In the first, flow is symmetrical either side of the mid-plane, at which point it is zero. If the 'apparatus' is put in motion from the outer

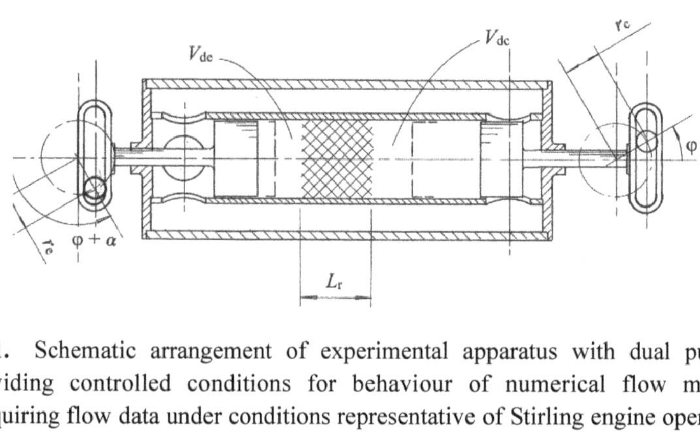

Fig. C.1. Schematic arrangement of experimental apparatus with dual purpose of (a) providing controlled conditions for behaviour of numerical flow model and (b) reacquiring flow data under conditions representative of Stirling engine operation.

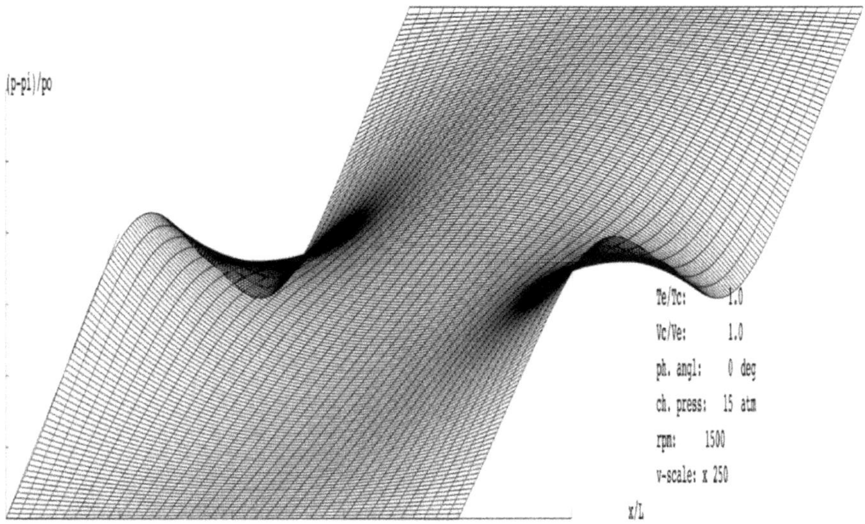

Fig. C.2. Start-up cycle for the conditions shown in the legend. Pressure scale amplified by the factor declared in the legend.

dead-centre positions, pressure swings left and right will be mirror images. There is no inertia term, so any departure from uniformity at the end of an initial cycle will reflect the fact that flow does not diffuse instantaneously. Figure C.2 shows evidence of this.

For the second case (phase angle = π) flow rates are generally higher, and the scale for Fig. C.3 has been adjusted accordingly. Flow has again been started from rest at outer dead-centre. There is an expectation of greater departure from the original uniformity at the end of the cycle, and Fig. C.3 confirms.

Case 3 ($\alpha = \pi/2$) might be thought as close to conditions in the functioning engine as the 'apparatus' could offer. Not so! The cycle distribution of mass can be adjusted to that of any corresponding 'Schmidt' case by setting κ to give a common value of the product κN_T in both cases (see the algebra of Chapter 4). Unless artificial means are applied to replicate the gradation of density and viscosity, cycle pressure history will not be representative. On the other hand, overall pressure swing has been isolated from flow rate (Fig. C.4). Wind-tunnel testing does not allow this. Moreover, regenerator theory has evolved around the assumption of uniform density. Identifying the potential of the Stirling engine calls for the use of relevant

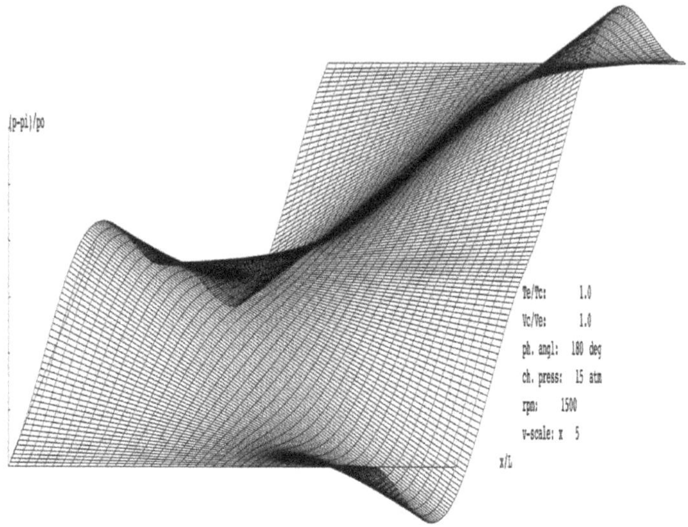

Fig. C.3. Initial cycle for the earlier conditions, except that phase angle is now π. Pressure scale amplification adjusted for the increased flow rates and corresponding greater pressure drop.

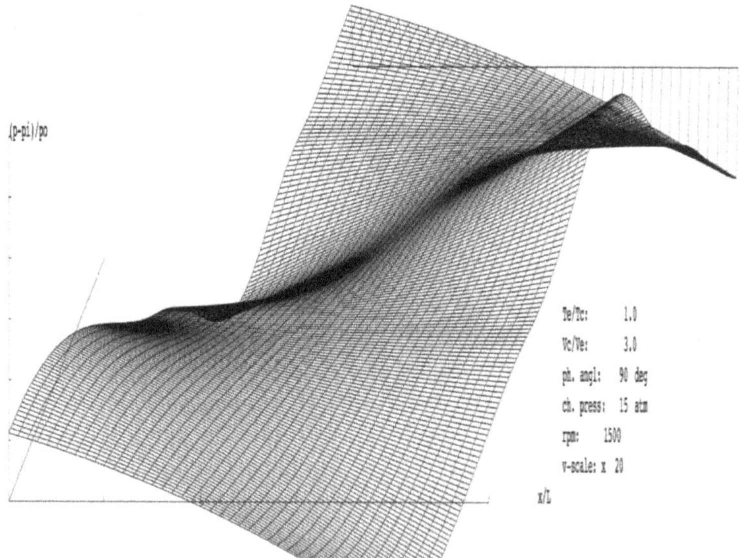

Fig. C.4. Start-up cycle for the earlier conditions, except that phase angle is now $\pi/2$ and $\kappa = 3$ in partial simulation of flow in the Stirling engine. (Note adjusted pressure scale.)

design tools. Use of the apparatus sketched in Fig. C.1 in conjunction with a suitable model of unsteady, compressible flow could be a step in the direction of superseding the data from incompressible, steady-flow tests which have been used indiscriminately and with indeterminate outcome to date.

Appendix D

Tentative Expansion of Steady-Flow Momentum Equation

D.1. Rationale

The troublesome equation is the steady-flow momentum equation:

$$\partial p/\partial x = -\frac{1}{2}\rho u^2 C_f / r_h sign(u). \tag{D.1}$$

The most worrying feature is in terms of the large number of one-dimensional flow models which have evidently embodied it with no reported difficulty.

The direct parallel between Eq. (D.1) and the one-dimensional steady thermal conduction equation $q' = -kA\mathrm{d}T/\mathrm{d}x$ suggests that a problem ought to *arise*! The latter equation is *never* used directly in tackling problems in unsteady heat flow, and to this extent reliance on Eq. (D.1) in the context of unsteady (compressible) fluid flow is misplaced.

One cannot argue against a chorus of successful cycle simulations, but one can attempt to understand. The avenue explored here is whether this author has missed a trick in failing to expand Eq. (D.1) in terms of time (or crank-angle). The algebra and corresponding numerical analysis present no difficulty. Whether algebra and physical reality amount to the same thing might be a topic for somebody's final year dissertation.

Except for the fact that wire-mesh regenerator and parallel exchanger require slightly different treatments on account of their distinct C_f–Re

correlations, expansion for compatibility with the present implicit formulation is straightforward. Moreover, since the thermodynamic properties of the formulation are p and T, a state relationship (e.g., the ideal gas equation) is required to eliminate ρ in favour of p and T, e.g., via $\rho = p/RT$. The move implies that Eq. (D.1) is valid for compressible flow.

A widely-used correlation for steady incompressible flow through a stack of wire screens is $C_f = c/Re + d$. Incorporating Reynolds number $Re = 4\rho|u|r_h/\mu(T)$:

$$\mathrm{d}p/\mathrm{d}x = -\frac{1}{2}\rho u^2[c\mu(T)/(4\rho\,|\,u\,|\,r_h) + d] \times u/|\,u\,|. \tag{D.2}$$

(Incompressible because real and obvious problems associated with compressibility were circumvented while acquiring the data.)

The expansion process is essentially differentiation, so a hint as to prospects is to be had by considering the outcome of differentiating by the chain rule, starting with $1/2\rho u^2$. This is merely a term chosen arbitrarily for its convenient dimensions in normalising the original experimental data, and the result is predictably unhelpful.

Pre-multiplying (or re-multiplying) through by the same term restores a degree of physical meaning — the linear sum of a viscous component and a momentum component:

$$p_{j+1} - p_{j-1} = -\frac{1}{2}(x_{j+1} - x_{j-1})u[c\mu(T)/(4r_h) + u \times d \times p/RT] \times u/|\,u\,|. \tag{D.3}$$

Change $\mathrm{d}f$ in a function $f(x, y)$ of two variables which varies continuously about the point of interest is:

$$\mathrm{d}f \approx \partial f/\partial x.\mathrm{d}x + \partial f/\partial y.\mathrm{d}y + \frac{1}{2}\partial^2 f/\partial x^2 \cdot \mathrm{d}x^2 + \partial^2 f/\partial y\partial x.\mathrm{d}x\mathrm{d}y + \frac{1}{2}\partial^2 f/\partial y^2.$$

Function f in this case is $p_{j+1} - p_{j-1}$. The expansion is the linear sum of the expansions of both about current values of $(x_{j+1} - x_{j-1})$, of p and of T (u is not a parameter of the present formulation). A specimen expansion represents both:

$$\partial p_{j+1}/\partial x = -\frac{1}{2}u[c\mu(T)/(4r_h) + udp/RT]u/|\,u\,|,$$

(NB $\partial p_{j-1}/\partial x = \partial p_{j+1}/\partial x$ etc.),

$$\partial p_{j+1}/\partial p = -\frac{1}{2}u(x_{j+1} - x_{j-1})(u^2 dp/RT)u/|u|,$$

$$\partial p_{j+1}/\partial T = +\frac{1}{2}u(x_{j+1} - x_{j-1})(u^2 dp/RT^2)u/|u|,$$

$$\partial^2 p_{j+1}/\partial p^2 = 0,$$

$$\partial^2 p_{j+1}/\partial T^2 = -2T^{-1}\partial p_{j+1}/\partial T,$$

$$\partial^2 p_{j+1}/\partial p\partial T = -p^{-1}\partial p_{j+1}/\partial p.$$

In the order in which they are entered into the matrix for solution, the unknowns $\mathrm{D}x_{j-1}$, $\mathrm{D}x_{j+1}$ etc., with their coefficients are as follows:

$$\partial p_{j-1}/\partial x \mathrm{D}x_{j-1} - \partial p_{j+1}/\partial x \mathrm{D}x_{j+1}\ \partial p_{j-1}/\partial T \mathrm{D}T_j \partial p_{j+1}/\partial x \mathrm{D}T_j \cdots$$

$$\partial p_{j-1}/\partial p \mathrm{D}p_{j-1}\ \partial p_{j+1}/\partial x \mathrm{D}p_{j+1} = \partial^2 p_{j-1}/\partial T^2 \mathrm{D}T_j^2 + \partial^2 p_{j+1}/\partial T^2 \mathrm{D}T_j^2$$
$$+ \partial^2 p_{j-1}/\partial p \partial T \mathrm{D}p \mathrm{D}T + \partial^2 p_{j+1}/\partial p \partial T \mathrm{D}p \mathrm{D}T.$$

The process is essentially the same as that of perturbation (Appendix C), except that the latter is stabilised by underlying values from the ideal reference cycle.

In the coded implementation, respective $\partial p_{j-1}/\partial x$ etc. are added to existing coefficients of the unknowns $\mathrm{D}x$, $\mathrm{D}T$, $\mathrm{D}p$. Corresponding normalised forms are used throughout — for example, $\mathrm{D}v_{j-1}$ for $\mathrm{D}x_{j-1}$.

D.2. Observation

The added 'sophistication' makes the numerical instabilities worse rather than better.

Stability is be expected only if an increase in local particle speed generates a steepened pressure gradient, which in turn feeds back so as to oppose the increase. The raw equation (D.1) clearly does not promise this, and the more radical form (Eq. (D.3)) evidently does not yield it.

The difficulty in coercing the steady flow momentum equation into the unsteady, compressible scenario appears to stem from the ambivalent u^2 — a view supported by the fact that eliminating u (as achieved by forming the diffusion equation) also eliminates the problem.

D.3. Deduction

This mini-study was prompted by problems encountered during attempts to formulate a cycle model in Lagrange coordinates which reconciled all three Conservation Laws (CL-3). As a 'test-bed' a CL-2 formulation in Euler coordinates promised access to the anomalies — and suggested a dead-end.

Progress on the original front will call for reconciling the diffusion formulation with Lagrange coordinates. An impression of the task may be had on noting that modelling thermal conduction effects along the direction of gas flow reduces to embodying the Euler diffusion equation moving at particle speed. This has already been tried and tested. Momentum conservation in diffusion terms does not coincide, and more is required than the obvious transformation $\mathrm{D}/\mathrm{d}t = \partial/\partial t + u\partial/\partial x$.

Appendix E

Bearing Selection

E.1. What Type?

Rotary bearings are of three principal types: (a) dry (unlubricated), (b) plain, hydrodynamic and (c) rolling-element. Satisfactory operation of a Stirling engine is crucially dependent on the appropriate choice.

So-called *dry* bearings involve a cylindrical shaft, usually of metal, running in a housing lined with a non-metallic (polymer or resin-based) material. Although the technology has made great advances, service life under the combination of high pressure (radial load divided by projected area) and high rotational speed is limited. At the other extreme (low bearing pressure), coefficient of friction is disproportionately high.

With the appropriate combination of radial clearance, radial load, surface finish, component rigidity, rpm, and lubricant viscosity and supply-rate, a plain, cylindrical bearing 'floats' on a film of lubricant — the hydrodynamic film. There is no metal-to-metal contact and the result is high load-carrying capacity and low friction torque. 'Bearings' consisting of a close-fitting pin in a drilled or reamed hole generally do not meet the conditions. The result is high friction and rapid wear.

The rolling-element bearing is fundamentally different: here the load is transmitted between shaft and housing at multiple points of rolling contact. Load-carrying capacity and limiting speeds are high and friction torque is low at all combinations of both. A rolling-element bearing for a particular application is almost never designed from first principles. Instead it is *selected* from a range of stock items (Figs. E.3 and E.4 below) according to an algorithm evolved from dealings between manufacturers and the International Standards Organization, ISO.

The rolling-element bearing may be the definitive example of sophisticated, high-precision mass-production. £5 for a small ball or roller-race buys an impressive amount of engineering. At the same time, it is an off-the-shelf component, readily available in a wide range of sizes, and to this extent as accessible as the nut-and-bolt, the O-ring and the circlip.

Races are available for supporting purely radial load, purely axial load or a combination of both. The rolling elements themselves can be spherical, parallel-cylindrical,[1] conical[2] or barrel-shaped according to the load-carrying function.

'Design' is a process of *identifying* from within a manufacturer's standard range the item which will withstand specified service load at specified angular speed for the required service life. Excess load/speed/life capacity has to be paid for in terms of outlay and/or bulk and/or weight and is undesirable. So it is usually the case that only one or two races from the range available are serious candidates for the intended role.

E.2. Definitions

F_r	radial component of load	(N)	m	exponent in life equation
F_a	axial component of load	(N)		= 3 for ball bearings
F_{eq}	equivalent effective radial load	(N)		= 10/3 for roller brgs
L_{10h}	rated useful life	(hours)	X	radial factor (–)
N	rotational speed	(*rpm*)	Y	axial factor (–)

(e a parameter tabulated against the individual bearing)

E.3. Equivalent Radial Load F_{eq}

For the purposes of applying the selection procedure, actual loading (which may be cyclic — rather than uniform — and may involve an axial component) is converted to an 'equivalent', purely radial load F_{eq}:

$$F_{eq} = X.F_r + Y.F_a \text{ [N]}.$$

The method of determining the X and Y factors depends on the type of bearing — ball or roller. It involves considering a parameter e, the numerical value of which is tabulated against the individual bearing:

[1] In reality slightly cambered.

[2] In reality slightly cambered.

***Taper-roller bearings*:**

$F_a/F_r \le e$: $X = 1.0$ $F_a/F_r > e$: $X = 0.0$

$Y = 0.0$ Y: tabulated against bearing

Deep-groove ball bearings: See the following table.

Parallel and needle-roller bearings take no axial load, so $X = 1.0$, $Y = 0.0$ for all sizes/loads.

A standard formula relates F_{eq} via dynamic load-carrying capacity C (tabulated in N against the individual bearing) and *rpm* to service life L_{10h} [hours].

L_{10h} is the life that 90% of a representative batch of nominally identical bearings can be expected to achieve or exceed. (Median life is 5-times L_{10h}).

$$\textbf{Rated life } L_{10h} = \frac{10^6}{60 \times rpm}(C/F_{eq})^m. \tag{E.1}$$

The advanced stage of a typical selection process can involve several iterations between the life equation and manufacturers' catalogues. A graph offers an advantage over the algebraic form: all possible solutions can be taken in at-a-glance. The nomograms which follow (Fig. E.1 for ball races, Fig. E.2 for rollers) offer a choice between the two ways of proceeding.

E.4. Factors *X* and *Y* for Deep-Groove Ball Bearings

		$F_a/F_r \le e$		$F_a/F_r > e$	
F_a/C_0	e	X	Y	X	Y
0.025	0.22	1	0	0.56	2.0
0.04	0.24	1	0	0.56	1.8
0.07	0.27	1	0	0.56	1.6
0.13	0.31	1	0	0.56	1.4
0.25	0.37	1	0	0.56	1.2
0.50	0.44	1	0	0.56	1.0

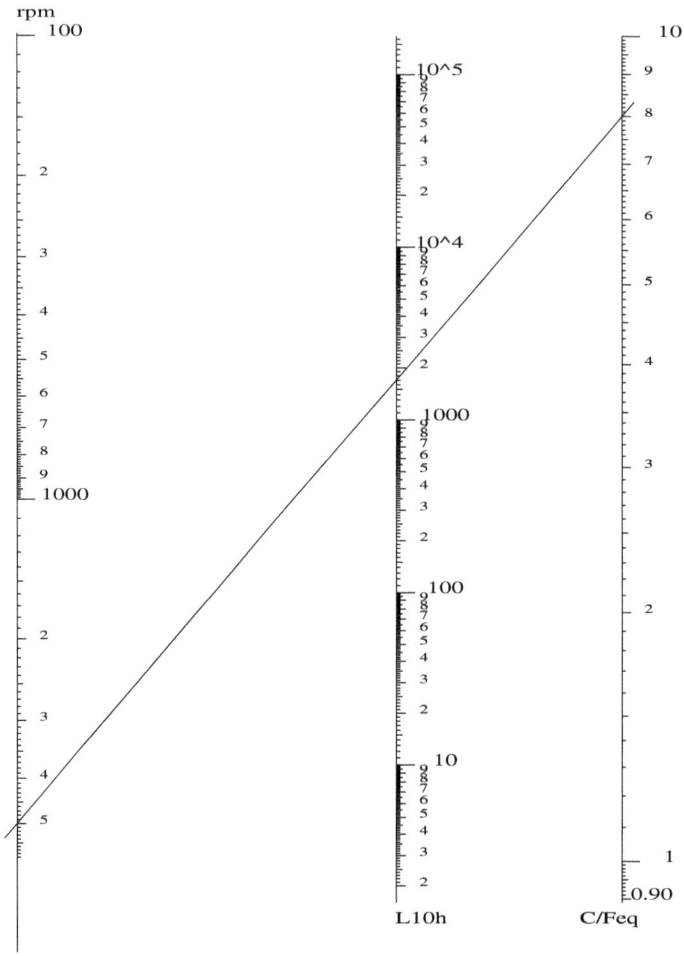

Fig. E.1. Nomogram display of life equation — ball races.

A colleague was exploring the selection procedure by 'reverse-engineering' the choice of needle-roller bearing at the big-end of the Honda CRF 450 motorcycle engine. Relevant details: bore 96 mm × 62.1 mm, compression ratio 11.5:1. Max torque at 6500 rpm, max power at 9000 rpm. Needle roller in question $C = 28.5$ kN (dimensions 35 × 42 × 20 mm).

Assuming peak combustion pressure of 750 psi and multiplying by piston area suggests peak gas load at the big-end of 36 kN and a value of

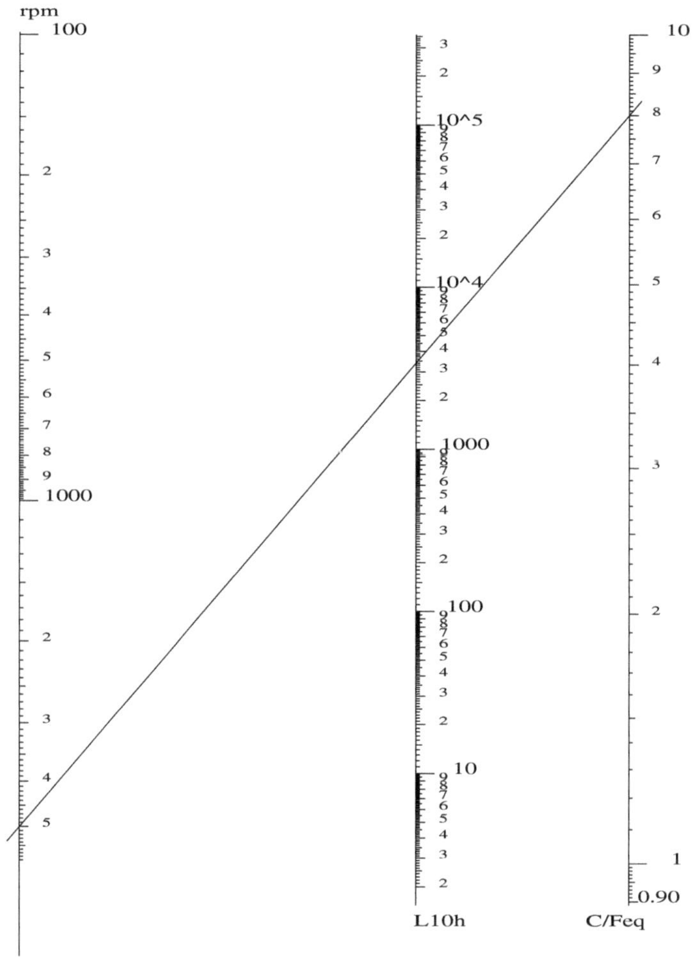

Fig. E.2. Nomogram display of life equation — roller races.

C/F_{eq} of 0.792. However, rated life from Eq. (E.1) is apparently less than one hour (0.815 hours).

It is necessary to enquire more deeply: Assuming combined mass of (piston + gudgeon-pin + connecting rod) = 0.5 kg then centrifugal force at inner dead-centre is 27.58 kN, giving a net (provisional) F_{eq} of 36 – 27.58 = 8.42 kN. The engine operates on the four-stroke cycle with one rotation loaded principally by centrifugal force. At dead-centres, this 27.58 kN alone exceeds the net 8.42 kN.

Deep groove ball bearings
single row
d **2,5–12** mm

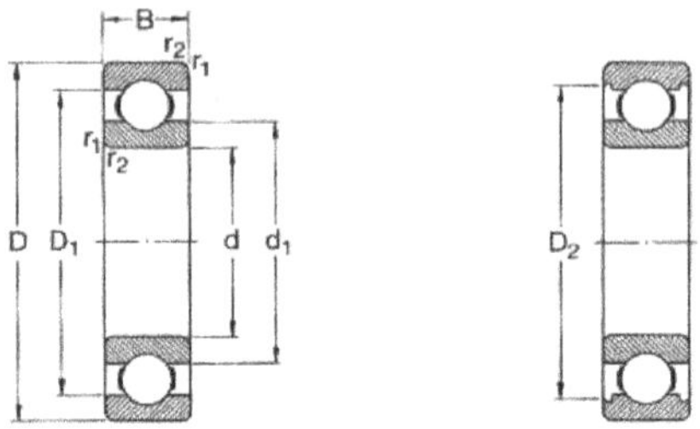

With full outer ring shoulders — With recessed outer ring shoulders

Principal dimensions			Basic load ratings		Fatigue load limit	Speed ratings		Mass	Designation
			dynamic	static		Lubrication grease	oil		
d	D	B	C	C_0	P_u				
mm			N		N	r/min		kg	–
2,5	8	2,8	319	106	4	67 000	80 000	0,0007	**60/2.5**
3	10	4	488	146	6	60 000	70 000	0,0015	**623**
4	9	2,5	540	180	7	63 000	75 000	0,0007	**618/4**
	12	4	806	280	12	53 000	63 000	0,0022	**604**
	13	5	975	305	14	48 000	56 000	0,0031	**624**
	16	5	1 110	380	16	43 000	50 000	0,0054	**634**
5	11	3	637	255	11	53 000	63 000	0,0012	**618/5**
	16	5	1 110	380	16	43 000	50 000	0,0050	**625**
	19	6	1 720	620	26	36 000	43 000	0,0090	**635**
6	13	3,5	884	345	15	48 000	56 000	0,0020	**618/6**
	19	6	1 720	620	26	36 000	43 000	0,0084	**626**
7	14	3,5	956	400	17	45 000	53 000	0,0022	**618/7**
	19	6	1 720	620	26	38 000	45 000	0,0075	**607**
	22	7	3 250	1 370	57	32 000	38 000	0,013	**627**
8	16	4	1 330	570	24	40 000	48 000	0,0030	**618/8**
	22	7	3 250	1 370	57	36 000	43 000	0,012	**608**
9	17	4	1 430	640	27	38 000	45 000	0,0034	**618/9**
	24	7	3 710	1 660	71	32 000	38 000	0,014	**609**
	26	8	4 620	1 960	83	28 000	34 000	0,020	**629**
10	19	5	1 380	585	25	36 000	43 000	0,0055	**61800**
	22	6	1 950	750	32	34 000	40 000	0,010	**61900**
	26	8	4 620	1 960	83	30 000	36 000	0,019	**6000**
	28	8	4 620	1 960	83	28 000	34 000	0,022	**16100**
	30	9	5 070	2 360	100	24 000	30 000	0,032	**6200**
	35	11	8 060	3 400	143	20 000	26 000	0,053	**6300**
12	21	5	1 430	670	28	32 000	38 000	0,0063	**61801**
	24	6	2 250	980	43	30 000	36 000	0,011	**61901**
	28	8	5 070	2 360	100	26 000	32 000	0,022	**6001**
	30	8	5 070	2 360	100	26 000	32 000	0,023	**16101**
	32	10	6 890	3 100	132	22 000	28 000	0,037	**6201**
	37	12	9 750	4 150	176	19 000	24 000	0,060	**6301**

186 **SKF**

Fig. E.3. Specimen ball bearings. From SKF General Catalogue 1989 with permission from SKF (UK) Ltd.

Drawn cup needle roller bearings
F_w **3–17** mm

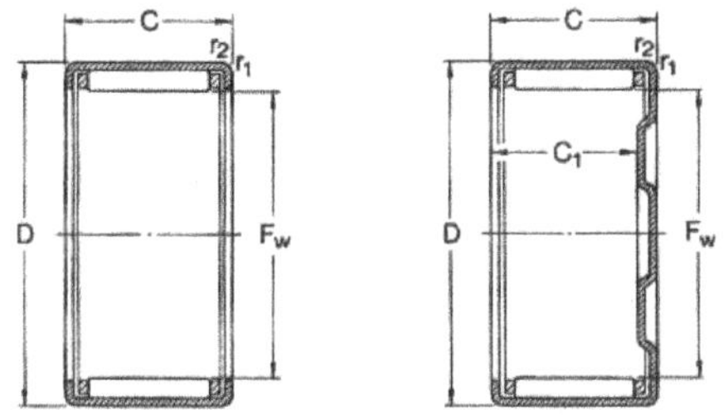

With open ends
series HK

With closed end
series BK

Principal dimensions			Basic load ratings dynamic	static	Fatigue load limit	Speed ratings Lubrication grease	oil	Masses With open ends	With closed end	Designations Bearings with open ends	closed end
F_w	D	C	C	C_0	P_u						
mm			N		N	r/min		g		–	
3	6,5	6	1 230	880	88	30 000	43 000	1	1	**HK 0306**	**BK 0306**
4	8	8	1 720	1 320	137	26 000	38 000	1,6	1,8	**HK 0408**	**BK 0408**
5	9	9	2 380	2 080	220	22 000	34 000	2	2,1	**HK 0509**	**BK 0509**
6	10	8	2 010	1 730	180	20 000	32 000	2,1	–	**HK 0608**	–
	10	9	2 810	2 700	285	20 000	32 000	2,5	2,6	**HK 0609**	**BK 0609**
7	11	9	3 030	3 050	325	18 000	28 000	2,6	2,9	**HK 0709**	**BK 0709**
8	12	8	2 700	2 750	285	17 000	26 000	2,7	3	**HK 0808**	**BK 0808**
	12	10	3 690	4 050	440	17 000	26 000	3	3,4	**HK 0810**	**BK 0810**
9	13	8	3 800	4 300	465	16 000	24 000	3	–	**HK 0908**	–
	13	10	4 130	4 800	530	16 000	24 000	4	4,3	**HK 0910**	**BK 0910**
	13	12	5 120	6 400	720	16 000	24 000	4,6	4,9	**HK 0912**	**BK 0912**
10	14	10	4 290	5 300	570	16 000	24 000	4,1	4,3	**HK 1010**	**BK 1010**
	14	12	5 390	6 950	780	16 000	24 000	4,8	5	**HK 1012**	**BK 1012**
	14	15	6 600	9 000	1 020	16 000	24 000	6	6,2	**HK 1015**	**BK 1015**
12	18	10	4 840	6 400	710	14 000	20 000	4,6	5,2	**HK 1210**	**BK 1210**
	18	12	6 270	7 350	850	14 000	20 000	9,1	10,5	**HK 1212**	**BK 1212**
13	19	12	6 600	8 000	915	13 000	19 000	9,5	11	**HK 1312**	**BK 1312**
14	20	12	6 820	8 650	980	12 000	18 000	10,5	12	**HK 1412**	**BK 1412**
15	21	12	7 650	9 500	1 080	11 000	17 000	11	13	**HK 1512**	**BK 1512**
	21	16	10 100	14 600	1 700	11 000	17 000	15	17	**HK 1516**	**BK 1516**
	21	22	13 000	20 000	2 280	11 000	17 000	20	22	**HK 1522**[1)]	**BK 1522**[1)]
16	22	12	7 370	9 800	1 120	10 000	16 000	12	14	**HK 1612**	**BK 1612**
	22	16	10 500	15 600	1 800	10 000	16 000	16	18	**HK 1616**	**BK 1616**
	22	22	12 800	19 600	2 240	10 000	16 000	22	24	**HK 1622**[1)]	**BK 1622**[1)]
17	23	12	7 650	10 600	1 200	10 000	16 000	12	15	**HK 1712**	**BK 1712**

[1)] double row

420 **SKF**

Fig. E.4. Specimen roller bearings. From SKF General Catalogue 1989 with permission from SKF (UK) Ltd.

This suggests *rpm* at peak torque (6500) as a more realistic design point. Peak inertia load reduces to 14.39 kN (negative relative to gas load at inner dead-centre) and net peak load becomes 36 – 14.39 = 21.61 kN. In terms of absolute values, peak loads are now more closely balanced. Refining the calculation would require an improved value for the mass of moving parts. This may be less critical than noting that the selection procedure recommended by SKF (1989) provides an allowance for fluctuating load. Applying this, $F_{eq} = (0 + 2 \times 21.61)/3$ — sufficiently close to 7 kN to be a job for mental arithmetic.

C/F_{eq} is now 28.5/7 — or about 4, giving L_{10h} (Eq. (E.1) or Fig. E.2) of 260 hours.

At 60 mph this adds up to 15,600 miles — and a short replacement interval. On the other hand, L_{10h} is the life which 90% of a sample of nominally-identical bearings will achieve or exceed by a factor of 5. On this basis the 'average' rider (if one can be identified) might look forward to 78,000 miles of rattle-free riding.

The example brigs out an observation of relevance to design calculations in general: it can help to have an idea of the outcome in advance.

Appendix F

UK Patent Application 2000428.9 *Improved Heat Rejection Provision for Stirling Engines*

F.1. A Word of Explanation

Grant of a patent follows a lengthy process involving scrutiny and 'searches' by the Patent Office. Submission of an 'application' merely gives the applicant 'priority' over his/her invention while matters run their course. The successful application is eventually published to the standard format of the country granting the patent rights. The description which follows is reproduced 'as filed', annotated, numbered and illustrated according to the requirements of the Intellectual Property Office, but obviously not to the format of published patents.

It appears below without the 'Claims'. It is these which give the legal weight. Paradoxically, they are also the aspect which a commercial competitor will ransack for potential vulnerabilities/loopholes.

Abstract

The Invention addresses a problem of thermal bottleneck in the coaxial, parallel Stirling engine by integrating tilting-piston technology from the field of reciprocating compressors with provision of extended surfaces or fins 3 integral with work piston crown and intercalating with corresponding fins 5 on the under-side of the displacer 4 to enhance convective heat transfer in the compression space by inducing shearing of the working fluid or 'squish' at the point in the cycle where such heat transfer would otherwise be of low intensity.

Figure F.1 to accompany Abstract.

Description

This Invention relates principally to un-pressurized Stirling engines employing air as working fluid, although it may find application to certain types of regenerative heat pump or refrigerator.

The essence of the Stirling engine is two spaces of variable volume in more or less free inter-communication via a regenerator, the latter being a porous matrix of solid parent material. One of the spaces is conveniently labelled the expansion space, the other the compression space. Where communication between either space and the regenerator is via a duct or ducts at nominally the temperature of the adjacent variable space, the combination of space and duct will, for present purposes, still be referred to as expansion space or compression space as appropriate. Either can also be appropriately called a 'working' space.

In the embodiment of the Stirling engine to which this Invention particularly relates, respective volume variations are achieved by piston-like members whose individual motions are coordinated by a kinematic mechanism of articulating links ultimately connecting to a conventional crankshaft.

More specifically, it applies to the parallel, coaxial configuration, known to those versed in the technology as the 'beta' layout, in which cylinder, work piston and displacer are aligned along a common axis. Description and discussion are facilitated by envisaging the machine positioned with cylinder axis vertical, having displacer located immediately above the work piston and actuated by a rod passing through a sliding seal in said piston from the crank mechanism below. This positions the expansion space as being between the enclosing cylinder head and the uppermost feature of the displacer, and the compression space as lying between the lowermost feature of the displacer and the uppermost of the work piston. The regenerative matrix sits within the body of the displacer, sharing the axial motion of the latter and offering the working gas 'straight-through' passage between the working spaces.

The Stirling engine differs from other reciprocating prime movers in a fundamental respect, illustrated by the manner in which waste energy from the cycle is rejected. Rather than expelling the working-fluid charge after it has been expanded, the charge is retained in entirety, and waste energy led away instead as heat through the enclosing member(s) in a sequence of thermal convection, thermal conduction and further thermal convection down the succession of respective temperature gradients.

The Stirling engine is an instance of a heat engine, and the maximum cycle work available from such an engine is the difference between heat input per cycle from the high-temperature source and heat rejected per cycle to the low-temperature sink. Improving performance is largely a matter of increasing that difference and/or of increasing the cycle rate at which input and output are achieved at that differential. Heat rate by the mechanisms at issue is proportional to temperature difference, to the degree of agitation of the gaseous elements and to the areas, interior and external, to which those elements are exposed. This reality exposes a 'thermal bottle-neck' at the compression end of the device, where it is not possible to increase temperature difference by arbitrarily lowering the temperature of the sink, since the latter is essentially the environment.

The commonest variant of the Stirling genre employs atmospheric air as working fluid. At its current state of development it is also the variant of least competitive performance – although also that which has most to gain from the benefits anticipated from the Invention. The fact that it operates un-pressurized allows it to function with crank mechanism unenclosed and with the external surfaces of the work piston thus exposed to the environment. This latter feature allows the piston to contribute to the heat-rejection path. It is a purpose of this Invention to exploit and to maximise that contribution.

The Invention builds on the fact that it is possible to enhance the heat rejection path by equipping the external surface of the work piston with integral fins. Suitably located and of suitable extent, they oscillate in the ambient air, generating beneficial forced thermal convection.

Potential for heat-rejection is further enhanced by equipping the interior face of the work piston with fins, optionally somewhat tapered, intercalating between similarly-tapered fins integral with the opposing, lower face of the displacer. If the two sets of fins embody similar degrees of taper, root to tip, then the volume enclosed on full engagement can be as close to zero as required.

The Invention beneficially addresses a residual problem: a design objective is to hold compression space volume to a minimum during the expansion phase. That minimum largely coincides with a period of minimal motion of the air relative to the adjacent metal surfaces. Minimal motion connotes minimal forced convective heat transfer. Superimposing a degree of relative motion between

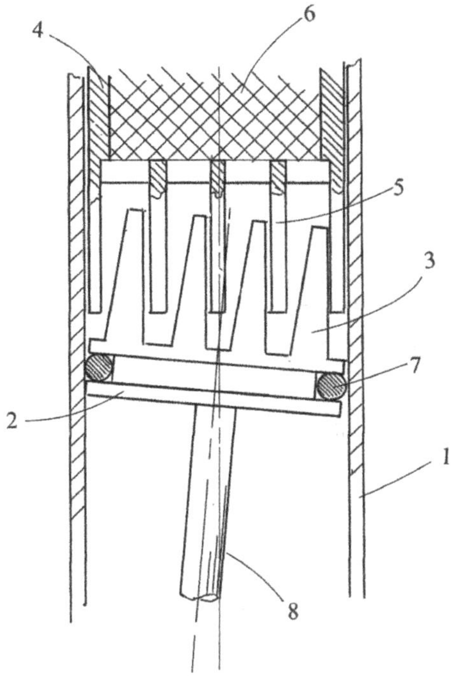

Fig. F.2

rotatably to the crank-pin of crankshaft 9, but connection can be to any suitable point on the drive mechanism. The latter does not form part of the invention, and indication at 9 of a conventional crankshaft does not define, characterise nor restrict the Invention: coaxial Stirling engines in general accept several alternative drive mechanisms, one of which is the subject of UK Patent Application 1906264.5, referred to here as being particularly suitable because its inbuilt kinematic flexibility offers a degree of control over the angular amplitude of piston tilt, and over the occurrence of tilt in relation to the extent of engagement of the two sets of fins.

There is a point in the cycle at which piston and displacer are both close to their respective uppermost positions with piston tilt biased either right or left. As the piston passes inner dead-centre, the angle of tilt reverses. There is a degree of displacement of the working gas from one end of the piston fins to the other, superimposed on a shearing effect on gas occupying the gaps between adjacent fins. The combination amounts to forced convection at a crucial stage in the compression phase of the working cycle.

The mode of operation just described might be thought of as the first mode – Mode 1. An alternative mode – Mode 2 – is indicated by Fig. F.2, in which both sets of interior fins, but not the external fins, have been re-positioned at right-angles to their orientation for Mode 1. A change in the degree of taper of one or both sets of fins accommodates the changed direction of tilt. The rocking motion now displaces gas occupying one side of any given fin to the opposite side. The resulting paddling action exposes any given element of gas to a large effective area of convecting surface. The action might be helpfully described by the term 'squish' borrowed from internal combustion engine parlance.

Operation in either mode can be thought of as increasing the agitation of the working fluid in the compression space to the benefit of convective heat transfer at the point in the cycle where such heat transfer would otherwise be of low intensity

Figure F.3 indicates means for guiding and sealing the displacer actuating rod. It shows, in cross-section through the principal axis of the machine, a portion of cylinder liner 1 and a relevant portion of the work-piston 2 with tilt exaggerated. Also shown is displacer actuating rod 11 and its guide bush 12. The peripheral sealing provision for the piston is again indicated by O-ring 7, but this does not preclude specification

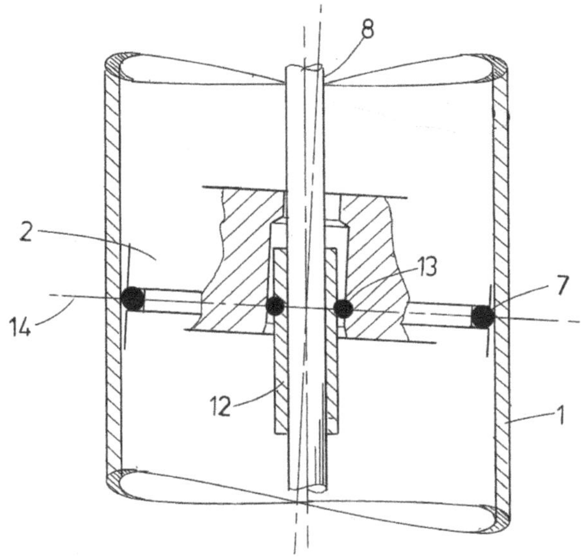

Fig. F.3

of a one-way lip seal where a degree of self-pressurisation is sought. Alternatively, sealing between piston and bore may be achieved by any means gleaned from the existing technology of the tilting-piston compressor.

Guide-bush 12 can advantageously be machined from a self-lubricating material. The proprietary ptfe-compound Rulon™ LD manufactured by the Dixon Corporation is a candidate. The bush is flexibly re-strained supported and aligned within the body of the piston by O-ring 13 accommodated in a groove internal to the piston body and in a counter-part groove in the bush itself. The former groove is preferably machined in the same plane as that for peripheral seal 7, as indicated by centre-line 14. Dimensions of respective grooves, and particularly radial depths, are chosen by calculation and/or experiment to provide pre-compression of O-ring 13 to give static sealing while offering the required angular compliance.

Alternatively, and particularly where short working-life is tolerable, or where the engine is constructed for demonstration purposes, sealing provision between piston and displacer actuating rod may be nothing more than a simple, close-fitting annulus of suitably compliant polymer material, such as that incorporated by the German company Leybold Heraeus in the Stirling engine which that Company markets for teaching purposes.

A variant of the Stirling engine in which the Invention promises particularly beneficial application is the parallel-coaxial version energised by solar energy focused onto an internal absorber via a transparent window of clear fused silica.

WHAT I CLAIM IS

Seven (7) itemised Claims. (Omitted from this Appendix — see **F.1 — A word of explanation**.)

Appendix G

UK Patent Application 2012230.5 *Internally-Focusing Thermal-Lag Engine*

Abstract

The Invention combines two concepts from the public domain to form a novel prime-mover. The first concept is the thermal-lag principle, well-known for its central role in the functioning of the pulse-tube cooler and the thermal-lag engine. The second is the internally-focusing principle, successfully demonstrated in the Stirling cycle machine to form an engine capable of operating from focused ambient sunlight.

The essential feature of the new engine is the window, indicated in Figure G.1 as element 13, made from clear fused quartz or clear fused silica or other transparent and heat-tolerant material. With an adequate flow of radiant energy collected by a mirror or a lens and focused through the window onto internal absorbing element 12, operation of the engine in terms of the internal cycle of interacting thermodynamic process is indistinguishable from that of the conventional, externally-heated thermal-lag counterpart. There is, however, the big and advantageous feature that heat at working temperature is released inside the

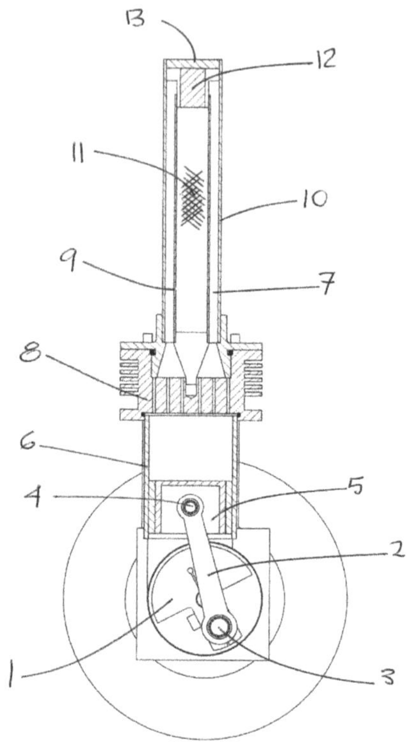

Fig. G.1

machine rather than requiring to be transferred by thermal conduction through the enclosing structure.

Figure G.1 accompanies the Abstract.

Internally-focusing thermal-lag engine

The Invention belongs in the field of solar energy capture and conversion. An outline definition is in terms of its own inventive step: two concepts, both from the public domain but not previously associated, are advantageously integrated into a novel combination.

One of the concepts is the thermal-lag engine. The other is the internally-focusing feature, previously demonstrated in conjunction with the Stirling engine.

The principle underlying operation of the thermal-lag engine is that of any heat-engine: an elastic substance (typically air) is compressed at low temperature. The temperature is by some means raised and the substance re-expanded. High-temperature expansion releases greater work than is absorbed by low-temperature compression. But for inevitable losses, the difference is work output per cycle.

While all other prime movers utilise valves or equivalent means to synchronise heat-exchange and volume-change events, the thermal-lag engine has only passive components, relying instead on the time-lag between creation of a temperature difference and corresponding flow of heat across that difference.

A representative thermal-lag engine consists of a rigid container closed by a piston, the motion of which may be controlled by a conventional crank and connecting-rod, or alternatively the coupling may be electro-mechanical. There are no valves or variable-area ports, and all internal features are in open communication. Actuation of the piston and corresponding volume brings about motion of the air relative to heated and cold portions of the solid enclosure.

Variants of the thermal-lag principle demonstrated to date rely for heating on thermal conduction through the solid enclose from a high-temperature external source, such as combusting hydrocarbon fuel – or, indeed, solar radiation focused externally. In the present invention, by contrast, a portion of the solid enclosure is an optical window of transparent, heat-resisting material such as clear fused quartz or clear fused silica. Thermal energy supply to the engine is solar radiation focused through the window onto a

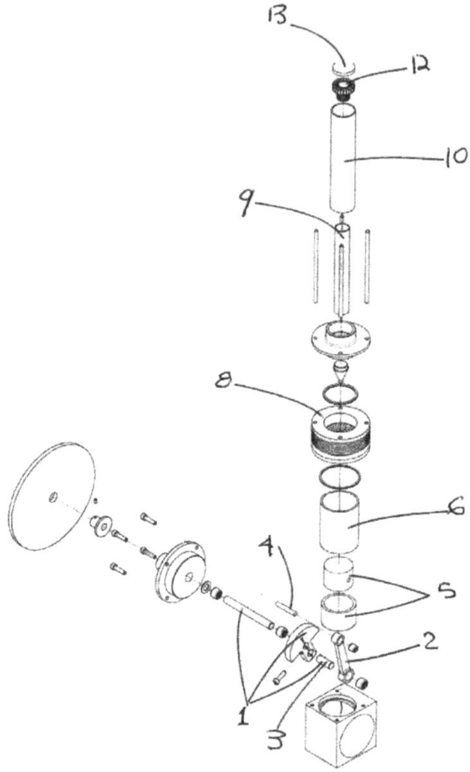

Fig. G.2

stationary receiving surface internal to the system.

The invention can be implemented using only two moving components – crankshaft and tilting (or rocking) piston. This may make it the most ruthlessly-simple thermal prime-mover ever disclosed. Indeed, it may be flippantly – but accurately and relevantly – thought of as requiring only two more components than the direct competition – the p-v solar panel. Thermal-lag engines in general are characterised by specific power which is low by comparison with that of the internal-combustion engine. However, the solar-powered variant in particular operates at zero penalty to the environment.

The Invention will now be described by reference to two figures, of which Fig. G.1 uses the conventions of engineering drawing to show the engine in cross-section through the axis of the vertical cylinder. Fig. G.2 shows the same engine in exploded-isometric form.

A built-up single-throw crankshaft is indicated by 1. It is conventionally supported in bearings preferably of rolling-element type. A conventional connecting-rod 2 locates rotatably on crank-pin 3, pivoting at its opposite end on gudgeon pin 4 located in piston 5. The piston is free to slide axially within liner 6, and may optionally be a two-part assembly as shown. The mutual fit forms a gas-tight seal while introducing minimal friction. A commercially-available combination which routinely reconciles these conflicting requirements is the carbon/graphite piston and glass cylinder of the 'snubber' or 'damper' of the scanning photo-copier.

The space of variable volume above the piston is in open communication with pulse-tube 7 via heat exchanger 8, the latter preferably being integral with exterior heat-dissipating means. The passage through the exchanger is preferably in the form of multiple slots or holes, number and dimensions being chosen in terms of the heat transfer capacity called for by the design and the corresponding pumping loss which can be tolerated.

In the configuration illustrated, the annular gap between the outside of regenerator housing 9 and cylindrical enclosure 10 is called the pulse-tube, here labelled 7. The latter is in open communication with regenerator 11 via channels through a further heat exchanger 12. The exchanger functions to receive, absorb and to pass on the heat of solar radiation focused on it via window 13. Alternatively, exchanger 12 may be omitted, and radiation concentrated directly onto the exposed end of regenerator 11. In the orientation of the figures this is the uppermost end. The regenerator is a matrix of porous and thermally-conducting but heat-tolerant material.

With inflow of thermal energy to receiving element established, the sequence of gas process interactions by which that energy is partially converted to work is identical to that of a conventionally-heated thermal lag engine. For completeness it will be outlined here. The working gas will be assumed to be air at nominally atmospheric pressure, although other combinations of gas and pressure may be viable – possibly potentially more advantageous.

For conciseness the temperature of receiving element 12 will be referred to as T_E. That of exchanger 8 will be referred to as T_C. The latter is nominally ambient temperature while T_E is higher, the exact value depending on the intensity of radiation incident on the absorber relative to the rate of heat converted to mechanical power during steady operation.

Features in Figs G.1 and G.2 which remain unnumbered are retained to show that the two concepts which combine to form the invention are capable of practical realisation.

With piston 5 initially stationary before start-up at the outer dead-centre position, the gas temperature distribution throughout the system is that of the immediately adjacent solid enclosure. Thermal conduction or 'heat-leak' along the walls of the enclosure ensures gradients of temperature from T_E in the vicinity of receiver 12 to T_C in the vicinity of exchanger 8. A parallel gradient exists over the axial length of the regenerator packing.

The crankshaft is now put into rotary motion in either direction. The piston accelerates towards inner dead centre with three-fold outcome: an overall rise in pressure in response to the reduction in total enclosed volume; a corresponding tendency for an overall temperature rise in sympathy (so-called 'adiabatic' component of temperature change); wholesale translation of the temperature distribution originally established in the air. Quantifying and displaying the effects by isolating them is not possible. However, it is sufficient for present purposes to note that the temperature of the gas locally no longer matches that of the solid enclosure with which it is in contact: in some regions it is higher, in others lower.

Passage of the piston through inner dead-centre is an interval of minimal motion, or of dwell, during which the temperature of the gas at any given location makes a partial re-adjustment to the temperature of its enclosing surface. The process might be aptly described as a partial 'relaxation' of temperature. Because displacement has been towards the generally hotter region of the machine, overall pressure rises further above the level reached by the preceding volume change alone.

The piston is now accelerating back towards outer dead-centre with simultaneous three-fold outcome: an overall reduction of pressure corresponding to increasing volume; a corresponding tendency for an overall fall in temperature ('adiabatic' component of temperature change); overall shift of the recently-modified gas temperature profile. Having started at an increased value, pressure during expansion remains higher than the corresponding point during the earlier compression. The excess per cycle is imparted to the crankshaft to be delivered as work observed or measured externally.

The approach to, and passage through, outer dead-centre amounts to a second dwell period, allowing time for the temperature profile displaced by expansion to relax somewhat towards that of the solid containing surfaces. Relaxation to the starting profile is incomplete, but a sufficient number of cycles at constant angular speed and constant heat input ensures eventual cyclic equilibrium whereby overall pressure and temperature distribution at the end of a cycle replicate those at which the cycle started.

The description is accompanied by two figures on separate sheets.

Appendix H

Mean Effective Temperature

Between adjacent particle paths j and j+1 gas temperature gradient $\partial T_g/\partial x$ is assumed linear. A finite difference form which tends to the 'exact' value for modest pressure gradient is:

$$dm = \frac{1}{2}(p_{j+1} + p_j)\,(v_{j+1} - v_j)\frac{\log_e(T_{j+1}/T_j)}{R\left(T_{j+1}/T_j - 1\right)}. \tag{H.1}$$

In Eq. (H.1), effective temperature T_{geff} is evidently:

$$T_{geff} = \frac{T_{j+1}/T_j - 1}{\log_e(T_{j+1}/T_j)}. \tag{H.2}$$

This does not linearise obviously on expanding to T_{gj} + DT_{gj}, T_{gj+1} + DT_{gj+1}.

***Formulation 1*:**
Intuitive assumption: for given $\partial T_g/\partial x$, T_{eff} is approximated by the sum of two components: the mean of T_j and T_{j+1}, viz, ½(T_{j+1} + T_j) and a factor C times the difference $T_{j+1} - T_j$:

$$T_{eff} \approx \frac{1}{2}(T_{j+1} + T_j) + C(T_{j+1} - T_j). \tag{H.3}$$

The right-hand side of Eq. (H.2) is expanded in terms of the standard series for $\log x/(x - 1)$ and equated to the right-hand side of Eq. (H.3). For compactness T_j/T_{j+1} is represented here by x:

$$\frac{1}{2}(1+x)+C(1-x)=\frac{1}{1-\frac{1}{2}(x-1)+1/3(x-1)^2-\cdots}. \tag{H.4}$$

Equation (H.4) is readily made explicit in C which, for given temperature ratio $x = T_j/T_{j+1}$ can be calculated to any desired degree of accuracy. The matter is of concern only within the regenerator, where a suitable value for x is available in terms of the temperature gradient of the matrix dT_w/dx.

Evaluating Eq. (H.4) over a range of x yields a relationship between $\log_{10}C$ and $\log_{10}(x - 1)$ which is almost linear, i.e., of the form:

$$\log_{10} C = m \log_{10}(x - 1) + c. \tag{H.5}$$

Over the range $1.01 < x < 1.4$ $m = 0.95477$ and $c = -1.15766$. With $x = T_j/T_{j+1} = 1.22$, for example, exact T_{geff} from Eq. (H.2) is 1.106357. At 1.11, the arithmetic mean $½(T_j + T_{j+1})$ is 0.33% in error. By contrast, the value using C from Eq. (H.5) in Eq. (H.3) is closer to exact T_{geff} at 1.106395 — an error of 0.0034%.

***Formulation 2*:**

For convenience, effective pressure $½(p_{j+1} + p_j)$ is represented by p and volume $(v_{j+1} - v_j)$ of mass element dm by V. The gas law applied to dm becomes

$$R\,dm\,T_j/(T_j/T_{j+1} - 1) = pV\log_e(T_j/T_{j+1}). \tag{H.6}$$

Expanding by a time step:

$$R\,dm(T_j + dT_j - T_{j+1} - dT_{j+1}) \approx (p + dp)(V + dV)\log_e[(T_j + dT_j)/(T_{j+1} + dT_{j+1})].$$

Using the identity $\log(a/b) = \log a - \log b$:

$$\text{LHS} \approx (p + dp)(V + dV)[\log_e(T_j/T_{j+1} + \log_e(1 + dT_j/T_j) - \log_e(1 + dT_{j+1}/T_{j+1})]. \tag{H.7}$$

Terms of the form dT_j/T_j are considerably smaller than unity, which allows expanding $\log_e(1+x)$ as $x - 1/2x^2 + 1/3x^3 - 1/4x^4 + \cdots$.

Multiplying out:

$$Rdm(T_j + dT_j - T_{j+1} - dT_{j+1}) \approx (pV + pdV + Vdp + dpdV)\cdots.$$

$$(\log_e(T_j/T_{j+1}) + dT_j/T_j - \frac{1}{2}(dT_j/T_j)^2 + \cdots - dT_{j+1}/T_{j+1} + \frac{1}{2}(dT_{j+1}/T_{j+1})^2 - \cdots.$$

Equation (H.6) is subtracted, leaving:

$$\begin{aligned}Rdm(dT_j - dT_{j+1}) \approx\ & pV\left(dT_j/T_j - \frac{1}{2}(dT_j/T_j)^2 + \cdots - dT_{j+1}/T_{j+1} + \text{hot}\right)\\ & + pdV\left(\log_e(T_j/T_{j+1}) + T_j/T_j - \frac{1}{2}(dT_j/T_j)^2 + \cdots - dT_{j+1}/T_{j+1} + \text{hot}\right)\\ & + Vdp\left(\log_e(T_j/T_{j+1}) + T_j/T_j - \frac{1}{2}(dT_j/T_j)^2 + \cdots - dT_{j+1}/T_{j+1} + \text{hot}\right)\\ & + dpdV\left(\log_e(T_j/T_{j+1}) + T_j/T_j - \frac{1}{2}(dT_j/T_j)^2 + \cdots - dT_{j+1}/T_{j+1} + \text{hot}\right).\end{aligned}$$

Higher-order terms (hot) $- \frac{1}{2}(dT_j/T_j)^2 + \frac{1}{3}(dT_j/T_j)^3 - \frac{1}{4}(dT_j/T_j)^4 + \cdots + \frac{1}{2}(dT_{j+1}/T_{j+1})^2 - \frac{1}{3}(dT_{j+1}/T_{j+1})^3 + \frac{1}{4}(dT_{j+1}/T_{j+1})^4 - \cdots$ are evaluated as far as justified by processor architecture (32/64-bit).

Occurrences of p and dp are replaced at coding time by $1/2p_{j+1} + 1/2p_j$ and by $1/2dp_{j+1} + 1/2dp_j$ respectively. V and dV are replaced by $v_{j+1} - v_j$ and by $dv_{j+1} - dv_j$, respectively.

Appendix I

Scaling by Dynamic Similarity as a Diagnostic Tool

I.1. Scaling and the Real World

The author's acquaintance with Geoff Vaizey began with a shared interest in Stirling engines. It continued after he switched his machining skills to scale steam locomotives. In due course, it resulted in an e-mail inviting a view on an anomaly which had come to light in experiments with coal firing of a $7\frac{1}{4}$ inch gauge Garratt steam locomotive. Having come across an account of the 'Rosebud' grate, Vaizey had characteristically lost no time in making one, installing and testing. There had been an unmistakable improvement in steaming, accompanied by a dramatic drop in coal consumption.

The e-mail evidently assumed that this author might be qualified to offer an explanation.

While the latter played for time with off-the-cuff thoughts around air–fuel ratio, stoichiometry, *NTU*, etc., Vaizey adopted the pragmatic approach, made a fire-door from ceramic glass and took a video of the furnace while driving the locomotive at full regulator loaded by the hand brake.[1]

The footage amply confirms a high intensity of burning.

An undeniable reality calls for a more disciplined search for the explanation! Overdue consultation of the internet revealed that Rosebud

[1] His video can be watched at: https://uk.video.search.yahoo.com/yhs/search?fr=yhs-bt-btyuser&hsimp=yhs-btyuser&hspart=bt&p=rosebud+grate+steam+loco#id=1&vid=922ba838d159e0e94a2e59add8301bda&action=click.

is a county in Montana, USA, newsworthy for its troubled coal-mining industry. The 'Rosebud' grate, with which some full-size US steam locomotives were retro-fitted, was designed to burn the low grade brown coal (lignite) from the mine of the same name. One account describes the grate as a flat plate perforated with an array of holes giving open-area ratio between 12% and 17% — compared with the 40% or so of the conventional slotted grate.

Figures for the calorific value (CV) of lignite vary between 10 and 20 MJ/kg, compared with 32 MJ/kg or thereabouts for good-quality steam coal (anthracite). This suggests that the air/fuel ratio allowed by the coal grate of the full-size locomotive would be well in excess of the ideal (stoichiometric) figure when burning lignite, the excess diluting and cooling the combustion gases. On the basis of the relative CVs, the air/fuel ratio when lignite is substituted should be 30% to 50% of that for coal. The figures are comparable to the ratios of open areas, Rosebud-to-conventional.

Vaizey's model locomotive, however, was burning the same fuel as the full-size counterpart, i.e., coal rather than lignite. If the underlying problem (air/fuel ratio) was the same, it was necessary to look further for an explanation.

Regardless of scale, the air supply is drawn by the intermittent partial vacuum in the smoke box. This in turn is caused when the combustion products leaving the fire-tubes are inducted into the blast pipe system by the pulse of escaping exhaust steam. The enquiry can be pursued by supposing *provisionally* that both model and full-size exhaust systems draw comparable vacuum. For present purposes, this is the pressure drop between entry at (constant) atmospheric pressure below the grate and instantaneous pressure in the smoke-box.

Air-flow rate through the grate and fire depends on what fraction of the total depression prevails across it. This is more difficult to estimate than the pressure drop Δp across the fire tubes — i.e., the drop between fire-box and smoke-box. So the approach is to calculate Δp and subtract from total depression. If the difference calculated for the model is significantly greater than that for the full-size case, then this is consistent with increased flow per unit area, i.e., with excess air — and with the phenomenon cured by Vaizey's Rosebud grate.

Details of the US locomotive which burned lignite on a Rosebud grate are not available. This enquiry proceeds tentatively using data from a UK coal-burning loco — an A1 class Peppercorn design. The specification of

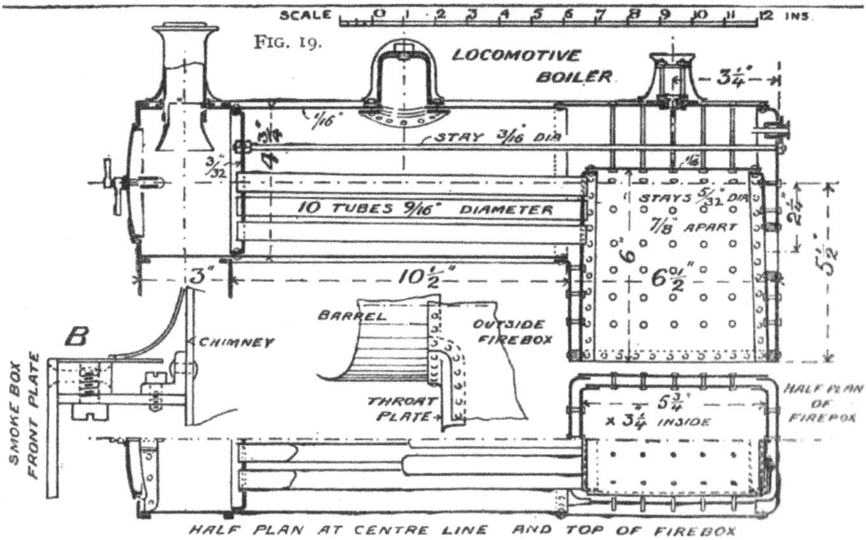

Fig. I.1. Boiler for model to 'one inch' scale (from Pearce (undated)).

Table I.1. Relevant dimensions of full-size and 1/12-linear-scale coal-burning boilers.

	Peppercorn A1	1/12 linear-scale model
No. of small-dia. tubes	121	10
No. of large-dia. (superheater tubes	43	–
I.d. smaller tubes	57 mm	14 mm
I.d. larger tubes	133 mm	–
Hydraulic radius r_h (water side)	23 mm	20.8 mm
Hydraulic radius r_h (flue side)	14.25 mm	3.57 mm
Length L between tube-plates	5170 mm	266 mm
L/r_h flue side	362	75
Grate area	4.65×10^6 mm^2	0.012×10^6 mm^2

NB The latter is not a model of Peppercorn A1. It is simply a boiler design in the 'model' or 'serious toy' category.

the counterpart model (Fig. I.1) was located in a venerable title by Pearce (undated except in terms of its price — 9 pre-decimalisation pence).

Table I.1 is noteworthy for entries for water-side hydraulic radii r_h — 23 mm and 20.8 mm, respectively. Scaling is normally in terms of dimensionless numbers, but r_h (with dimensions) has a special significance as ratio of wetted area to wetted volume. All other things being equal, the same numerical value is expected between non-flow systems of different sizes — but fulfilling the same function, as is the case here.

The flow phenomenon of interest (intermittent through the fire-tubes) is even more unsteady than that in the Stirling engine. In both cases, however, the steady-flow pipe equation is the only symbolic model available: $\Delta p = -1/2\rho u^2 Cf \times L/r_h$ where r_h is the flue-side value. The choice of correlation between Cf and Re matters little if the same fit is to be used between full-size and small-scale. (Cylindrical tubes in both cases!). Using the Blasius model for the turbulent regime:

$$Cf \approx 16/Re + 0.079/Re^{0.25}. \tag{I.1}$$

Assuming ideal gas behaviour for the flue gases, i.e., $\rho = p/RT$, recalling $Re = 4\rho u r_h/\mu$, substituting and multiplying through by $1/2\rho u^2$

$$\Delta p|_{\text{tube}} \approx -\left\{\left(16\mu u/r_h + \left[\frac{1}{2}(p/RT)^{3/4} u^{7/4} u^{1/4}\right] / r_h^{1/4} 2\sqrt{2}\right\} \times L/r_h. \tag{I.2}$$

Do not be daunted by this equation: Using it in the present context could not be more simple: p, R, T, μ being equal (as provisionally assumed) and the r_h, being more or less equal (as calculated — Table I.1), pressure drop depends only upon L/r_h and u. Numerical values are available for respective L/r_h, leaving only the relative values of the two u to be estimated.

At common T and p, steaming rate in m^3/s is proportional to boiler volume and thus to L^3. Consumption in m^3/s by the cylinders is proportional to $L^2 \times$ piston speed. The latter is proportional to the peripheral speed of the driving wheels, suggesting a *scale speed*:

$$\text{speed}_{\text{model}} = \text{speed}_{\text{full-size}} \times L_{\text{model}}/L_{\text{full-size}}. \tag{I.3}$$

According to Eq. (I.3) the 90 mph maximum speed of the Black Five locomotive scales to $90/12 = 7\frac{1}{2}$ mph in the $\frac{1}{12}$ model — somewhat less than twice walking pace.

Under assumptions to this point (which include $p_{\text{model}} = p_{\text{full-size}}$) exhaust pressures are the same. This means that the smoke-box depression which can be drawn when the exhaust expands through the linearly-scaled blast-pipe is the same between full- and model-size.

Flue gas volume rate uA is proportional to steaming rate, and the latter is proportional (via a constant $k(t)$ with units of t^{-1}) to boiler volume (L^3):

$$u_{\text{model}}L^2_{\text{model}} = k(t)L^3_{\text{model}},\ u_{\text{full-size}}L^2_{\text{full-size}} = k(t)L^3_{\text{full-size}}.$$

Dividing and transposing:

$$u_{\text{model}} = L_{\text{model}}/L_{\text{full-size}}. \tag{I.4}$$

The value of Eq. (I.4) for the present case is $\frac{1}{12}$ or 0.0833. (This chimes with the picture of forced convective heat transfer in the parallel duct, since a particle of combustion gas in the model now takes the same time to traverse length of the fire-tube as does a particle in the full-size locomotive.)

Substituting values of L/r_h from Table 1 makes it unnecessary to evaluate the terms in u of Eq. (I.3) to see that $\Delta p_{\text{model}}/\Delta p_{\text{full-size}}$ is fractional. This is consistent with disproportionately-high air-flow per unit area in a model fitted with a scale grate. In turn this is consistent with the anticipated cooling of the combustion gases by air-flow exceeding the stoichiometric requirement.

The restriction introduced by the Rosebud grate promises reduced airflow. And according to Vaizey's experiment, endorsed by the current scaling exercise, it does just that!

Appendix J

A Note on Computer-Generated Diagrams (Including Nomograms)

One of many luxuries of an appointment to a university faculty of engineering is unrestricted access to computing resources, hardware, software — and to specialist advice!

Continuing research and publishing on leaving that environment means selecting — and financing — replacements. Thus it is that, in the course of turning out 6 titles over 19 years, this writer has gone through a total of 6 personal computers — one per title (by chance!). In terms of years per computer — slightly over 3 — the ratio is somewhat less unfavourable, but does not reflect the associated outlay on software. At each replacement it has been necessary for one reason or another (licensing agreements, incompatibility) to invest in a different Fortran compiler, or in a different development environment/'studio' or, in one case, in both.

In the most recent fatality, irremediable crashes of the hard drives of two computers within days of each other galvanised action on two fronts: (a) abandonment of reliance on two machines for mutual back-up and (b) evaluation of actual graphics requirements. The result has been a 'working man's' solution which may appeal to others who, like the writer, are not qualified to the level of systems analyst, and who are daunted by the prospect of £2000+ for an Intel™ compiler and a further £600 for a Winteracter™ GUI Fortran toolset.

The plotting requirement boiled down to being able to drive a pen in a straight line segment from x_1, y_1 to x_2, y_2, either marking a line in the process or simply changing position. The ability to annotate with alphanumeric characters would be a bonus.

These basic facilities are features of PostScript — apparently an accompaniment to any self-respecting operating system — which recognises and processes a text file with extension .ps. All that remains is to generate a .ps file from within a Fortran program.

A first step is to open a file with suitable destination and .ps extension. Why not call it 'file':

```
OPEN(10, FILE = 'Desktop/file.ps')
```

The first WRITE to file.ps should generate the 4-character .ps instruction %!PS. This can be achieved in Fortran by

```
WRITE(10,*) '%!PS'
```

For convenience of debugging, a second (text) file can be opened with a view to duplicating writes to file.ps, thus creating a mirror image for quick inspection. Call the destination file 'mirror':

```
OPEN(11, FILE = 'Desktop/mirror.txt')
```

A Fortran sub-program with suitable arguments adds instructions to *file.ps* and to *mirror.txt* through formatted WRITEs:

```
    SUBROUTINE GRDRAW(X, Y)
    WRITE(10,100)X, Y
    WRITE(11,100)X, Y
    RETURN
100 FORMAT(1X, E10.4, 1X, E10.4, 1X, 'lineto')
    END
```

The WRITE(10, instruction adds a line of .ps code to *file.ps* (and mirrors the instruction in file *mirror.txt*).

```
26.5214E+02 18.0021E+02 lineto
```

A sub-routine to achieve the corresponding position-shift could be

```
      SUBROUTINE GRMOVE(X, Y)
      WRITE(10,100)X, Y
      WRITE(11,100)X, Y
      RETURN
  100 FORMAT(1X, E10.4, 1X, E10.4, 1X, 'moveto')
      END
```

file.ps needs to be properly closed. A FORTRAN sequence which achieves this is:

```
      SUBROUTINE WINDUP
      WRITE(10,*) '0 setgray'
      WRITE(10,*) 'stroke'
      WRITE(10,*) 'showpage'
      CLOSE(10)
      RETURN
      END
```

It would be possible to get by with a repertoire of fewer than a dozen PostScript instructions. The *Tutorial and Cook-Book* (Adobe Systems Inc., 1985) gives the full range and examples of its use.

Mac users are now at a potential advantage: a powerful and sophisticated development environment is available as a free download from AppStore. Early releases of X-Code apparently came integrated with compilers for a number of high-level coding languages. The author's iMac runs on OX 10.11.6 (el Capitan). The contemporary version of X-Code was v8.1 (8B62). It would be possible to spend a career-lifetime mastering the facilities of X-Code, and a further career-lifetime getting the value-for-money evidently on offer. With a view to retaining a feeling of control over proceedings, this author downloaded the GNU open-source Fortran compiler. (Finding out whether X-Code recognises or over-rides it threatened a great deal of work. Anderson's authoritative *Step into X-Code* 2006 was studied without locating the answer.) In this case, the OS-X and X-Code jointly achieved one or the other.

Compilation, link and execution can be managed in Unix commands entered via a *console* window. On the top bar click the sequence *Go/utilities/terminal* to produce a window as shown at Fig. J.1.

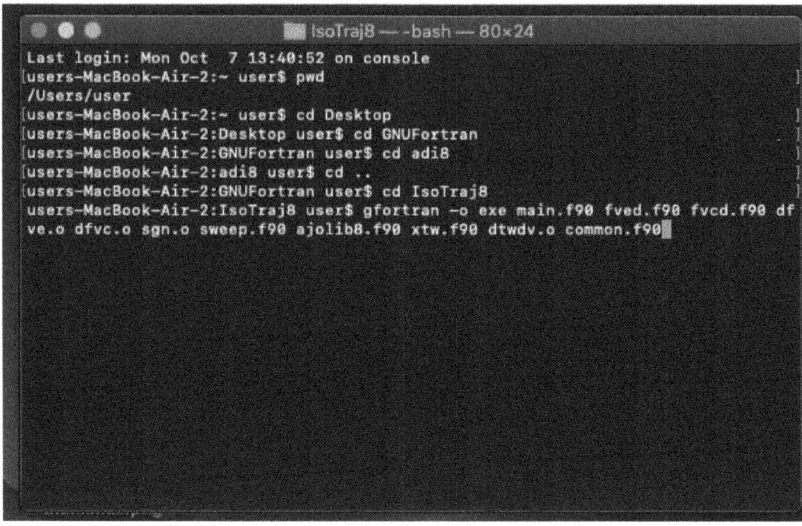

Fig. J.1. Specimen Unix instructions.

The Unix repertoire required is:

pwd: Prints the current directory

cd (followed by the name of a destination directory at the next level in the tree)

Moves to the nominated directory

cd .. reverts to the previous directory

The screenshot shows the path to folder IsoTraj8 which in this case contains the Fortran source code to be processed (edited and/or compiled and/or linked and/or executed). The final two lines on the console (*gfortran -o exe*, etc.) are an instruction to invoke GNUFortran, to compile files having extension .f90, to link pre-compiled files (extension .o), to produce re-locatable binary file and to link to shortcut icon *exe*. A viable program will execute on double-clicking on *exe*. A window appears showing progress.

The string of source file names can be stored in a .txt file within the project folder and copied and pasted into the console window.

The following command-line example generates a file of executable code and displayed an EXE icon as short-cut.

gfortran -o exe main.f90 fve.o fvc.o dfve.o dfvc.o patheq.f90 gaslaw.f90 SetAdi.o SetIso.o sweep.f90 scan.f90 sgn.o fdpop.o fmu.o ajolib8.f90 xtw.o dtwdv.o common.f90

A final step is to run the .ps file. Simply double-click on the icon. All this (as they say) for free.

Appendix K

Unsteady Gas Dynamics of Musk/Virgin HyperLoop

This facsimile reproduction of the original report retains coverage of aspects skipped under the priorities of Chapter 25.

Allan J Organ

The original study (1) suggests that aerodynamic design is being approached first and foremost in terms of quasi-steady, incompressible flow. It is equally arguable is that this is first and foremost a problem in <u>unsteady, compressible</u> flow.

HyperLoop operation is depicted here in terms of a pair of close-fitting vehicles in a tube of 3.4 m diameter. Attention focuses on the interval between the cars, assumed to accelerate, coast and to decelerate in synchronism, i.e., with separation distance L constant and equal to the 50 km. of a representative journey stage. Assuming isentropic conditions and applying the Method of Characteristics (MoC) yields a pictorial display of the interactions between the left-running and right-running waves.

Shock initiation and propagation differ markedly from the steady-flow picture. Operation at reduced pressure promises no benefit in terms of adiabatic temperature rise $\Delta T = T_\infty\{(p/p_\infty)^{(\gamma-1)/\gamma} - 1\}$ *nor in the viscous contribution* $\mu\partial u/\partial y$ *to aerodynamic drag.*

K.1. Notation

a	local, instantaneous acoustic speed — $\sqrt{\gamma RT}$ in an ideal gas	m/s
a_∞	value of a at ambient local temperature T_∞	m/s
D	diameter	m
g	acceleration due to gravity	m/s^2
L	stage length	m
M	Mach number u/a_∞	–
p	absolute pressure	Pa
R	specific gas constant	J/kgK
t	time	s
T	local, instantaneous temperature	K
T_∞	environment or reference temperature	K
u	local, instantaneous particle velocity	m/s
x	distance coordinate	m
α	acceleration factor — decimal fraction of g	–
γ	isentropic index, or specific heat ratio	–
ρ	density — p/RT	kg/m^3

Special Characters

I, II indicate right-running and left-running wave respectively

K.2. Background

Projectiles in ducts set up travelling waves — pressure and rarefaction waves — and are subject to the resulting interactions. Suitable assumptions allow numerical solution of the defining equations by the Method of Characteristics (MoC). Graphical depiction of the wave interactions is intrinsic to the integration scheme, and a quasi-pictorial relief of pressure as a function of time and location emerges automatically. Computation is in terms of local, instantaneous acoustic speed *ratio* a/a_∞ and Mach number u/a_∞, leaving computed pressure ratio p/p_∞ independent of base pressure p_∞. To this extent the MoC applies without reservation to the low-pressure environment proposed for HyperLoop.

K.3. Assumptions

(1) Before being set in motion, air in the tube is at the design pressure of 100 Pa (0.001 atm.). It behaves throughout as an ideal gas, viz, as $p/\rho = RT$.
(2) Cars ride in a straight, parallel, horizontal tube of diameter D. A pair of cars is considered, accelerating, coasting and decelerating in synchronism, i.e., at constant separation distance L, where L is stage distance. (Alternative motion profiles can be modelled with no adjustment to the method *per se*.)
(3) Flow provoked by car motion is one-dimensional ('slab' flow).
(4) To the point of shock formation, processes are isentropic.
(5) The front-mounted fan of the original proposal is disregarded, being absent from descriptions of the later 'HyperLoop One' (2).

K.4. Integration Scheme in Brief

The equations describing one-dimensional, unsteady, compressible flow of a gas are hyperbolic, of second-order and non-linear. Accounts of transformation to 'characteristics' coordinates abound (see, e.g., 3, 4), allowing the derivation to taken for granted here — other than to summarise:

(1) Characteristic directions are loci in the physical ($t - x$) plane of discontinuities in the *gradients* of fluid properties p and T (and thus of $a = \sqrt{\gamma RT}$). They can equally be thought of as disturbances in fluid properties p and a propagating at speeds $u + a$ (right-running) and $u - a$ (left-running).
(2) Right-running waves are distinguished by subscript I (Roman one), left-running by II. Respective directions in the $x - t$ plane (the Mach plane or 'physical' lane) are $dx/dt_{\mathrm{I}} = u + a$ (right-running) and $dx/dt_{\mathrm{II}} = u - a$ (left-running).
(3) Changes of state along I and II characteristic directions in 'state' ($a - u$) coordinates are $da_{\mathrm{I}} = -1/2(\gamma - 1)du_{\mathrm{I}}$ (right-running) and $da_{\mathrm{II}} = +1/2(\gamma - 1)du_{\mathrm{II}}$ (left-running).

Figure K.1 illustrates a single integration step (unit process). This amounts to simultaneous solution between the state plane and the physical plane for point 3, given respective values of a, u, x, and t for points 1 and 2. Numerical work is usually carried out in terms of normalized variables

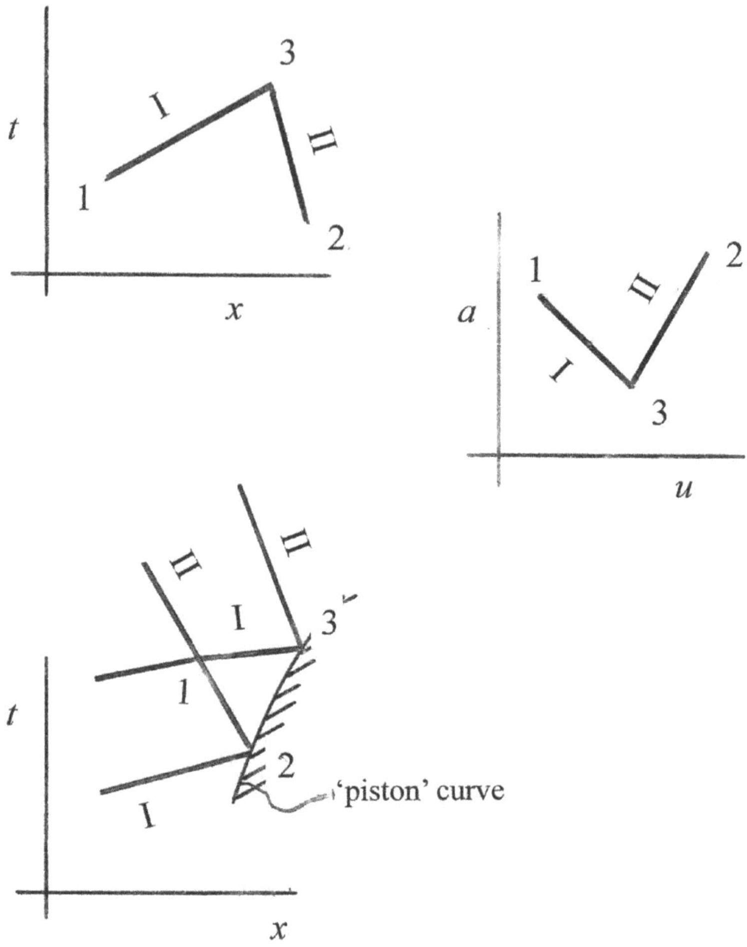

Fig. K.1. Graphical equivalent of unit process (integration step). Upper left: physical ($t - x$) or Mach plane. Centre right: state ($a - u$) plane. Lower: Mach plane — right-hand boundary.

u/a_∞, a/a_∞, x/L and $t/(L/a_\infty)$. Appendix I offers the algebraic equivalent of the graphical construction.

K.5. Boundary Conditions (Capsule Motion Profile)

Solution proceeds by applying the unit process in a left-right sweep (i.e., along a I-characteristic) starting from x and u values at the left boundary.

The right boundary (trailing face of the leading capsule) is reached at time t, calling for values of x and u at that t.

Under present assumptions, all that is required to start the cycle of left-right integration sweeps is x_{LH} as a function of t_{LH}. Corresponding u can be acquired by numerical differentiation as $[x(t + \Delta t) - x(t - \Delta t)]/2\Delta t$, where Δt is some suitably small value of t (independent of integration interval dt). At the right-hand (RH) boundary, $x_{\mathrm{RH}} = x(t_{\mathrm{RH}})$ is calculated at t_{RH} for the left-hand boundary, but increased by separation distance L. Velocity u_{RH} corresponding to x_{RH} is found by calculating x_{LH} at $t = t_{\mathrm{RH}}$.

HyperLoop operation is specified in terms of acceleration (fraction of g) and coast speed (fraction of M). The MoC functions in terms of x and u at the boundaries. Appendix II has the conversion algebra.

K.6. Specimen Wave Diagrams

Interpretation of the specimen wave diagrams calls for two items of background:

The numerical equivalent of shock formation is a Mach-line attempting to overtake the Mach line ahead. At the point of coalescence, pressure gradient is infinite. The thickness of the real shock is too small to model, but processes within are known to be irreversible. If the construction is allowed to press ahead across the discontinuity it becomes a wave in which the crest overlaps the slower-moving base. The robustness of a properly-formulated MoC means that breaking of the virtual wave is delayed. The construction nevertheless becomes chaotic eventually, to this extent mimicking the ocean-wave analogy.

Illustrations offered here called for a minimum of 200 Mach lines in both (I and II) directions covering the 50 km station-to-station distance of the virtual test-section.

Plotting all Mach lines leads to an image which is too dense for the printing process. Instead, every nth Mach line is represented, where n is 3, 4 or 5 as appropriate. The reduced number faithfully outlines the hidden construction.

In all diagrams the right-hand boundary marks progress, lower- to upper-right, of the trailing face of the leading vehicle. The left boundary represents the leading face of the trailing vehicle. In each case, initial acceleration causes the lead vehicle to trail a rarefaction wave. The leading face of the trailing vehicle propagates a pressure wave — both reversed on deceleration.

Postponing shock formation requires holding acceleration to 0.025g m/s^2 and peak velocity to M 0.2. Figure K.2 shows the wave interactions as they eventually coalesce into a shock (overlapping wave at upper right).

More ambitious operating conditions of M 0.3 and 0.3 g m/s^2 lead to shock formation early into the coast phase. Figure K.3 is the Mach plane construction for the early part of the journey. Shock formation is recognised as the point of initial crossing of successive I-waves (*cf.* incipient breaking of an ocean wave).

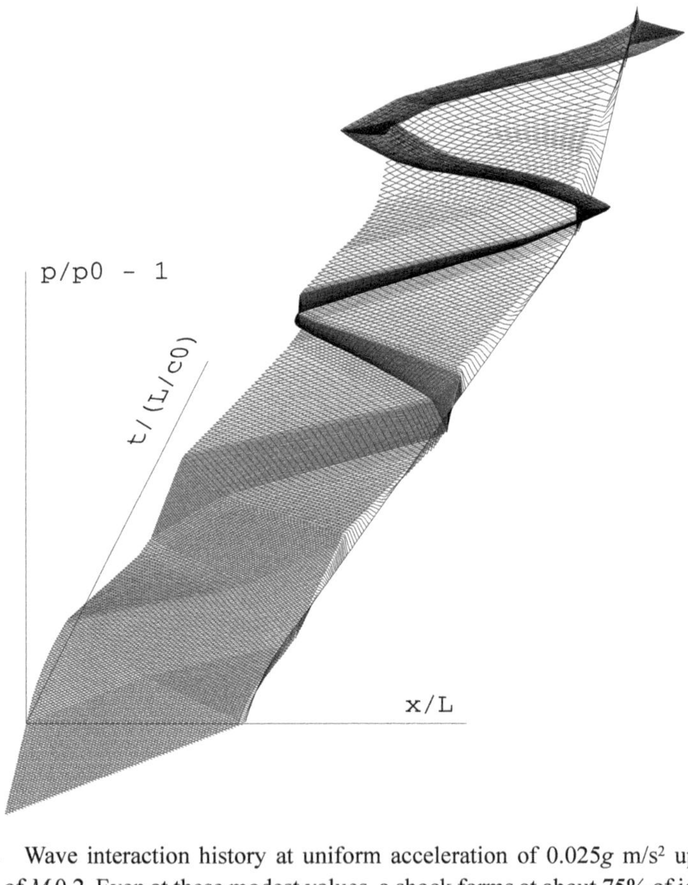

Fig. K.2. Wave interaction history at uniform acceleration of 0.025g m/s^2 up to peak car speed of M 0.2. Even at these modest values, a shock forms at about 75% of journey — i.e., well before start of the deceleration phase.

Because the integration sweep is from lower left to upper right, the construction relating to the right-hand car is further advanced in time than that for the left. Beyond that time, the left-hand boundary outline no longer represents the progress of the left-hand car.

Figure K.3(b) is the pressure relief corresponding to Fig. K.3(a). Neither car reaches its destination before formation of the 'virtual' shock interrupts the sequence.

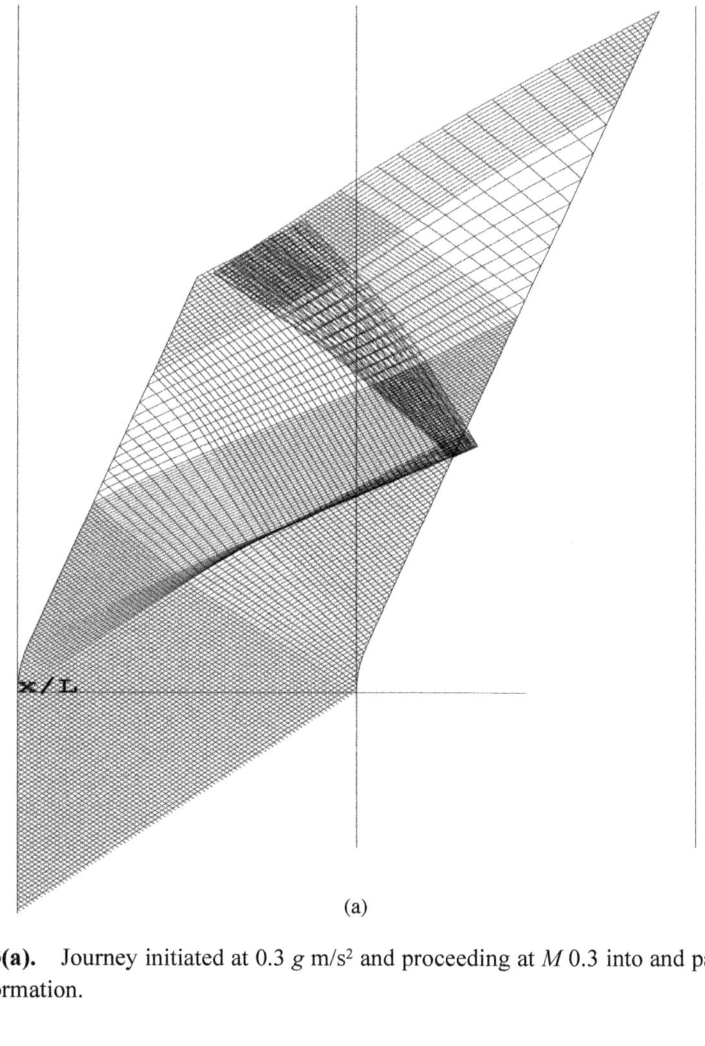

(a)

Fig. K.3(a). Journey initiated at 0.3 g m/s^2 and proceeding at M 0.3 into and past initial shock formation.

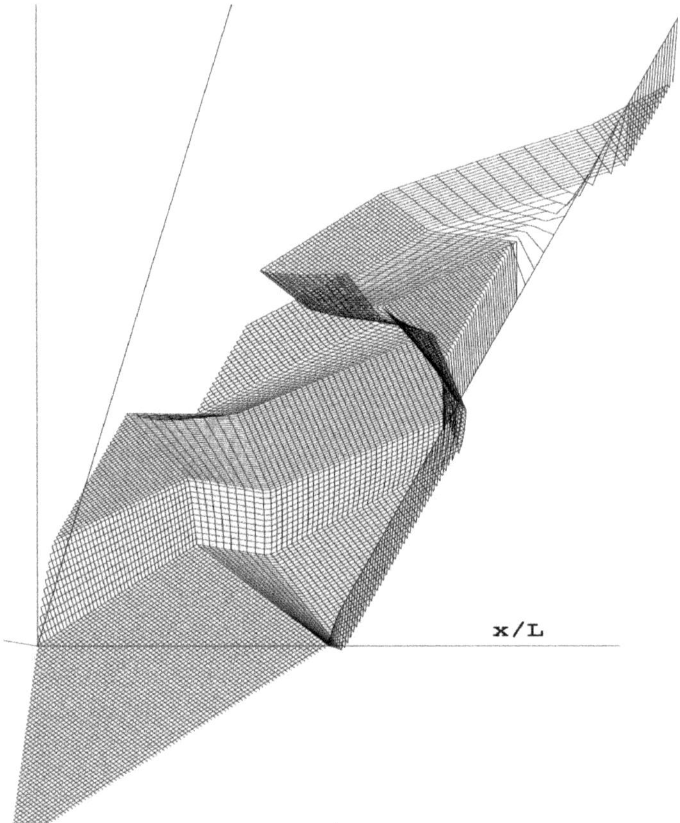

(b)

Fig. K.3(b). Pressure relief corresponding to the Mach plane construction of Fig. K.3(a). Note the deep depression propagating leftward at start-up corresponding to the rarefaction wave trailed by the leading (right-hand) car.

When acceleration and coast speed are set close to design target values of 0.5 *g* and *M* 0.9 a shock forms well before attainment of coast speed. Figure K.4(a) shows the pressure surface generated to the point of formation of the virtual shock. This writer has yet to master art of hidden-line elimination, and the attempt at rendering in perspective for this extreme operating condition is less revealing than the corresponding Mach plane in honest two dimensions shown at Fig. K.4(b).

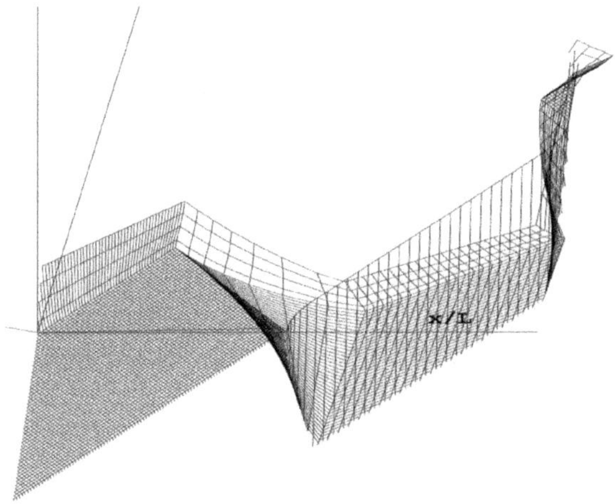

(a)

Fig. K.4(a). Pressure surface for start-up at 0.5 *g* to *M* 0.9. The shock (overlapping Mach lines) has already developed part-way through the acceleration phase.

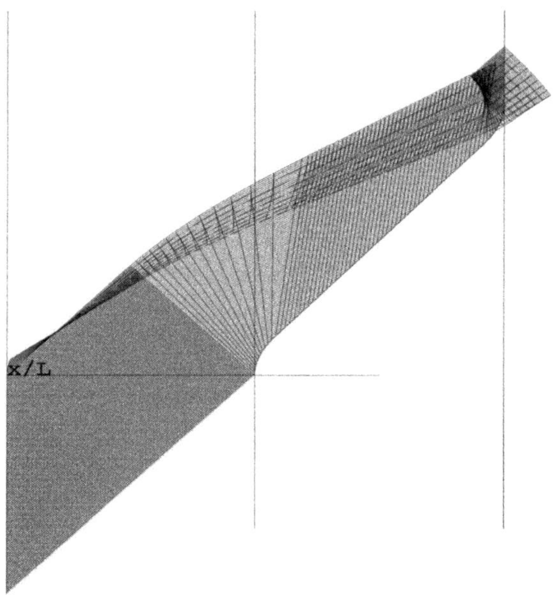

(b)

Fig. K.4(b). Initial stage of Mach plane construction corresponding to Fig. K.4(a). Shock wave building from left is interrupted by deep rarefaction wave outlined by the strong expansion fan. (Only the first few Mach lines of the left-hand envelope of the construction coincide with forward motion of the car.)

K.7. Discussion

Reports of initial tests of HyperLoop One suggest that they have not focused on unsteady compressibility effects. It would therefore be unsurprising if no such phenomena had been noted.

Whether these are eventually important depends largely upon whether or not flow around the capsule chokes as Mach number increases: choking would go some way to causing the capsule to behave as the close-fitting piston of the one-dimensional flow supposed here. Matters should become clearer as test speeds are attained above the $M\,0.3$ or so recently claimed.

Matters which may need to be taken into account during HyperLoop development are that stagnation temperatures to be accommodated depend not on absolute pressure p_∞, but on local pressure *ratio* p/p_∞, viz, as $T_{stag}/T_\infty = (p_{stag}/p_\infty)^{(\gamma-1)/\gamma}$ that p/p_∞ – *and thus* T_{stag}/T_∞ — can exceed counterpart values based on the assumption of steady-state that coefficient of dynamic viscosity μ [Pas] is independent of pressure. Drag due to skin friction is $A_w\mu\partial u/\partial y$, where A_w is wetted area. All things remaining equal, operation at low pressure has done nothing to attenuate the viscous component of aerodynamic drag.

K.8. References

1 Anon https://www.tesla.com/sites/default/files/blog_images/hyperloop-alpha.pdf (Last accessed 10 Feb. 2018).

2 Anon http://www.spacex.com/sites/spacex/files/hyperloop_alpha-20130812.pdf. (Last accessed 10 Feb. 2018).

3 Shapiro A H. 1954 *Dynamics and Thermodynamics of Compressible Fluid Flow*. Ronald, New York, v II.

4 Abbott M B. 1966 *An Introduction to the Method of Characteristics*. Thames and Hudson, London.

Appendix I — Algebra of the Integration Process

General unit process

State equations 1a and 1b are applied to the finite intervals between known points 1 and 2 and point 3 of so-far unknown states:

$$u_3 - u_1 = -2(a_3 - a_1)/(\gamma - 1), \qquad \text{(A-I.1a)}$$

$$u_3 - u_2 = +2(a_3 - a_2)/(\gamma - 1). \qquad \text{(A-I.1b)}$$

Equations (A-I.1a) and (A-I.1b) are linear in the two unknowns a_3 and u_3 and elementary manipulation leads to

$$a_3 = \frac{1}{2}(a_1 + a_2) - \frac{1}{4}(\gamma - 1)(u_2 - u_1). \qquad \text{(A-I.2a)}$$

From which u_3 follows as:

$$u_3 = -2(a_3 - a_1)/(\gamma - 1) + u_1. \qquad \text{(A-I.2b)}$$

Mean wave velocities v_{I} and v_{II} are

$$\underline{v}_{\mathrm{I}} = \frac{1}{2}(u_1 + u_3) + \frac{1}{2}(a_1 + a_3); \qquad \underline{v}_{\mathrm{II}} = \frac{1}{2}(u_2 + u_3) - \frac{1}{2}(a_2 + a_3).$$

Substituting into the Mach plane relationships:

$$t_3 = (-x_1 + x_2 + \underline{v}_{\mathrm{I}} t_1 - \underline{v}_2 t_2)/(\underline{v}_1 - \underline{v}_2), \qquad \text{(A-I.3)}$$

$$x_3 = (t_3 - t_1)\underline{v}_1 + x_1. \qquad \text{(A-I.4)}$$

With $a_{\mathrm{I}} = 1/2(a_1 + a_3)$ for mean acoustic speed between 1 and 3:

$$p_3 = p_1 - \gamma^{1/2}(u_3 - u_1)/\underline{a}_{\mathrm{I}}. \qquad \text{(A-I.5)}$$

Specimen boundary condition (right-hand)

Along path-line 2 – 3 (trailing face of leading vehicle) average $\underline{u} = (x_3 - x_2)/(t_3 - t_2)$. Inverting:

$$x_3 - x_2 = \frac{1}{2}\{u_2 + u_3(t_3)\}(t_3 - t_2). \qquad \text{(A-I.6)}$$

Location x_3 is prescribed in function of t, so the Mach plane equation along the I-characteristic direction can be written:

$$x_3(t_3) - x_1 = \left\{\frac{1}{2}(u_1 + u_3(t_3) + \frac{1}{2}(a_1 + a_3)\}(t_3 - t_1)\right\}. \qquad \text{(A-I.7a)}$$

Pending a definitive value for unknown t_3, the left-hand side is approximated:

$$x_3(t_3) - x_2 \approx (u_1 + a_1)(t_3 - t_1). \tag{A-I.7b}$$

The right-hand side of Eq. (A-I.7b) is expanded. Subtracting from Eq. (A-I.4) eliminates unknown $x_3(t_3)$:

$$t_3 \approx \{-x_2 + x_1 + u_2 t_2 - u_1 t_1 - a_1 t_1\}/(u_2 - u_1 - a_1). \tag{A-I.8}$$

Using this t_3 in Eq. (A-I.6) yields an improved value of x_3:

$$x_3 \approx x_2 + \frac{1}{2}\{u_2 + u_3(t_3)\}(t_3 - t_2). \tag{A-I.9}$$

Changes of state along path-line 1 – 3 are in terms of Eq. (A-I.2b), transposed here as:

$$a_3 \approx \frac{1}{2}(\gamma - 1)(u_1 - u_3(t_3)) + a_1. \tag{A-I.10}$$

Abbreviating the right-hand side of Eq. (A-I.7a) to *RHS*:

$$t_3 \approx (x_3 - x_1)/RHS + t_1. \tag{A-I.11}$$

The improved value of t_3 is fed into Eq. (A-I.9). The sequence convergences rapidly.

Appendix II — Conversion of HyperLoop Journey Profile to MoC Boundary Conditions

Initial acceleration αg m/s^2 is at uniform fraction α of g, dropping to zero when u reaches coast speed cM (c fractional). Deceleration rate is set here as (minus) αg m/s^2, using the same value for α. However, with minor modification the algebra accepts any other value.

Time taken t_{acc} and distance x_{acc} travelled to attain coast speed are from the simple equations of motion with initial velocity u_0 set to zero, viz, from $x = u_0 t + 1/2\alpha g t^2$ and from $u = u_0 + \alpha g t$:

$$t_{acc} = cM/\alpha g. \tag{AII.1}$$

$$x_{acc} = \frac{1}{2}\alpha g t_{acc}^2. \tag{AII.2}$$

With distance to decelerate equal in this case to x_{acc}, coast distance x_{coast} and coast time t_{coast} are respectively:

$$x_{coast} = L - 2x_{acc}. \tag{AII.3}$$

$$t_{coast} = (L - 2x_{acc})/cM. \tag{AII.4}$$

The relationships are used to define the path of the left-hand boundary of the Mach plane (leading face of trailing car):

$$0 \leq t \leq t_{acc}: x = \frac{1}{2}\alpha g t^2,$$

$$t_{acc} \leq t \leq t_{acc} + t_{coast}: \qquad x = x_{acc} + (t - t_{acc})\, cM.$$

Abbreviating $t - (t_{acc} + t_{coast})$ to Δt:

$$t_{acc} + t_{coast} \leq t \leq 2t_{acc} + t_{coast}: \qquad x = L - x_{acc} + cM\Delta t - \frac{1}{2}\alpha g \Delta t^2,$$

$$t \geq t_{acc} + 2t_{coast}: \qquad x = L.$$

Velocity u at the right-hand boundary (trailing face of leading car) is this same function of t. Lapsed distance is equal at any given t to $L + x$ of the trailing (left-hand) car.

Bibliography

Adler A A 1946 Variations with Mach number of static and total pressure through various screens. NACA Wartime Report L-23. (Originally Confidential Bulletin L5F-28.)

Adobe Systems Inc. 1985 *PostScript Language — Tutorial and Cookbook.* Addison-Wesley, Menlo Park, CA.

Airpot Corp. (undated) Company website Airpot Corporation, 35 Lois Street, Norwalk, CT 06851 USA.

Anderson F 2006 *Step into X-Code*. Pearson Education Inc., Boston MA.

Annand W J D 1953 The resistance to air flow of wire gauzes. *J. Roy. Aero Soc.* 57 pp. 141–146.

Anon 1986 Woven wire cloth. Commercial catalogue of G Bopp and Co., Metallgewebe- und Drahtwarenfabrik, Bachmannweg 21, 8046 Zürich, Switzerland.

Anon 1994 Glacier Industrial Bearings. Designer's Handbook. The Glacier Metal Co. Ltd, Riccarton, Kilmarnock, Strathclyde UK.

Anon 1998 *Specification Tables for Industrial Wire Cloth.* Haver and Boeker, D-59299 Oelde, Germany.

Anon 2018 *Rolling Bearings*. SKF Group, Bedford UK.

Bomford N 2004 Personal communication (3 Jan).

Brundrett E and Baines D 1964 Production and diffusion of vorticity in duct flow. *J. Fluid. Mech.* 19 p. 375.

Cornell W G 1958 Losses in flow normal to plane screens. Paper 57-F-19 presented at ASME Fall Meeting, Hartford, Conn. Sept. 23–25, 1957 and published in *Trans. ASME* 80, pp. 791–799.

Crema L *et al.* 2011 *Development of a Pellet Boiler with Stirling Engine for m-CHP Domestic Application.* Energy, Sustainability and Society, Springer.

Eckert E R G 1959 *Heat and Mass Transfer*. McGraw-Hill Book Co, New York and Drake R M.

FCA 2018 (Financial Conduct Authority) Carbon credit trading scams. fica.org. uk. Site updated 17/05/2018.

Feulner P (2003, 2005) Personal communication.

Finkelstein T 1960a Optimization of volume ratio and phase angle for Stirling engines. Paper 118C, SAE Annual Meeting, Detroit, Michigan, Jan. 11–15.

Finkelstein T 1960b Generalized thermodynamic analysis of Stirling engines. Paper 118B, SAE Annual Meeting, Detroit, Michigan, Jan. 11–15.

Finkelstein T and Eibling J A 1961 Internally-focusing solar power systems. Paper 61-WA-297, *Proc. ASME Winter Annual Meeting*, New York, 26 Nov–1st Dec.

Gleick J 1987 *Chaos*. Abacus, London.

Gunther M 2011 Advanced CSP Teaching Materials. Deutsches Zentrum fur Luft und Raumfahrt e.V. Chapter 2 Solar Radiation.

Hargreaves C 1991 *The Philips Stirling Engine*. Elsevier Science Publishers BV, Amsterdam and New York.

Hausen H 1929 Wärmeausausch in Regeneratoren. Zeitschrift des Vereins deutscher Ingenieure **73**, p. 432.

Heraeus Conamic (undated) Company website. Clear fused quartz products.

Hinch E J 2020 *Think Before You Compute. A prelude to Computational Fluid mechanics*. Cambridge University Press in their series *Texts in Applied Mathematics*.

Hodgman C D (Ed-in-Chief) 1954 *Mathematical Tables from Handbook of Chemistry and Physics*. Chemical Rubber Co., Cleveland OH.

Ibrahim M B and Tew R C Jr. 2012 *Stirling Converter Regenerators*. CRC Press, Boca Raton.

Jakob M 1957 *Heat Transfer* Vol II. John Wiley & Sons, New York and Chapman and Hall, London.

Karabulut H *et al.* 2009 An experimental study on the development of a beta-type Stirling engine for low and moderate heat sources. Applied Energy **86** pp 68–73.

Kays W M and London A L 1964 *Compact Heat Exchangers*. McGraw-Hill Inc., New York.

Kays W M 1966 *Convective Heat and Mass Transfer*. McGraw-Hill Inc, New York

Kleemann M 1993 Regenerative Energiequellen. 2nd edition. Springer. Berlin.

Kolin I 1991 *Stirling Motor — history, theory, practice*. Inter-University Centre, Dubrovnik.

Lee K 1976 The Stirling cycle with adiabatic compression and expansion. MSc. diss., University of Calgary.

Lennard-Jones J E 1924 On the determination of molecular fields. *Proc. Roy Soc.* **106 A**, p. 463.

MacKay D J C 2009 *Sustainable Energy — Without the Hot Air*. UIT, Cambridge, UK.

Martini W 1983 *Stirling Engine Design Manual*. Ed II. (Possibly 'self-published' — no copyright or publisher claimed.)

Newport Co (undated) Company web site: https://www.newport.com.

NMiTE 2019 https://nmite.ac.uk.

Organ A J 1962 An appraisal of the 'hot-air' engine. Final year dissertation, Department of Mechanical Engineering, University of Birmingham.

Organ A J 1996 The great Stirling cycle simulation fraud. Proc. StirlingForum Ecos Consult, Westerbreite 7, D-49084, Germany, pp. 245–269.

Organ A J 1997 *The Regenerator and the Stirling Engine*. Wiley 1997, Chichester, UK.

Organ A J 2005 *Stirling and Pulse-Tube Cry-Coolers*. Originally Professional Engineering Publishing, London and Bury St Edmunds. Acquired by Wiley, Chichester, UK.

Organ A J 2007 *The Air Engine — Stirling Cycle Power for a Sustainable Future*; Woodhead, Cambridge UK and CRC Boca Raton USA. Title subsequently acquired by Elsevier.

Organ A J 2014 *Stirling Cycle Engines — Inner Workings and Design*. Wiley, 2014, Chichester, UK.

Organ A J 2014 *The Bridge — A Love Affair*. CreateSpace. Available *via* Amazon.

Organ A J 2016a *A re-appraisal of the Stirling Engine*. FTRR, Inverness.

Organ A J 2016b An algorithm to guide in situ regenerator testing. *Proc. 17th Int. Stirling Engine Conf.*, Northumbria University, Newcastle upon Tyne, 24–26 Aug. pp. 395–405.

Organ A J 2016c Confronting the realities of the Stirling engine. *Proc. 17th Int. Stirling Engine Conf.*, Northumbria University, Newcastle upon Tyne, 24–26 Aug. pp. 406–424.

Organ A J 2019 Comprehensive adiabatic cycle analysis avoiding integration *Proc. 17th Int. Stirling engine conf.*, Northumbria University, Newcastle upon Tyne, 24–26 Aug. pp 425–436.

Organ A J 2019 In search of the efficient, affordable, environment-friendly 150 W air engine. (Awaiting publication.)

Organ A J 2019 *A planar mechanism affording improved working fluid utilisation in the Stirling engine*. Patent application #1906264.5.

Pearce E L (undated) *Model Boiler Making*. Percival Marshall and Co., London.

Pinker R A and Herbert M V. 1967 Pressure loss associated with compressible flow through square-mesh wire gauzes. *Proc. Inst. Mech. Engrs. (J. Mech. Eng. Sci.)* 9, pp. 11–23.

Qvale E B and Smith J L 1968 A mathematical model for steady operation of Stirling-type engines. *ASME J. Eng. Power*. **91**, pp. 109–112.

Rampen W 2015 'Energy storage our salvation?' Inaugural lecture, University of Edinburgh (June).

Rios P A 1969 An analytical and experimental investigation of the Stirling Cycle. PhD diss., MIT.

Schmidt G 1871 Theorie der Lehmann'schen kalorischen Maschine. Zeitschrift des Vereins Deutscher Ingenieure. Band XV, Heft 1 (Jan.).

Shapiro A H 1954 *Thermodynamics and Gas Dynamics of Compressible Fluid Flow*. **II** Ronald, New York.

Sher D 2019 Lung-inspired 3D printed heat exchanger could make carbon-dioxide power plants a reality. https://www.3dprintingmedia.network/lung-inspired-3d-printed-heat-exchanger/ 03 July.

Sier R 1999 *Hot-Air, caloric and Stirling Engines*. L A Mair, Chelmsford, UK.

Stirling R 1816 Patent 4081 *Improvements for diminishing the consumption of fuel, and in particular an engine capable of being applied to the moving of machinery on a principle entirely new*.

Su C C 1986 An enquiry into the mechanism of pressure drop in the regenerator of the Stirling cycle machine. PhD. diss., Dept. Engineering, Cambridge (CUED).

Tailer P 1993 Thermal lag machine. US Patent US5414997A.

Tobias S A 1969 High-speed metal forming with PetroForge machines. Proc 2nd. Int. Conf., Centre for High Energy-rate Forming, Estes Park, 1.

UHI 2019 https://www.lews.uhi.ac.uk/courses/.

Urieli I and Berchowitz D 1984 *Stirling Cycle Engine Analysis*. Adam Hilger, Bristol.

Vaizey G 2020 Personal communications via e-mail.

van Rijn C 2016 Personal communication by e-mail 03 May.

Walker G 1980 *Stirling Engines*. Oxford University Press.

White F M 1979 *Fluid Mechanics*. McGraw-Hill Kogakusha, Tokyo, Japan.

Zaremba H 2019 How the military will spark the death of fossil fuels. OilPrice.com.

Index

H

I

J

K

L

W

X

Z

www.ingramcontent.com/pod-product-compliance
Ingram Content Group UK Ltd.
Pitfield, Milton Keynes, MK11 3LW, UK
UKHW020025100525
458400UK00007B/32

9 781800 611047